Merchants of Doubt

For Embere,
With best wishes
Naomi

MERCHANTS OF DOUBT

How a Handful of Scientists
Obscured the Truth on Issues from
Tobacco Smoke to Global Warming

NAOMI ORESKES
and
ERIK M. CONWAY

BLOOMSBURY PRESS
New York • Berlin • London

Published by Bloomsbury Press, New York

All papers used by Bloomsbury Press are natural, recyclable products made from wood grown in well-managed forests. The manufacturing processes conform to the environmental regulations of the country of origin.

LIBRARY OF CONGRESS CATALOGING-IN-PUBLICATION DATA

Oreskes, Naomi.
Merchants of doubt : how a handful of scientists obscured the truth on issues from tobacco smoke to global warming / Naomi Oreskes and Erik M. Conway.—1st U.S. ed.
p. cm.
Includes bibliographical references and index.
ISBN: 978-1-59691-610-4 (alk. paper hardcover)
1. Scientists—Professional ethics.   2. Science news—Moral and ethical aspects.
3. Democracy and science.   I. Conway, Erik M., 1965–   II. Title.
Q147.O74 2010
174'.95—dc22
2009043183

First U.S. edition 2010

3 5 7 9 10 8 6 4

Typeset by Westchester Book Group
Printed in the United States of America by Quad/Graphics Fairfield

To Hannah and Clara
It's in your hands now.

This generation has altered the composition of the atmosphere on a global scale through . . . a steady increase in carbon dioxide from the burning of fossil fuels.
—Lyndon Johnson
Special Message to Congress, 1965

The trouble with Americans is that they haven't read the minutes of the previous meeting.
—Adlai Stevenson

Contents

Introduction

Ben Santer is the kind of guy you could never imagine anyone attacking. He's thoroughly moderate—of moderate height and build, of moderate temperament, of moderate political persuasions. He is also very modest—soft-spoken, almost self-effacing—and from the small size and nonexistent décor of his office at the Lawrence Livermore National Laboratory, you might think he was an accountant. If you met him in a room with a lot of other people, you might not even notice him.

But Santer is no accountant, and the world has noticed him.

He's one of the world's most distinguished scientists—the recipient of a 1998 MacArthur "genius" award and numerous prizes and distinctions from his employer—the U.S. Department of Energy—because he has done more than just about anyone to prove the human causes of global warming. Ever since his graduate work in the mid-1980s, he has been trying to understand how the Earth's climate works, and whether we can say for sure that human activities are changing it. He has shown that the answer to that question is yes.

Santer is an atmospheric scientist at the Lawrence Livermore National Laboratory's Model Diagnosis and Intercomparison Project, an enormous international project to store the results of climate models from around the globe, distribute them to other researchers, and compare the models, both with real-world data and with each other. Over the past twenty years, he and his colleagues have shown that our planet *is* warming—and in just the way you would expect if greenhouse gases were the cause.

Santer's work is called "fingerprinting"—because natural climate variation leaves different patterns and traces than warming caused by greenhouse gases. Santer looks for these fingerprints. The most important one involves two parts of our atmosphere: the troposphere, the warm blanket closest to the Earth's surface, and the stratosphere, the thinner, colder part above it. Physics tells us that if the Sun were causing global warming—as some skeptics continue to insist—we'd expect both the troposphere and the stratosphere to warm, as heat comes into the atmosphere from outer space. But if the warming is caused by greenhouse gases emitted at the surface and largely trapped in the lower atmosphere, then we expect the troposphere to warm, but the stratosphere to cool.

Santer and his colleagues have shown that the troposphere is warming and the stratosphere is cooling. In fact, because the boundary between these two atmospheric layers is in part *defined* by temperature, that boundary is now moving upward. In other words, the whole structure of our atmosphere is changing. These results are impossible to explain if the Sun were the culprit. It shows that the changes we are seeing in our climate are not natural.

The distinction between the troposphere and the stratosphere became part of the Supreme Court hearing in the case of *Massachusetts et al. v. the EPA,* in which twelve states sued the federal government for failing to regulate carbon dioxide as a pollutant under the Clean Air Act. Justice Antonin Scalia dissented, arguing that there was nothing in the law to require the EPA to act—but the honorable justice also got lost in the science, at one point referring to the stratosphere when he meant the troposphere. A lawyer for Massachusetts replied, "Respectfully, Your Honor. It is not the stratosphere. It's the troposphere." The justice answered, "Troposphere, whatever. I told you before I'm not a scientist. That's why I don't want to deal with global warming."[1]

But we all have to deal with global warming, whether we like it or not, and some people have been resisting this conclusion for a long time. In fact, some people have been attacking not just the message, but the messenger. Ever since scientists first began to explain the evidence that our climate was warming—and that human activities were probably to blame—people have been questioning the data, doubting the evidence, and attacking the scientists who collect and explain it. And no one has been more brutally—or more unfairly—attacked than Ben Santer.

THE INTERGOVERNMENTAL PANEL on Climate Change (IPCC) is the world's leading authority on climate issues. Established in 1988 by the World Meteorological Organization and the United Nations Environment Program, it was created in response to early warnings about global warming. Scientists had known for a long time that increased greenhouse gases from burning fossil fuels could cause climate change—they had explained this to Lyndon Johnson in 1965—but most thought that changes were far off in the future. It wasn't until the 1980s that scientists started to worry—to think that the future was perhaps almost here—and a few mavericks began to argue that anthropogenic climate change was actually already under way. So the IPCC was created to evaluate the evidence and consider what the impacts would be if the mavericks were right.

In 1995, the IPCC declared that the human impact on climate was now "discernible." This wasn't just a few individuals; by 1995 the IPCC had grown to include several hundred climate scientists from around the world. But *how* did they know that changes were under way, and how did they know they were caused by *us*? Those crucial questions were answered in *Climate Change 1995: The Science of Climate Change*, the Second Assessment Report issued by the IPCC. Chapter 8 of this report, "Detection of Climate Change and Attribution of Causes," summarized the evidence that global warming really was caused by greenhouse gases. Its author was Ben Santer.

Santer had impeccable scientific credentials, and he had never before been involved in even the suggestion of impropriety of any kind, but now a group of physicists tied to a think tank in Washington, D.C., accused him of doctoring the report to make the science seem firmer than it really was. They wrote reports accusing him of "scientific cleansing"—expunging the views of those who did not agree.[2] They wrote reports with titles like "Greenhouse Debate Continued" and "Doctoring the Documents," published in places like *Energy Daily* and *Investor's Business Daily*. They wrote letters to congressmen, to officials in the Department of Energy, and to the editors of scientific journals, spreading the accusations high and wide. They pressured contacts in the Energy Department to get Santer fired from his job. Most public—and most publicized—was an op-ed piece published in the *Wall Street Journal*, accusing Santer of making the alleged changes to "deceive policy makers and the public."[3] Santer *had* made changes to the report, but not to deceive anyone. The changes were made in response to review comments from fellow scientists.

Every scientific paper and report has to go through the critical scrutiny of other experts: peer review. Scientific authors are required to take reviewers' comments and criticisms seriously, and to fix any mistakes that may have been found. It's a foundational ethic of scientific work: no claim can be considered valid—not even *potentially* valid—until it has passed peer review.

Peer review is also used to help authors make their arguments clearer, and the IPCC has an exceptionally extensive and inclusive peer review process. It involves both scientific experts and representatives of the governments of the participating nations to ensure not only that factual errors are caught and corrected, but as well that all judgments and interpretations are adequately documented and supported, and that all interested parties have a chance to be heard. Authors are required either to make changes in response to the review comments, or to explain why those comments are irrelevant, invalid, or just plain wrong. Santer had done just that. He had made changes in response to peer review. He had done what the IPCC rules required him to do. He had done what *science* requires him to do. Santer was being attacked for being a good scientist.

Santer tried to defend himself in a letter to the editor of the *Wall Street Journal*—a letter that was signed by twenty-nine co-authors, distinguished scientists all, including the director of the U.S. Global Change Research Program.[4] The American Meteorological Society penned an open letter to Santer affirming that the attacks were entirely without merit.[5] Bert Bolin, the founder and chairman of the IPCC, corroborated Santer's account in a letter of his own to the *Journal*, pointing out that accusations were flying without a shred of evidence, and that the accusers had not contacted him, nor any IPCC officers, nor any of the scientists involved to check their facts. Had they "simply taken the time to familiarize [themselves] with IPCC rules of procedure," he noted, they would have readily found out that no rules were violated, no procedures were transgressed, and nothing wrong had happened.[6] As later commentators have pointed out, no IPCC member nation ever seconded the complaint.[7]

But the *Journal* only published a portion of both Santer and Bolin's letters, and two weeks later, they gave the accusers yet another opportunity to sling mud, publishing a letter declaring that the IPCC report had been "tampered with for political purposes."[8] The mud stuck, and the charges were widely echoed by industry groups, business-oriented newspapers and magazines, and think tanks. They remain on the Internet today. If you Google "Santer IPCC," you get not the chapter in question—much less the whole IPCC report—but instead a variety of sites that repeat the 1995 accusations.[9] One site even asserts (falsely) that Santer admitted that he had "adjusted the data to make it fit with political policy," as if the U.S. government even *had* a climate policy to adjust the data to fit. (We didn't in 1995, and we still don't.)[10]

The experience was bitter for Santer, who spent enormous amounts of time and energy defending his scientific reputation and integrity, as well as trying to hold his marriage together through it all. (He didn't.) Today, this normally mild-mannered man turns white with rage when he recalls these events. Because no scientist starts his or her career expecting things like this to happen.

Why didn't Santer's accusers bother to find out the facts? Why did they continue to repeat charges long after they had been shown to be unfounded? The answer, of course, is that they were not interested in finding facts. They were interested in fighting them.

A FEW YEARS later, Santer was reading the morning paper and came across an article describing how some scientists had participated in a program, organized by the tobacco industry, to discredit scientific evidence linking tobacco to cancer. The idea, the article explained, was to "keep the controversy alive."[11] So long as there was doubt about the causal link, the tobacco industry would be safe from litigation and regulation. Santer thought the story seemed eerily familiar.

He was right. But there was more. Not only were the tactics the same, the people were the same, too. The leaders of the attack on him were two retired physicists, both named Fred: Frederick Seitz and S. (Siegfried) Fred Singer. Seitz was a solid-state physicist who had risen to prominence during World War II, when he helped to build the atomic bomb; later he became president of the U.S. National Academy of Sciences. Singer was a physicist—in fact, the proverbial rocket scientist—who became a leading figure in the development of Earth observation satellites, serving as the first director of the National Weather Satellite Service and later as chief scientist at the Department of Transportation in the Reagan administration.[12]

Both were extremely hawkish, having believed passionately in the gravity of the Soviet threat and the need to defend the United States from it with high-tech weaponry. Both were associated with a conservative think tank in Washington, D.C., the George C. Marshall Institute, founded to defend Ronald Reagan's Strategic Defense Initiative (SDI or "Star Wars"). And both had previously worked for the tobacco industry, helping to cast doubt on the scientific evidence linking smoking to death.

From 1979 to 1985, Fred Seitz directed a program for R. J. Reynolds Tobacco Company that distributed $45 million to scientists around the country for biomedical research that could generate evidence and cultivate experts to be used in court to defend the "product." In the mid-1990s, Fred Singer coauthored a major report attacking the U.S. Environmental Protection

Agency over the health risks of secondhand smoke. Several years earlier, the U.S. surgeon general had declared that secondhand smoke was hazardous not only to smokers' health, but to anyone exposed to it. Singer attacked this finding, claiming the work was rigged, and that the EPA review of the science—done by leading experts from around the country— was distorted by a political agenda to expand government control over all aspects of our lives. Singer's anti-EPA report was funded by a grant from the Tobacco Institute, channeled through a think tank, the Alexis de Tocqueville Institution.[13]

Millions of pages of documents released during tobacco litigation demonstrate these links. They show the crucial role that *scientists* played in sowing doubt about the links between smoking and health risks. These documents—which have scarcely been studied except by lawyers and a handful of academics—also show that the same strategy was applied not only to global warming, but to a laundry list of environmental and health concerns, including asbestos, secondhand smoke, acid rain, and the ozone hole.

Call it the "Tobacco Strategy." Its target was science, and so it relied heavily on scientists—with guidance from industry lawyers and public relations experts—willing to hold the rifle and pull the trigger. Among the multitude of documents we found in writing this book were *Bad Science: A Resource Book*—a how-to handbook for fact fighters, providing example after example of successful strategies for undermining science, and a list of experts with scientific credentials available to comment on any issue about which a think tank or corporation needed a negative sound bite.[14]

IN CASE AFTER CASE, Fred Singer, Fred Seitz, and a handful of other scientists joined forces with think tanks and private corporations to challenge scientific evidence on a host of contemporary issues. In the early years, much of the money for this effort came from the tobacco industry; in later years, it came from foundations, think tanks, and the fossil fuel industry. They claimed the link between smoking and cancer remained unproven. They insisted that scientists were mistaken about the risks and limitations of SDI. They argued that acid rain was caused by volcanoes, and so was the ozone hole. They charged that the Environmental Protection Agency had rigged the science surrounding secondhand smoke. Most recently—over the course of nearly two decades and against the face of mounting evidence—they dismissed the reality of global warming. First they claimed there was none, then they claimed it was just natural variation, and then they claimed that even if it was happening and it was our fault, it didn't matter because we could just adapt to it. In case after case, they steadfastly denied the existence of scientific agreement, even though they, themselves, were pretty much the only ones who disagreed.

A handful of men would have had no impact if no one paid any attention, but people did pay attention. By virtue of their earlier work in the Cold War weapons programs, these men were well-known and highly respected in Washington, D.C., and had access to power all the way to the White House. In 1989, to give just one example, Seitz and two other players in our story, physicists Robert Jastrow and William Nierenberg, wrote a report questioning the evidence of global warming.[15] They were soon invited to the White House to brief the Bush administration. One member of the Cabinet Affairs Office said of the report: "Everyone has read it. Everyone takes it seriously."[16]

It wasn't just the Bush administration that took these claims seriously; the mass media did, too. Respected media outlets such as the *New York Times*, the *Washington Post*, *Newsweek*, and many others repeated these claims as if they were a "side" in a scientific debate. Then the claims were repeated again and again and again—as in an echo chamber—by a wide range of people involved in public debate, from bloggers to members of the U.S. Senate, and even by the president and the vice president of the United States. In all of this, journalists and the public never understood that these were *not* scientific debates—taking place in the halls of science among active scientific researchers—but misinformation, part of a larger pattern that began with tobacco.

This book tells the story of the Tobacco Strategy, and how it was used to attack science and scientists, and to confuse us about major, important issues affecting our lives—and the planet we live on. Sadly, Ben Santer's story is not unique. When scientific evidence mounted on the depletion of stratospheric ozone, Fred Singer challenged Sherwood Rowland—the Nobel laureate and president of the American Association for the Advancement of Science who first realized that certain chemicals (CFCs) could destroy stratospheric ozone. When a graduate student named Justin Lancaster tried to set the record straight on Roger Revelle's views in the face of the claim that Revelle had changed his mind about global warming, he became the defendant in a libel lawsuit. (Lacking funds to defend himself, Lancaster was forced to settle out of court, leaving both his personal and professional life in tatters.)[17]

Fred Seitz and Fred Singer, both physicists, were the most prominent and persistent scientists involved in these campaigns. William Nierenberg and Robert Jastrow were physicists, too. Nierenberg was a one-time director of the distinguished Scripps Institution of Oceanography and member of Ronald Reagan's transition team, helping to suggest scientists to serve in important positions in the administration. Like Seitz, he had helped to build the atomic bomb, and later was associated with several Cold War weapons programs and laboratories. Jastrow was a prominent astrophysicist, successful popular author, and director of the Goddard Institute for Space Studies, who had long been involved with the U.S. space program. These men had no particular expertise in environmental or health questions, but they did have power and influence.

Seitz, Singer, Nierenberg, and Jastrow had all served in high levels of science administration, where they had come to know admirals and generals, congressmen and senators, even presidents. They had also dealt extensively with the media, so they knew how to get press coverage for their views, and how to pressure the media when they didn't. They used their scientific credentials to present themselves as authorities, and they used their authority to try to discredit any science they didn't like.

OVER THE COURSE of more than twenty years, these men did almost no original scientific research on any of the issues on which they weighed in. Once they had been prominent researchers, but by the time they turned to the topics of our story, they were mostly attacking the work and the reputations of others. In fact, on every issue, they were on the wrong side of the scientific consensus. Smoking does kill—both directly and indirectly. Pollution does cause acid rain. Volcanoes are not the cause of the ozone hole. Our seas are rising and our glaciers are melting because of the mounting effects of greenhouse gases in the atmosphere, produced by burning fossil fuels. Yet, for years the press quoted these men as experts, and politicians listened to them, using their claims as justification for inaction. President George H. W. Bush once even referred to them as "my scientists."[18] Although the situation is now a bit better, their views and arguments continue to be cited on the Internet, on talk radio, and even by members of the U.S. Congress.[19]

Why would scientists dedicated to uncovering the truth about the natural world deliberately misrepresent the work of their own colleagues? Why would they spread accusations with no basis? Why would they refuse to correct their arguments once they had been shown to be incorrect? And why did the press continue to quote them, year after year, even as their claims were shown, one after another, to be false? This is the story we are about to tell. It is a story about a group of scientists who fought the scientific evidence and spread confusion on many of the most important issues of our time. It is a story about a pattern that continues today. A story about fighting facts, and merchandising doubt.

CHAPTER 1

Doubt Is Our Product

O N MAY 9, 1979, A GROUP OF tobacco industry executives gathered to hear about an important new program. They had been invited by Colin H. Stokes, the former chairman of R. J. Reynolds, a company famous for pioneering marketing, including the first cigarette advertisements on radio and television ("I'd walk a mile for a Camel"). In later years, Reynolds would be found guilty of violating federal law by appealing to children with the character Joe Camel (which the Federal Trade Commission compared to Mickey Mouse), but the executives had not come to hear about products or marketing. They had come to hear about science. The star of the evening was not Stokes, but an elderly, balding, bespectacled physicist named Frederick Seitz.

Seitz was one of America's most distinguished scientists. A wunderkind who had helped to build the atomic bomb, Seitz had spent his career at the highest levels of American science: a science advisor to NATO in the 1950s, president of the National Academy of Sciences in the 1960s, president of the Rockefeller University—America's leading biomedical research institution—in the 1970s. In 1979, Seitz had just retired, and he was there to talk about one last job: a new program, which he would run on behalf of R. J. Reynolds, to fund biomedical research at major universities, hospitals, and research institutes across the country.

The focus of the new program was degenerative diseases—cancer, heart disease, emphysema, diabetes—the leading causes of death in the United States. And the project was huge: $45 million would be spent over the next six years. The money would fund research at Harvard, the universities of

Connecticut, California, Colorado, Pennsylvania, and Washington, the Sloan-Kettering Institute, and, not surprisingly, the Rockefeller University.[1] A typical grant was $500,000 per year for six years—a very large amount of money for scientific research in those days.[2] The program would support twenty-six different research programs, plus six young investigators on "RJR research scholarships," in the areas of chronic degenerative disease, basic immunology, the effect of "lifestyle modes" on disease.[3]

Seitz's role was to choose which projects to fund, to supervise and monitor the research, and to report progress to R. J. Reynolds. To determine the project criteria—what types of projects to fund—he enlisted the help of two other prominent colleagues: James A. Shannon and Maclyn McCarty.

Shannon was a physician who pioneered the use of the antimalaria drug Atabrine during World War II. Atabrine was effective, but had lousy side effects; Shannon figured out how to deliver the drug without the sickening side effects, and then administered the program that delivered it to millions of troops throughout the South Pacific, saving thousands from sickness and death.[4] Later, as director of the National Institutes of Health from 1955 to 1968, he transformed the NIH by convincing Congress to allow them to offer grants to university and hospital researchers. Before that, NIH funds were spent internally; very little money was available to American hospitals and universities for biomedical research. Shannon's external grant program was wildly popular and successful, and so it grew—and grew. Eventually it produced the gargantuan granting system that is the core of the NIH today, propelling the United States to leadership in biomedical research. Yet, for all this, Shannon never won a Nobel Prize, a National Medal of Science, or even a Lasker Award—often said to be biology's next best thing to the Nobel Prize.

Maclyn McCarty similarly had a fabulously successful career without being fabulously recognized. Many people have heard of James Watson and Francis Crick, who won the Nobel Prize for deciphering the double helix structure of DNA, but Watson and Crick did not prove that DNA carried the genetic information in cells. That crucial first step had been done a decade earlier, in 1944, by three bacteriologists at the Rockefeller University—Oswald Avery, Maclyn McCarty, and Colin MacLeod. In an experiment with pneumonia bacteria, they showed that benign bacteria could be made virulent by injecting them with DNA from virulent strands. You could change the nature of an organism by altering its DNA—something we take for granted now, but a revolutionary idea in the 1940s.

Perhaps because Avery was a quiet man who didn't trumpet his discovery, or perhaps because World War II made it difficult to get attention for any discovery without immediate military relevance, Avery, McCarty, and McLeod got relatively little notice for their experiment. Still, all three had distinguished scientific careers and in 1994 McCarty won the Lasker Award. But in 1979, McCarty was definitely underappreciated.

So it is perhaps not surprising that when Shannon and McCarty helped Seitz to develop their criteria for judging proposals, they sought projects that took a different perspective from the mainstream, individuals with unusual or offbeat ideas, and young investigators in their "formative stages" who lacked federal support.[5] One funded study examined the impact of stress, therapeutic drugs, and food additives (like saccharin) on the immune system. Another explored the relation between "the emotional framework and the state of . . . the immune system . . . in a family of depressed patients." A third asked whether the "psychological attitude of a patient can play a significant role in determining the course of a disease."[6] Projects explored the genetic and dietary causes of atherosclerosis, possible viral causes of cancer, and details of drug metabolism and interactions.

Two scientists in particular caught Seitz's personal attention. One was Martin J. Cline, a professor at UCLA who was studying the lung's natural defense mechanisms and was on the verge of creating the first transgenic organism.[7] Another was Stanley B. Prusiner, the discoverer of prions—the folded proteins that cause mad cow disease—for which he later won the Nobel Prize in Physiology or Medicine.[8]

All of the chosen studies addressed legitimate scientific questions, some that mainstream medicine had neglected—like the role of emotions and stress in somatic disease. All the investigators were credentialed researchers at respected institutions.[9] Some of the work they were doing was pathbreaking. But was the purpose simply to advance science? Not exactly.

Various R. J. Reynolds documents discuss the purpose of Seitz's program. Some suggest that supporting research was an "obligation of corporate citizenship." Others note the company's desire to "contribute to the prevention and cure of diseases for which tobacco products have been blamed." Still others suggest that by using science to refute the case against tobacco, the industry could "remove the government's excuse" for imposing punitive taxes.[10] (In 1978, smokers paid over a billion and a half dollars in cigarette excise taxes in the United States and abroad—taxes that had been raised in part in response to the scientific evidence of its harms.)

But the principal goal, stressed by Stokes to his advisory board that day in May and repeated in scores of industry documents, was to develop "an extensive body of scientifically, well-grounded data useful in defending the industry against attacks."[11] No doubt some scientists declined the offer of industry funding, but others accepted it, presumably feeling that so long as they were able to do science, it didn't really matter who paid for it. If any shareholders were to ask why company funds were being used to support basic (as opposed to applied) science, they could be told that the expenditure was "fully justified on the basis of the support it provides for defending the tobacco industry against fundamental attacks on its business."[12] The goal was to fight science with science—or at least with the gaps and uncertainties in existing science, and with scientific research that could be used to deflect attention from the main event. Like the magician who waves his right hand to distract attention from what he is doing with his left, the tobacco industry would fund distracting research.

In a presentation to R. J. Reynolds' International Advisory Board, and reviewed by RJR's in-house legal counsel, Stokes explained it this way: The charges that tobacco was linked to lung cancer, hardening of the arteries, and carbon monoxide poisoning were unfounded. "Reynolds and other cigarette makers have reacted to these scientifically unproven claims by intensifying our funding of objective research into these matters."[13] This research was needed because the case against tobacco was far from proven.

"Science really knows little about the causes or development mechanisms of chronic degenerative diseases imputed to cigarettes," Stokes went on, "including lung cancer, emphysema, and cardiovascular disorders." Many of the attacks against smoking were based on studies that were either "incomplete or . . . relied on dubious methods or hypotheses and faulty interpretations." The new program would supply new data, new hypotheses, and new interpretations to develop "a strong body of scientific data or opinion in defense of the product."[14] Above all, it would supply witnesses.

By the late 1970s, scores of lawsuits had been filed claiming personal injury from smoking cigarettes, but the industry had successfully defended itself by using scientists as expert witnesses to testify that the smoking-cancer link was not unequivocal. They could do this by discussing research that focused on other "causes or development mechanisms of chronic degenerative diseases imputed to cigarettes."[15] The testimony would be particularly convincing if it were their *own* research. Experts could supply reasonable doubt, and who better to serve as an expert than an actual scientist?

The strategy had worked in the past, so there was no reason to think it would not continue to work in the future. "Due to favorable scientific testimony," Stokes boasted, "no plaintiff has ever collected a penny from any tobacco company in lawsuits claiming that smoking causes lung cancer or cardiovascular illness—even though one hundred and seventeen such cases have been brought since 1954."[16]

In later years, this would change, but in 1979 it was still true. No one had collected a penny from the tobacco industry, even though scientists had been certain of the tobacco-cancer link since the 1950s (and many had been convinced before that).[17] Every project Reynolds funded could potentially produce such a witness who could testify to causes of illness other than smoking. Prusiner's work, for example, suggested a disease mechanism that had nothing to do with external causes. A prion, Seitz explained, could "take over in such a way that it over-produces its own species of protein and . . . destroys the cell," in "the manner in which certain genes . . . can be stimulated to over-produce cell division and lead to cancer."[18] Cancer might just be cells gone wild.

Cline's research suggested the possibility of preventing cancer by strengthening the cell's natural defenses, which in turn suggested that cancer might just be a (natural) failure of those defenses. Many of the studies explored other causes of disease—stress, genetic inheritance, and the like—an entirely legitimate topic, but one that could also help distract attention from the industry's central problem: the overwhelming evidence that tobacco killed people. Tobacco caused cancer: that was a fact, and the industry knew it. So they looked for some way to deflect attention from it. Indeed, they had known it since the early 1950s, when the industry first began to use science to fight science, when the modern era of fighting facts began. Let us return, for a moment, to 1953.

DECEMBER 15, 1953, was a fateful day. A few months earlier, researchers at the Sloan-Kettering Institute in New York City had demonstrated that cigarette tar painted on the skin of mice caused fatal cancers.[19] This work had attracted an enormous amount of press attention: the *New York Times* and *Life* magazine had both covered it, and *Reader's Digest*—the most widely read publication in the world—ran a piece entitled "Cancer by the Carton."[20] Perhaps the journalists and editors were impressed by the scientific paper's dramatic concluding sentences: "Such studies, in view of the corollary clinical data relating smoking to various types of cancer, appear urgent. They may not only result in furthering our knowledge of carcinogens, but in promoting some practical aspects of cancer prevention."

These findings shouldn't have been a surprise. German scientists had shown in the 1930s that cigarette smoking caused lung cancer, and the Nazi government had run major antismoking campaigns; Adolf Hitler forbade smoking in his presence. However, the German scientific work was tainted by its Nazi associations, and to some extent ignored, if not actually suppressed, after the war; it had taken some time to be rediscovered and independently confirmed.[21] Now, however, American researchers—not Nazis—were calling the matter "urgent," and the news media were reporting it.[22] "Cancer by the carton" was not a slogan the tobacco industry would embrace.

The tobacco industry was thrown into panic. One industry memo noted that their salesmen were "frantically alarmed."[23] So industry executives made a fateful decision, one that would later become the basis on which a federal judge would find the industry guilty of conspiracy to commit fraud—a massive and ongoing fraud to deceive the American public about the health effects of smoking.[24] The decision was to hire a public relations firm to challenge the scientific evidence that smoking could kill you.

On that December morning, the presidents of four of America's largest tobacco companies—American Tobacco, Benson and Hedges, Philip Morris, and U.S. Tobacco—met at the venerable Plaza Hotel in New York City. The French Renaissance chateau-style building—in which unaccompanied ladies were not permitted in its famous Oak Room bar—was a fitting place for the task at hand: the protection of one of America's oldest and most powerful industries. The man they had come to meet was equally powerful: John Hill, founder and CEO of one of America's largest and most effective public relations firms, Hill and Knowlton.

The four company presidents—as well as the CEOs of R. J. Reynolds and Brown and Williamson—had agreed to cooperate on a public relations program to defend their product.[25] They would work together to convince the public that there was "no sound scientific basis for the charges," and that the recent reports were simply "sensational accusations" made by publicity-seeking scientists hoping to attract more funds for their research.[26] They would not sit idly by while their product was vilified; instead, they would create a Tobacco Industry Committee for Public Information to supply a "positive" and "entirely 'pro-cigarette'" message to counter the anti-cigarette scientific one. As the U.S. Department of Justice would later put it, they decided "to deceive the American public about the health effects of smoking."[27]

At first, the companies didn't think they needed to fund new scientific research, thinking it would be sufficient to "disseminate information on hand." John Hill disagreed, "emphatically warn[ing] . . . that they should . . .

sponsor additional research," and that this would be a long-term project.[28] He also suggested including the word "research" in the title of their new committee, because a pro-cigarette message would need science to back it up.[29] At the end of the day, Hill concluded, "scientific doubts must remain."[30] It would be his job to ensure it.

Over the next half century, the industry did what Hill and Knowlton advised. They created the "Tobacco Industry Research Committee" to challenge the mounting scientific evidence of the harms of tobacco. They funded alternative research to cast doubt on the tobacco-cancer link.[31] They conducted polls to gauge public opinion and used the results to guide campaigns to sway it. They distributed pamphlets and booklets to doctors, the media, policy makers, and the general public insisting there was no cause for alarm.

The industry's position was that there was "no proof" that tobacco was bad, and they fostered that position by manufacturing a "debate," convincing the mass media that responsible journalists had an obligation to present "both sides" of it. Representatives of the Tobacco Industry Research Committee met with staff at *Time, Newsweek, U.S. News and World Report, BusinessWeek, Life,* and *Reader's Digest,* including men and women at the very top of the American media industry. In the summer of 1954, industry spokesmen met with Arthur Hays Sulzburger, publisher of the *New York Times*; Helen Rogers Reid, chairwoman of the *New York Herald Tribune*; Jack Howard, president of Scripps Howard Newspapers; Roy Larsen, president of Luce Publications (owners of *Time* and *Life*); and William Randolph Hearst Jr. Their purpose was to "explain" the industry's commitment to "a long-range . . . research program devoted primarily to the public interest"— which was needed since the science was so unsettled—and to stress to the media their responsibility to provide a "balanced presentation of all the facts" to ensure the public was not needlessly frightened.[32]

The industry did not leave it to journalists to seek out "all the facts." They made sure they got them. The so-called balance campaign involved aggressive dissemination and promotion to editors and publishers of "information" that supported the industry's position. But if the science was firm, how could they do that? *Was* the science firm?

The answer is yes, but. A scientific discovery is not an event; it's a process, and often it takes time for the full picture to come into clear focus. By the late 1950s, mounting experimental and epidemiological data linked tobacco with cancer—which is why the industry took action to oppose it. In private, executives acknowledged this evidence.[33] In hindsight it is fair to say—and science historians *have* said—that the link was already established beyond a reasonable doubt. Certainly no one could honestly say that science showed that smoking was safe.

But science involves many details, many of which remained unclear, such as why some smokers get lung cancer and others do not (a question that remains incompletely answered today). So some scientists remained skeptical. One of them was Dr. Clarence Cook Little.

C. C. Little was a renowned geneticist, a member of the U.S. National Academy of Sciences and former president of the University of Michigan.[34] But he was also well outside the mainstream of scientific thinking. In the 1930s, Little had been a strong supporter of eugenics—the idea that society should actively improve its gene pool by encouraging breeding by the "fit" and discouraging or preventing breeding by the "unfit." His views were not particularly unusual in the 1920s—they were shared by many scientists and politicians including President Theodore Roosevelt—but nearly everyone abandoned eugenics in the '40s when the Nazis made manifest where that sort of thinking could lead. Little, however, remained convinced that essentially all human traits were genetically based, including vulnerability to cancer. For him, the cause of cancer was genetic weakness, not smoking.

In 1954, the tobacco industry hired Little to head the Tobacco Industry Research Committee and spearhead the effort to foster the impression of debate, primarily by promoting the work of scientists whose views might be useful to the industry. One of these scientists was Wilhelm C. Hueper, chief of the Environmental Cancer Section at the National Cancer Institute. Hueper had been a frequent expert witness in asbestos litigation where he sometimes had to respond to accusations that a plaintiff's illnesses were caused not by asbestos, but by smoking. Perhaps for this reason, Hueper prepared a talk questioning the tobacco-cancer link for a meeting in São Paulo, Brazil. When the Tobacco Industry Research Committee learned about it, they contacted Hueper, who agreed to allow them to promote his work. Hill and Knowlton prepared and delivered a press release, with copies of Hueper's talk, to newspapers offices, wire services, and science and editorial writers around the country. They later reported that "as a result of the distribution [of the press release] in the U.S.A., stories questioning a link between smoking and cancer were given wide attention, both in headlines and stories."[35] *U.S. News and World Report* practically gushed, "Cigarettes are now gaining support from new studies at the National Cancer Institute."[36]

Little's committee prepared a booklet, *A Scientific Perspective on the Cigarette Controversy*, which was sent to 176,800 American doctors.[37] Fifteen thousand additional copies were sent to editors, reporters, columnists, and members of Congress. A poll conducted two years later showed that "neither the press nor the public seems to be reacting with any noticeable fear or alarm to the recent attacks."[38]

The industry made its case in part by cherry-picking data and focusing on unexplained or anomalous details. No one in 1954 would have claimed that everything that needed to be known about smoking and cancer was known, and the industry exploited this normal scientific honesty to spin unreasonable doubt. One Hill and Knowlton document, for example, prepared shortly after John Hill's meeting with the executives, enumerated fifteen scientific questions related to the hazards of tobacco.[39] Experiments showed that laboratory mice got skin cancer when painted with tobacco tar, but not when left in smoke-filled chambers. Why? Why do cancer rates vary greatly between cities even when smoking rates are similar? Do other environmental changes, such as increased air pollution, correlate with lung cancer? Why is the recent rise in lung cancer greatest in men, even though the rise in cigarette use was greatest in women? If smoking causes lung cancer, why aren't cancers of the lips, tongue, or throat on the rise? Why does Britain have a lung cancer rate four times higher than the United States? Does climate affect cancer? Do the casings placed on American cigarettes (but not British ones) somehow serve as an antidote to the deleterious effect of tobacco? How much is the increase in cancer simply due to longer life expectancy and improved accuracy in diagnosis?[40]

None of the questions was illegitimate, but they were all disingenuous, because the answers were known: Cancer rates vary between cities and countries because smoking is not the only cause of cancer. The greater rise in cancer in men is the result of latency—lung cancer appears ten, twenty, or thirty years after a person begins to smoke—so women, who had only recently begun to smoke heavily, would get cancer in due course (which they did). Improved diagnosis explained some of the observed increase, but not all: lung cancer was an exceptionally rare disease before the invention of the mass-marketed cigarette. And so on.

When posed to journalists, however, the loaded questions did the trick: they convinced people who didn't know otherwise that there was still a lot of doubt about the whole matter. The industry had realized that you could create the impression of controversy simply by asking questions, even if you actually knew the answers and they didn't help your case.[41] And so the industry began to transmogrify emerging scientific consensus into raging scientific "debate."[42]

The appeal to journalistic balance (as well as perhaps the industry's large advertising budget) evidently resonated with writers and editors, perhaps because of the influence of the Fairness Doctrine. Under this doctrine, established in 1949 (in conjunction with the rise of television), broadcast journalists were required to dedicate airtime to controversial issues of public concern in a balanced manner.[43] (The logic was that broadcasts licenses were a scarce resource, and therefore a public trust.) While the doctrine did not formally apply to print journalism, many writers and editors seem to have applied it to the tobacco question, because throughout the 1950s and well into the 1960s, newspapers and magazines presented the smoking issue as a great debate rather than as a scientific problem in which evidence was rapidly accumulating, a clear picture was coming into focus, and the trajectory of knowledge was clearly against tobacco's safety.[44] Balance was interpreted, it seems, as giving equal weight to both sides, rather than giving *accurate* weight to both sides.

Even the great Edward R. Murrow fell victim to these tactics. In 1956, Hill and Knowlton reported on a conference held with Murrow, his staff, and their producer, Fred Friendly:

> The Murrow staff emphasized the intention to present a coldly objective program with every effort made to tell the story as it stands today, with *special effort toward a balanced perspective and concrete steps to show that the facts still are not established and must be sought by scientific means such as the research activities the Tobacco Industry Research Committee will support.*[45]

Balance. Cold objectivity. These were Murrow's trademarks—along with his dangling cigarette—and the tobacco industry exploited them both. Murrow's later death from lung cancer was both tragic and ironic, for during World War II Murrow had been an articulate opponent of meretricious balance in reporting. As David Halberstam has put it, Murrow was not ashamed to take the side of democracy, and felt no need to try to get the Nazi perspective or consider how isolationists felt. There was no need to "balance Hitler against Churchill."[46]

Yet Murrow fell prey to the tobacco industry's insistence that their self-interested views should be balanced against independent science. Perhaps, being a smoker, he was reluctant to admit that his daily habit was deadly and reassured to hear that the allegations were unproven. Roger Ferger, publisher of the *Cincinnati Enquirer*, evidently felt that way, as he wrote a bread-and-butter note for his copy of the *Scientific Perspective* pamphlet: "I have been a smoker for some forty-five years and I am still a pretty healthy specimen."[47] It was certainly comforting to be told that the jury was still out.

Editors, however, might eventually be expected to notice if the only support for industry claims came from obscure conferences in Brazil. No doubt realizing this, the industry sought links with mainstream medicine, funding research projects at leading medical schools related to cancer pathology, diagnosis, and distribution, and potentially related diseases such as coronary heart disease. In 1955, the industry established a fellowship program to support research by medical degree candidates: seventy-seven of seventy-nine medical schools agreed to participate.[48] (Industry documents don't tell which two declined; perhaps they were affiliated with religious denominations that eschewed smoking.) The industry also sought to develop good relations with members of the National Cancer Institute and American Heart Association by inviting their representatives to board meetings.[49] Building on his success, in 1957 the Tobacco Industry Research Committee published 350,000 copies of a new pamphlet, *Smoking and Health*, mostly sent to doctors and dentists.[50]

By the end of the 1950s, the tobacco industry had successfully developed ties with doctors, medical school faculty, and public health authorities across the country. In 1962, when U.S. Surgeon General Luther L. Terry established an Advisory Committee on Smoking and Health, the tobacco industry made nominations, submitted information, and ensured that Dr. Little "established lines of communication" with the committee.[51] To ensure that the panel was "democratically" constituted, the surgeon general invited nominations from the tobacco industry, as well as from the Federal Trade Commission (who would become involved if restrictions were placed on tobacco advertising). To ensure that the panel was unbiased, he excluded anyone who had publicly expressed a prior opinion. One hundred and fifty names were put forward, and the tobacco industry was permitted to veto anyone they considered unsuitable.[52]

Despite these concessions, the 1964 report was not favorable to the tobacco industry.[53] Historian Allan Brandt recounts how half the members of the panel were smokers, and by the time their report was ready, most of them had quit.[54] For those close to the science, this was no surprise, because the evidence against smoking had been steadily mounting. In 1957, the U.S. Public Health Service had concluded that smoking was "the principal etiological factor in the increased incidence of lung cancer."[55] In 1959, leading researchers had declared in the peer-reviewed scientific literature that the evidence linking cigarettes and cancer was "beyond dispute."[56] That same year, the American Cancer Society had issued a formal statement declaring that "cigarette smoking is the major causative factor in lung cancer."[57] In 1962, the Royal College of Physicians of London had declared that "cigarette smoking is a cause of cancer and bronchitis and probably contributes to . . . coronary heart disease," a finding that was prominently reported in *Reader's Digest* and *Scientific American*. Perhaps most revealingly, the tobacco industry's own scientists had come to the same conclusion.

As University of California professor Stanton Glantz and his colleagues have shown in their exhaustive reading of tobacco industry documents, by the early 1960s the industry's own scientists had concluded not only that smoking caused cancer, but also that nicotine was addictive (a conclusion that mainstream scientists came to only in the 1980s, and the industry would continue to deny well into the 1990s).[58] In the 1950s, manufacturers had advertised some brands as "better for your health," implicitly acknowledging health concerns.[59] In the early 1960s, Brown and Williamson's in-house scientists conducted their own experiments demonstrating that tobacco smoke caused cancer in laboratory animals, as well as experiments showing the addictive properties of nicotine. In 1963, the vice president of Brown and Williamson concluded, presumably with reluctance, "We are, then, in the business of selling nicotine, an additive drug." Two years later, the head of research and development for Brown and Williamson noted that industry scientists were "unanimous in their opinion that smoke is . . . carcinogenic."[60] Some companies began secretly working on a "safe" cigarette, even while the industry as a whole was publicly denying that one was needed.

It's one thing for scientists to report something in peer-reviewed journals, however, and another for the country's doctor in chief to announce it publicly, loud and clear. The 1964 surgeon general's report, *Smoking and Health*, did just that. Based on review of more than seven thousand scientific studies and testimony of over one hundred and fifty consultants, the landmark report was written by a committee—in this case selected from nominations provided by the U.S. Food and Drug Administration, the Federal Trade Commission, the American Medical Association, and the Tobacco

Institute—but its conclusions were unanimous.[61] Lung cancer in the twentieth century had reached epidemic proportions, and the principal cause was not air pollution, radioactivity, or exposure to asbestos. It was to-bacco smoking. Smokers were ten to twenty times more likely to get lung cancer than nonsmokers. They were also more likely to suffer from em-physema, bronchitis, and heart disease. The more a person smoked, the worse the effects.

Terry realized that the report's release would be explosive, so when he gathered two hundred reporters into the State Department for a two-hour briefing, the auditorium doors were locked for security.[62] The report was released on a Saturday to minimize impact on the stock market, but it was still a bombshell. Nearly half of all adult Americans smoked—many men had picked up the habit while serving their country during World War II or in Korea—and the surgeon general was telling them that this pleasur-able habit, at worst a mild vice, was killing them. The government not only allowed this killing, but promoted and profited from it: the federal govern-ment subsidized tobacco farming, and tobacco sales were an enormous source of both federal and state tax revenues. To argue that tobacco killed people was to suggest that our own government both sanctioned and prof-ited from the sale of a deadly product. In hindsight, calling it the biggest news story of 1964 seems insufficient; it was one of the biggest news sto-ries of the era.[63] One tobacco industry PR director concluded that the ciga-rette business was now in a "grave crisis."[64] They did not sit idly by.

Immediately, they redoubled their effort to challenge the science. They changed the name of the Tobacco Industry Research Council to the Coun-cil for Tobacco Research (losing the word "industry" entirely), and severed their relations with Hill and Knowlton. They resolved that the new organi-zation would be wholly dedicated to health research, and not to "industry technical or commercial studies."[65] They "refined" the approval and review process for grants, intensifying their search for "experts" who would af-firm their views.

Given the evidence produced in their own laboratories, the industry might have concluded that the "debate" game was up. The PR director for Brown and Williamson suggested that perhaps the time had come to back off "assurances, denial of harm, and similar claims."[66] Others suggested identifying the hazardous components in cigarette smoke and trying to re-move them, or adopting voluntary warning labels.[67] In 1978, the Liggitt Group—makers of L&Ms, Larks, and Chesterfields—filed a patent appli-cation for a technique to reduce the "tumorigenicity" of tobacco. (Tumorigenicity is the tendency of something to generate tumors, so this was an implicit acknowledgement that tobacco did indeed cause tumors, as one newspaper realized.)[68]

The cigarette manufacturers did not give up. Rather, they resolved to fight harder. "A steady expansion in our program of scientific research into tobacco use and health has convinced us of the need for more permanent organizational machinery," one press release concluded. The industry had already given more than $7 million in research funds to 155 scientists at more than one hundred American medical schools, hospitals, and laboratories; now it would give even more.[69] When Congress held hearings in 1965 on bills to require health warnings on tobacco packages and advertisements, the tobacco industry responded with "a parade of dissenting doctors," and a "cancer specialist [who warned] against going off 'half cocked' in the controversy."[70]

Sometimes further research muddies scientific waters, as additional complications are uncovered or previously unrecognized factors are acknowledged. Not so with smoking. When a new surgeon general reviewed the evidence in 1967, the conclusions were even starker.[71] Two thousand more scientific studies pointed emphatically to three results, enumerated on the report's first page: One, smokers lived sicker and died sooner than their nonsmoking counterparts. Two, a substantial portion of these early deaths would not have occurred if these people had never smoked. Three, were it not for smoking "practically none" of the early deaths from lung cancer would have occurred. Smoking killed people. It was as simple as that. Nothing had been learned since 1964 that brought into question the conclusions of the earlier report.[72]

How did the industry respond to this? More denial. "There is no scientific evidence that cigarette smoking causes lung cancer and other disease," Brown and Williamson insisted.[73]

In 1969, when the Federal Communications Commission voted to ban cigarette advertising from television and radio, Clarence Little insisted that there was "no demonstrated causal relationship between smoking or [sic] any disease."[74] Publicly, the industry supported the advertising ban, because under the Fairness Doctrine health groups were getting free anti-smoking advertisements on television, and these were having an effect.[75] Privately, however, the Tobacco Research Council sent materials to the liquor industry suggesting that it would be the next target.[76] In fact, the FCC had disavowed any such intentions, declaring in their own press release, "Our action is limited to the unique situation and product; we . . . expressly

disclaim any intention to so proceed against other product[s]."[77] But the tobacco industry sought to foster the anxiety that controlling tobacco advertising was the first step down a slippery slope to controlling advertising of
all sensitive products.

Despite industry fears, the U.S. Congress did not ban or even limit sales
of tobacco, but it did require warning labels. The American people now
knew that smoking was dangerous. And the danger wasn't just cancer. A
host of ailments had been clearly linked to smoking: bronchitis, emphysema, coronary heart disease, hardening of the arteries, low birth weight in
infants, and many more. As the 1960s came to a close, the numbers of
Americans who smoked had declined significantly. By 1969, the number
of adult Americans who smoked was down to 37 percent. By 1979 it would
fall to 33 percent—among doctors it would fall to 21 percent—and the *New
York Times* would finally stop quoting tobacco industry spokesmen to provide "balance."[78]

While smoking had declined, industry profits had not. In 1969, R. J.
Reynolds reported net revenues of $2.25 billion. Despite the mounting polit-
ical pressure to control tobacco sales and discourage tobacco use, Reynolds's
directors reported records for sales, revenues, and earnings, and the contin-
uation of its seventy-year record of uninterrupted dividends to its stockhold-
ers. "Tobacco," they concluded, "remains a good business."[79] Protecting that
business—against regulation, punitive taxes, FDA control, and, especially,
lawsuits—became a growing concern.[80]

Although 125 lawsuits related to health impairment were filed against
the tobacco industry between 1954 and 1979, only nine went to trial, and
none were settled in favor of the plaintiffs.[81] Still, industry lawyers were increasingly concerned, in part because their insistence that the debate was
still open was contradicted not just by academic science, but by their own
internal company documents. To cite just one example: in 1978, the minutes from a British American Tobacco Company research conference concluded that the tobacco-cancer link "has long ceased to be an area for
scientific controversy."[82] (Brown and Williamson lawyers recommended
the destruction or removal of documents that spoke to this point.)[83]

How could the industry possibly defend itself when the vast majority of
independent experts agreed that tobacco was harmful, and their own documents showed that they knew this? The answer was to continue to market
doubt, and to do so by recruiting ever more prominent scientists to help.

Collectively the industry had already spent over $50 million on biomed-
ical research. Individual tobacco companies had invested millions more—

bringing the total to over $70 million. By the mid-1980s, that figure had exceeded $100 million. One industry document happily reported that "this expenditure exceeds that given for research by any other source except the federal government."[84] Another noted that grants had been distributed to 640 investigators in 250 hospitals, medical schools, and research institutions.[85] The American Cancer Society and American Lung Association in 1981 devoted just under $300,000 to research; that same year, the tobacco industry gave $6.3 million.[86] It was time to do even more.

In the 1950s, the tobacco industry had enlisted geneticist C. C. Little— a member of the U.S. National Academy of Sciences—to lend credibility to their position. This time they went one step better: they enlisted Dr. Frederick Seitz—the balding man introduced to Reynolds executives in 1979—a former *president* of the Academy.[87]

Seitz was part of the generation of bright young men whose lives were transformed by the Manhattan Project, catapulted into positions of power and influence on the basis of brainpower. Before World War II, physics was a fairly obscure discipline; nobody expected to become rich, famous, or powerful through a career in physics. But the atomic bomb changed all that, as hundreds of physicists were recruited by the U.S. government to build the most powerful weapon ever known. After the war, many of these physicists were recruited to build major academic departments at elite universities, where they frequently also served as consultants to the U.S. government on all kinds of issues—not just weapons.

Seitz's link to the atomic bomb was even closer than most. A solid-state physicist, he had trained under Eugene Wigner at Princeton, the man who, along with colleague Leo Szilard, convinced Albert Einstein to send his famous letter to Franklin Roosevelt urging him to build the atomic bomb. Later, Wigner won the Nobel Prize for work in nuclear physics; Seitz was Wigner's best and most famous student.

From 1939 to 1945, Seitz had worked on a variety of projects related to the war effort, including ballistics, armor penetration, metal corrosion, radar, and the atomic bomb. He also managed to complete a textbook published in 1940, *The Modern Theory of Solids*—widely acknowledged as the definitive textbook of its day on solid-state physics—and a second volume, *The Physics of Metals*, in 1943. He also found time to consult for the DuPont Corporation.

In 1959, Seitz became science advisor to NATO and from there moved into the highest echelons of American science and policy. From 1962 to 1969, he served as president of the National Academy of Sciences and as ex officio member of the U.S. President's Science Advisory Committee. In 1973, he received the National Medal of Science from President Richard Nixon. As Academy president, he developed an interest in biology, and in 1968 became president of the Rockefeller University—American's preeminent biomedical research center. In 1979 he went to work for R. J. Reynolds.

It's obvious why R. J. Reynolds would have wanted a man of Seitz's credentials on their team, but why would Seitz want to work for R. J. Reynolds?[88] Speaking to the industry executives in 1979, Seitz stressed the debt of gratitude he felt to Reynolds for the funding they had supplied his institution. Rockefeller was one of the universities that the tobacco industry had long funded, and Seitz put it this way:

> About a year ago, when my period as President of the Rockefeller University was nearing its end, [I was] asked if I would be willing to serve as advisor to the Board of Directors of R. J. Reynolds Industries, as it developed its program on the support of biomedical research related to degenerative diseases in man—a program which would enlarge upon the work supported through the consortium of tobacco industries. Since . . . R. J. Reynolds had provided very generous support for the biomedical work at The Rockefeller University, I was more than glad to accept.[89]

Reynolds *had* been generous to Rockefeller. In 1975 they had established the R. J. Reynolds Fund for the Biomedical Sciences and Clinical Research, with a grant of $500,000 per year for five years, with an additional $300,000 in year one to endow the R. J. Reynolds Industries Postdoctoral Fellowship "to make possible permanent recognition of RJR's assistance."[90]

There was a bit more to it than gratitude. Seitz also harbored an enormous grudge against the scientific community that he once led. Over the years, Seitz had come to view the scientific community as fickle, even irrational. As president of the National Academy, he had become "keenly aware how quickly, and irrationally, the mood of the membership of an organization can change. I could become highly unpopular almost overnight because of some seemingly trivial issue."[91]

Seitz was particularly unpopular for his support of the Vietnam War, which increasingly isolated him from colleagues on the President's Science Advisory Committee, who by the early 1970s had concluded not only that the war was a morass, but that they, like the rest of America, had been lied to about its progress.[92] As the 1970s drew to a close, Seitz also parted company with scientific colleagues on questions of nuclear preparedness. The scientific community generally supported arms limitations talks and treaties, and rejected as impossible the idea of achieving permanent technology superiority. Seitz, on the other hand, was committed to a muscular military strengthened by the most technologically advanced weaponry. He never rejected the idea of achieving American political superiority through superior weaponry, an idea that most colleagues had abandoned, but which would continue to crop up and cause conflict in the 1980s.

Above all, Seitz, like his mentor Eugene Wigner (a Hungarian refugee), was ardently anti-Communist. (Wigner in later years lent his support to Reverend Sun Myung Moon's Unification Church, evidently feeling that any enemy of Communism was his friend.)[93] Seitz's support for aggressive weapons programs was a reflection of this anti-Communism, but the feeling went further. As president of the Academy, Seitz had been a strong supporter of Taiwan, developing exchange programs with Taiwanese scientists as a counterbalance to the influence of "red" China. Exchange programs with Taiwanese scientists was an idea that most colleagues found reasonable enough, but in later years Seitz's anti-Communism would seem to lose a sense of proportion, as he increasingly defended anything that private enterprise did, and attacked anything with the scent of Socialism.[94]

Seitz justified his increasing social and intellectual isolation by blaming others. American science had become "rigid," he insisted, his colleagues dogmatic and closed-minded. The growing competition for federal funds stifled creativity, and discouraged work that didn't fall into clean disciplinary categories. This, perhaps, was the most important basis for his connection with the tobacco industry, as he explained in a presentation to Reynolds's International Advisory Board: "From time to time, [there are] exceptional cases where the ever-growing rigidity of the support provided by the federal government excludes the support of an important program in the hands of a distinguished and imaginative investigator."[95] Seitz would welcome the role of being the arbiter of who these distinguished and imaginative investigators were, and his judgment was not necessarily bad. Witness his support for Stanley Prusiner.

Seitz, however, did not simply want to support creative science. He was also angry at what he saw as an increasingly antiscience and antitechnology attitude in American life. He accepted the industry argument that attacks on the use of tobacco were "irrational," and that "independent" science was needed to "sift truth from fiction" (although independent from whom was never made clear).[96] Seitz saw irrationality everywhere, from the attack on tobacco to the "attempt to lay much of the blame for cancer upon industrialization."[97] After all, the natural environment was hardly carcinogen-free, he noted, and even "the oxygen in the air we breathe . . . plays a role in radiation-induced cancer."[98] (Oxygen, like most elements, has a radioactive version—oxygen-15—although it is not naturally occurring.)[99]

Seitz believed passionately in science and technology, both as the cause of modern health and wealth and the only means for future improvements, and it infuriated him that others didn't see it his way. In his memoir, he confidently proclaimed his faith in technology, insisting that "technology is continuously devising procedures to protect our health and safety and the natural beauty and resources of our world."[100]

While in his own mind a staunch defender of democracy, Seitz had an uneasy time with the masses. Environmentalists, he felt, were Luddites who wanted to reverse progress. His academic colleagues were ingrates who failed to appreciate what science and technology had done for them. Democracy as a whole had an uncertain relation to science, Seitz noted, and higher culture in general. Popular culture was a morass—Seitz despised Hollywood—and he wondered with more than a trace of bitterness whether the "culminating struggle to create free and open societies" would culminate in the "triumph of the ordinary." Seitz did not help build the atomic bomb to make the world safe for action-adventure films.[101]

These attitudes all help to explain how and why Seitz would have been willing to work for the tobacco industry. And there is one more important piece of the puzzle. Like C. C. Little before him, Seitz was something of a genetic determinist (perhaps because he was loath to admit that environmental hazards related to technology might cause serious health harms, or perhaps because he just saw the science that way). In his memoir, he attributed the early death of his friend William Webster Hansen, co-inventor of the klystron (important in the development of radar) to "a genetic defect leading to emphysema," but this interpretation is highly unlikely.[102]

Medical experts believe that emphysema is almost invariably caused by environmental assaults. The Aetna insurance company concludes that up to 90 percent of cases are caused by smoking and most of the rest to other airborne toxins; only 1 percent of cases are attributable to a rare genetic defect.[103] Hansen's case was strange, because he died so young—only thirty-nine—so perhaps he did have a genetic defect, but his disease could also have been caused by inhaling the beryllium he used in his research.[104]

Beryllium—a heavy metal—is well-known to be extraordinarily toxic; in later years the U.S. federal government would compensate workers exposed to beryllium in the nation's nuclear weapons programs.[105] Seitz clearly had trouble accepting that Hansen's exposure to beryllium could have been the cause of his early death.[106]

Given these various views—hawkish, superior, technophilic, and communophobic—Seitz may well have felt more comfortable in the company of conservative men from the tobacco industry (who perhaps shared his political views) than with his mostly liberal academic colleagues (who generally did not). Over the years, he had spent a good deal of time in corporate America, first as a physicist at General Electric in the 1930s, and then, for thirty-five years during his academic career, as a consultant to DuPont. He was also a member of the Bohemian Grove, an exclusive men's club in San Francisco, which in those days counted among its members Secretary of Defense Caspar Weinberger, as well as many executives of California banks, oil companies, and military-industrial contractors. (One former president of Caltech recalls that he joined Bohemian Grove because the trustees of his institution insisted it was important, but as a liberal and a Jew he never felt comfortable there.)[107]

Seitz no doubt also enjoyed the perks he received while working for the tobacco industry, such as flying to Bermuda with his wife when the Reynolds Advisory Committee met there in November 1979, as well as the heady feeling of distributing money to researchers that he had handpicked.[108] Given his views that genetic weakness was the crux of disease susceptibility, and that modern science had become narrow-minded, Seitz may well have honestly believed that tobacco was being unfairly attacked, and that Reynolds money could do some real good. But we know from tobacco industry documents that the criteria by which he chose projects for funding were not purely scientific.

By May 1979, Seitz had made commitments for over $43.4 million in research grants. During this time, he corresponded frequently with H. C. Roemer—R. J. Reynolds's legal counsel—discussing with him which particular projects they planned to fund and why; all press releases regarding the research program had to be cleared by the legal department.[109] It's not normal for granting agencies to consult legal counsel on each and every grant they make, so this connection alone might suggest a criterion related to legal liability. But we don't have to speculate, because industry documents tell us so: "Support [for scientific research] over the years has produced a number of authorities upon whom the industry could draw for expert testimony in court suits and hearings by governmental bodies."[110] The industry wasn't just generating reasonable doubt; it was creating friendly witnesses—witnesses that could be called on in the future.

One of these witnesses was Martin J. Cline, who had earlier caught Seitz's attention. Cline was one of the most famous biomedical researchers in the United States. Chief of the Division of Hematology-Oncology at UCLA's medical school, he had created the world's first transgenic organism: a genetically modified mouse. In 1980, however, he was censured by UCLA and the National Institutes of Health for an unapproved human experiment injecting bone marrow cells that had been altered with recombinant DNA into two patients with a hereditary blood disorder.[111] Cline was found to have misrepresented the nature of the experiment to hospital authorities, telling them that the experiment did not involve recombinant DNA.[112] He later admitted that he had performed the experiments, but claimed that he did it because he believed it would work. Cline lost nearly $200,000 in research grants and was forced to resign his position as division chief, although he was permitted to stay on as a professor of medicine.[113]

Many years later—in 1997—Cline was deposed in the case of *Norma R. Broin et al. v. Philip Morris*.[114] (Broin was a nonsmoking flight attendant who contracted lung cancer at the age of thirty-two, and sued—along with her husband and twenty-five other flight attendants—charging that their illnesses were caused by secondhand smoke in airline cabins, and the tobacco industry had suppressed information about its hazards.)[115] In the deposition Cline acknowledged that he had been a witness in two previous trials, one in which he testified that a plaintiff's cancer was not caused by exposure to toxic fumes, and another in which he testified that a plaintiff's leukemia was not caused by exposure to radiation. He had also served as a paid consultant in a previous tobacco litigation case, had given seminars to a law firm representing the tobacco industry, and had served on a so-called Scientific Advisory Board for R. J. Reynolds. (The scientists that Seitz supported were also sometimes called upon as an advisory group, attending periodic meetings to offer "advice and criticism." One letter suggested that they might also act as an advocacy group—although this was later struck out.)[116]

When asked point blank in the Norma Broin case, "Does cigarette smoking cause lung cancer?" attorneys for Philip Morris objected to the "form of the question."[117] When asked, "Does direct cigarette smoking cause lung cancer?" the attorneys objected on the grounds that the question was "irrelevant and immaterial." When finally instructed to answer, Cline was evasive.

> Cline: Well, if by "cause" you mean a population base or epidemiologic risk factor, then cigarette smoking is related to certain types of lung cancer. If you mean: In a particular individual is the cigarette smoking the cause of his or her cancer? Then . . . it is difficult to say "yes" or "no." There is no evidence.[118]

When asked if a three-pack-a-day habit might be a *contributory* factor to the lung cancer of someone who'd smoked for twenty years, Cline again answered no, you "could not say [that] with certainty . . . I can envision many scenarios where it [smoking] had nothing to do with it." When asked if he was paid for the research he did on behalf of the tobacco industry, he acknowledged that the tobacco industry had supplied $300,000 per year over ten years—$3 million—but it wasn't "pay," it was a "gift."[119]

What Cline said about cancer was technically true: current science does not allow us to say with certainty that any one particular person's lung cancer—no matter how much she smoked—was caused by smoking. There are always other possibilities. The science *does* tell us that a person with a twenty-year, three-pack-a-day habit who has lung cancer most probably got that cancer from smoking, because other causes of lung cancer are very rare. If there's no evidence that the woman in question was ever exposed to asbestos or radon, or smoked cigars or pipes, or had prolonged occupational exposure to arsenic, chromium, or nickel, then we could say that her lung cancer was almost certainly caused by her heavy smoking. But we couldn't say it for sure. In scientific research, there is always doubt. In a lawsuit we ask, Is it *reasonable* doubt? Ultimately, juries began to say no, but it took a long time, in large part because of witnesses like Martin Cline, witnesses that the industry had cultivated by supporting their research. Reynolds supported scientists, and when the need arose they were available to support Reynolds.

Stanley Prusiner would have been an even better witness for the industry—his work on prions was groundbreaking and his reputation untarnished—and his name did appear on a list of potential witnesses in the 2004 landmark federal case against the tobacco industry: *U.S. vs. Philip Morris et al.*[120] (He evidently did not testify; available documents do not indicate why.) The industry was finally found guilty under the RICO Act (Racketeer Influenced and Corrupt Organizations).[121] In 2006, U.S.

district judge Gladys Kessler found that the tobacco industry had "devised and executed a scheme to defraud consumers and potential consumers" about the hazards of cigarettes, hazards that their own internal company documents proved they had known about since the 1950s.[122]

But it took a long time—just about half a century—to get to that point. Along the way the tobacco industry won many of the suits that were brought against it. Juries, of course, were much more likely to believe scientific experts than industry executives—especially scientists who appeared to be independent—and neither Cline nor Prusiner ever worked "directly" for the tobacco industry; many of the funds were channelled through law firms.[123] External research could also help bolster the industry's position that the public should decide for themselves. "We believe any proof developed should be presented fully and objectively to the public and that the public should then be allowed to make its own decisions based on the evidence," they had argued, seemingly reasonably.[124] The problem was that public had no way to know that this "evidence" was part of an industry campaign designed to confuse. It was, in fact, part of a criminal conspiracy to commit fraud.

Cline and Prusiner were reputable scientists, so one might ask, Didn't they have a right to be heard? In later years Seitz and his colleagues would often make this claim, insisting that they deserved equal time, and their ability to invoke the Fairness Doctrine to obtain time and space for their views in the mainstream media was crucial to the impact of their efforts. Did they deserve equal time?

The simple answer is no. While the idea of equal time for opposing opinions makes sense in a two-party political system, it does not work for science, because science is not about opinion. It is about evidence. It is about claims that can be, and have been, tested through scientific research—experiments, experience, and observation—research that is then subject to *critical review by a jury of scientific peers*. Claims that have not gone through that process—or have gone through it and failed—are not scientific, and do not deserve equal time in a scientific debate.

A scientific hypothesis is like a prosecutor's indictment; it's just the beginning of a long process. The jury must decide not on the elegance of the indictment, but on the volume, strength, and coherence of the evidence to support it. We rightly demand that a prosecutor provide evidence—abundant, good, solid, consistent evidence—and that the evidence stands up to the scrutiny of a jury of peers, who can take as much time as they need.

Science is pretty much the same. A conclusion becomes established not when a clever person proposes it, or even a group of people begin to discuss it, but when the jury of peers—the community of researchers—reviews the evidence and concludes that it is sufficient to accept the claim. By the 1960s, the scientific community had done that with respect to tobacco. In contrast, the tobacco industry was never able to support its claims with evidence, which is why they had to resort to obfuscation. Even after decades and tens of millions of dollars spent, the research they funded failed to supply evidence that smoking was really OK. But then, that was never really the point of it anyway.

THE TOBACCO INDUSTRY was found guilty under the RICO statute in part because of what the Hill and Knowlton documents showed: that the tobacco industry knew the dangers of smoking as early as 1953 and conspired to suppress this knowledge. They conspired to fight the facts, and to merchandise doubt.

But it took a long time for those facts to emerge, and the doubt to be dispelled. For many years, the American people did continue to think that there was reasonable doubt about the harms of smoking (and some still do). While hazard labels were strengthened, it was not until the 1990s that the industry began to lose cases in courts. And although the FDA sought to regulate tobacco as an addictive drug in the early 1990s, it was not until 2009 that the U.S. Congress finally gave them the authority to do so.[125]

One reason the industry's campaigns were successful is that not everyone who smokes gets cancer. In fact, most people who smoke will not get lung cancer. They may suffer chronic bronchitis, emphysema, heart disease, or stroke, and they may suffer cancer of the mouth, uterus, cervix, liver, kidney, bladder, or stomach. They may develop leukemia, suffer a miscarriage, or go blind. The children of women who smoke are much more likely to be low birth weight babies than the children of women who don't, and to suffer high rates of sudden infant death syndrome. Today, the World Health Organization finds that smoking is the known or probable cause of twenty-five different diseases, that it is responsible for five million deaths worldwide every year, and that half of these deaths occur in middle age.[126] By the 1990s, most Americans knew that smoking was generally harmful, but as many as 30 percent could not tie that harm to specific disease. Even many doctors do not know the full extent of tobacco harms, and nearly a quarter of poll respondents still doubt that smoking is harmful at all.[127]

Industry doubt-mongering worked in part because most of us don't really understand what it means to say something is a cause. We think it means that if A causes B, then if you do A, you will get B. If smoking causes cancer, then if you smoke, you will get cancer. But life is more complicated than that. In science, something can be a *statistical* cause, in the sense that that if you smoke, you are much more *likely* to get cancer. Something can also be a cause in the everyday sense of being an occasion for something—as in "the cause of the quarrel was jealousy."[128] Jealousy does not always cause quarrels, but it very often does. Smoking does not kill everyone who smokes, but it does kill about half of them.

Doubt-mongering also works because we think science is about facts—cold, hard, definite facts. If someone tells us that things are uncertain, we think that means that the science is muddled. This is a mistake. There are always uncertainties in any live science, because science is a process of discovery. Scientists do not sit still once a question is answered; they immediately formulate the next one. If you ask them what they are doing, they won't tell you about the work they finished last week or last year, and certainly not what they did last decade. They will tell you about the new and uncertain things they are working on *now*. Yes, we know that smoking causes cancer, but we still don't fully understand the mechanism by which that happens. Yes, we know that smokers die early, but if a particular smoker dies early, we may not be able to say with certainty how much smoking contributed to that early death. And so on.

Doubt is crucial to science—in the version we call curiosity or healthy skepticism, it drives science forward—but it also makes science vulnerable to misrepresentation, because it is easy to take uncertainties out of context and create the impression that *everything* is unresolved. This was the tobacco industry's key insight: that you could use *normal* scientific uncertainty to undermine the status of actual scientific knowledge. As in jujitsu, you could use science against itself. "Doubt is our product," ran the infamous memo written by one tobacco industry executive in 1969, "since it is the best means of competing with the 'body of fact' that exists in the minds of the general public."[129] The industry defended its primary product—tobacco—by manufacturing something else: doubt about its harm. "No proof" became a mantra that they would use again in the 1990s when attention turned to secondhand smoke. It also became the mantra of nearly every campaign in the last quarter of the century to fight facts.

For tobacco is not the end of our story. It is just the beginning. In the years to come various groups and individuals began to challenge scientific evidence that threatened their commercial interests or ideological beliefs. Many of these campaigns involved the strategies developed by the tobacco industry, and some of them involved the same people. One of these people was Frederick Seitz.

As the industry campaign to defend tobacco was reaching the end of its course—and the claim that smoking's harms were unproven became harder to say with a straight face—Seitz moved on to other things. One of these was to found the George C. Marshall Institute, created to challenge scientists' conclusions in a whole new arena—strategic defense. When that debate was over, they would turn to the environment. Seitz had railed about scientific colleagues who made "simplified, dramatic statements" to capture public attention, rather than remaining "sober," yet in the later years of his life, he would do exactly that when discussing the ozone hole, global warming, and other environmental threats.[130]

The tobacco road would lead through Star Wars, nuclear winter, acid rain, and the ozone hole, all the way to global warming. Seitz and his colleagues would fight the facts and merchandise doubt all the way.

CHAPTER 2

Strategic Defense, Phony Facts, and the Creation of the George C. Marshall Institute

THE TOBACCO INDUSTRY WAS HAPPY to have a man of Frederick Seitz's scientific stature on their side, but by the late 1980s, Seitz was aligning himself with men of increasingly extreme views. Often, they were scientists in their twilight years who had turned to fields in which they had no training or experience, such as Walter Elsasser, a geophysicist who argued that biology as a science was a dead end because of "the unfathomable complexity" of organisms, a view that even a sympathetic biographer described as "ignored by most biologists and attacked by some."[1] Many colleagues thought Elsasser had become irrational, and some began to think the same of Seitz. In August 1989, one tobacco industry executive recommended against soliciting his further advice: "Dr. Seitz is quite elderly and not sufficiently rational to offer advice."[2]

But Seitz had found other allies, and by the mid-1980s a new cause: rolling back Communism. He did this by joining forces with several fellow physicists—old cold warriors who shared his unalloyed anti-Communism—to support and defend Ronald Reagan's Strategic Defense Initiative. SDI (Star Wars to most of us) was rejected by most scientists as impractical and destabilizing, but Seitz and his colleagues began to defend it by challenging the scientific evidence that SDI would not work and promoting the idea that the United States could "win" a nuclear war.

Seitz's hawkish politics had deep roots. Like nearly every American physicist of his generation, he grew up alongside the nuclear weapons programs, watching the national security state build his science as his science helped to build the national security state. Nearly all scientists who participated in the atomic bomb program felt they had done the right thing, given the frightening prospect of a German atomic bomb, and with it a German victory. That changed after the war, however, as many retreated from weapons work. By the 1950s, the arms race had turned many academic physicists into arms control advocates, and by the 1960s, Vietnam had turned many more into outright doves. Seitz would have none of that. As president of the National Academy of Sciences during the 1960s, Seitz had been disgusted by colleagues' antiwar activities, and had opposed the arms control efforts of the Johnson, Nixon, and Ford administrations as well as Nixon's policy of détente—the U.S.-Soviet effort to move toward more peaceful relations. Détente was about finding ways to coexist peacefully with the Soviet Union; Seitz found that morally repugnant, believing that the Soviets would use disarmament to achieve military superiority and conquer the West.

Seitz's strident anti-Communism was shared at influential foreign policy think tanks. These included the Hoover Institution (originally founded as the Hoover War Library, dedicated to promoting the "ideas that define a free society"), the Hudson Institute (founded by the military strategist Herman Kahn during the mid-1970s), and the Heritage Foundation (established in 1973 to promote conservative ideas).[3] These organizations and their allies in Congress fostered an assault on détente. By the end of the decade, they had destroyed the idea of peaceful coexistence, justifying a major new arms buildup during the Reagan years. This attack was mounted in very similar ways to the effort to protect tobacco: opponents of détente cast doubt on the official intelligence assessments prepared by the Central Intelligence Agency and created an alternative body of "facts"—which often weren't. They planted their claims in American minds by using large-scale publicity campaigns in the mass media, campaigns that relied on the demand for equal time for their views.

During the late 1970s and early 1980s, when the campaign first got under way, Seitz was still mainly focused on tobacco, while two close colleagues—physicist Edward Teller, father of the hydrogen bomb, and astrophysicist Robert Jastrow, founder of the Goddard Institute for Space Studies—led the way on strategic defense. Initially, their argument was a political one: that détente was naïve—a latter-day version of appeasement. They argued that Soviet capabilities were far greater than we knew, and that it was essential to continue to maintain and even expand our nuclear weapons stockpile. They defended the concept of SDI based on the scientific capability of the United States to build an effective defensive system against incoming Soviet missiles, and insisted that if war broke out the United States could win.

Then astronomer Carl Sagan and his colleagues threw a spanner in the works, arguing that any exchange of nuclear weapons—even a modest one—could plunge the Earth into a deep freeze that would devastate the whole planet. If that were true, then no nuclear war was winnable. The SDI lobby decided to attack the messenger, first attacking Sagan himself, and then attacking science generally. Just as the tobacco industry had created an institute to foster its claims, so did they: the George C. Marshall Institute, promoting "science for better public policy," with Frederick Seitz as the founding chairman of the board.[4]

The Birth of Team B

The right-wing attack on détente began in the final year of the Ford administration. In 1976, opponents of détente convinced a new Central Intelligence Agency director, George H. W. Bush, to support an "independent" analysis of Soviet capabilities and intentions. The idea was promoted by the President's Foreign Intelligence Advisory Board, which counted among its members Edward Teller. One of the most hawkish physicists ever to serve the U.S. government, Teller asserted that "student demonstrators and radical administrations at MIT and Stanford had wiped out military R&D," leaving the United States short of scientists ready and able to build the next generation of nuclear weapons.[5] It was only a matter of time before the Soviets surpassed us technologically, probably just a few years. While he was not an intelligence expert, and never had been, Teller thought he understood the Soviet threat better than the CIA, insisting that the agency's estimates greatly downplayed its true magnitude. A "competitive" reevaluation by the right people could set the record straight.[6]

The Central Intelligence Agency publishes National Intelligence Estimates (NIEs) to assess threats to the United States, but the CIA is not America's only intelligence agency; in the 1970s there were about a dozen. The NIEs were composed jointly, with various agencies supplying input and reviewing the estimates before publication. The result of this joint drafting and review process was an approved text, a compromise representing the best judgment that could earn general agreement. Any strong disagreement between agencies was recorded in the footnotes, so the estimates acknowledged dissent, even while they focused on producing a consensus.

As a member of the President's Foreign Intelligence Advisory Board, Teller was part of the reviewing process, and had seen the top secret draft of the CIA's 1975 estimate of Soviet capability. He didn't like it. He thought that it greatly understated the threat in three crucial areas. First, the CIA had concluded that Soviet ICBMs (intercontinental ballistic missiles) were not very accurate, so a Soviet "first strike" would leave plenty of U.S. capability to destroy the U.S.S.R. on the rebound. Second, the CIA believed that the vast Soviet air defense system would provide little protection against low-flying American bombers; our bombers would still be able to get through. Third, the CIA did not think that the Soviets were able to locate American submarines.[7] "The Soviets currently do not have an effective defense against the U.S. submarine force," the estimate stated plainly, and the CIA didn't imagine that this would change in the next decade.[8] In each of these areas the CIA had concluded that the United States was in the stronger position.

Teller didn't believe it. Moreover, he thought the entire process of trying to achieve a statement of likely Soviet capabilities was wrongheaded: what was needed was a bald statement of the worst-case scenario for which we had to prepare. One official put it this way: "Intelligence officers should deliberately try to shape policy by calling attention to the worst things the Soviets could do in order to stimulate appropriate countermeasure responses by the U.S. Government."[9] To ensure U.S. safety, we had to be alarmed, and Teller wanted the most alarming statement possible.

Teller got his wish for an independent threat assessment when a public conflict erupted between the CIA and the Defense Intelligence Agency over Soviet military expenditures. The Defense Intelligence Agency believed the U.S.S.R. spent twice as much on its military as the CIA thought—about 15 percent of its gross national product, as compared with only 6–8 percent of the U.S. GNP. The conservative press picked up the argument, using it to suggest that the Soviet Union had embarked on a vast military expansion. But this was very misleading, because the two agencies *agreed* on the numbers of Soviet soldiers, tanks, missiles, and aircraft; what they disagreed on was how much it all *cost*. The Defense Intelligence Agency thought the Soviet forces cost twice what CIA thought, implying that the Soviet military economy was half as efficient.[10]

If the Defense Intelligence Agency's analysis was correct, then the Soviet Union was a *weaker* adversary, not a stronger one; it took them twice as much money to achieve the same level of military preparedness. In fact, the Soviets were weaker still, because the United States had a vastly larger economy: 6 percent of our GDP was far more than 15 percent of the U.S.S.R.'s. Ours was a better, stronger system; we were literally getting more bang for each buck. Yet it was easy to take the Defense Intelligence Agency's claim out of context and argue that the United States was falling behind. So there was strong political pressure on the CIA to allow an independent analysis.

In June, CIA director Bush approved the formation of three independent review panels, each chartered to review a different aspect of the Soviet threat. One panel reviewed Soviet missile accuracy, and a second reviewed Soviet air defense capabilities, as Teller and his fellow critics wanted. The third area—submarine warfare capability—was blocked by the navy, which didn't want to release information about its submarines, so the third panel was chartered to review Soviet "Strategic Objectives" instead.[11]

The members of these panels came to be known as "Team B." While they were supposed to provide an objective review of the NIE, their composition ensured otherwise: the membership was composed entirely of foreign policy hawks who already believed that the CIA was underplaying the Soviet threat. Harvard historian Richard Pipes chaired the Strategic Objectives Panel, selecting the rest of the panel with assistance from Richard Perle—later assistant secretary of defense for Ronald Reagan (and later still an architect of the second Iraq War). Other panelists included Paul Nitze, one of the original architects of American Cold War foreign policy in the Truman administration, Lt. General Daniel O. Graham, the originator of the concept of "High Frontier," a space-based weapons program, and Paul Wolfowitz, a rising star in "neoconservative" circles who would later serve as deputy secretary of defense in the administration of George W. Bush.[12] Teller and Perle served as reviewers.

On every front, the panel cast the Soviet effort in the most alarming possible light. The Strategic Objectives Panel argued that the Soviets were only interested in détente to give them breathing space during which to achieve their real objective, "global Soviet hegemony."[13] They sought not a sufficiency of nuclear weapons to sustain deterrence (the U.S. strategy), but quantitative and qualitative strategic superiority, enabling them to fight and win any kind of war, including a nuclear one, essential to their goal of conquest.[14] Once the Soviets had achieved strategic superiority, they would use it, and they would achieve that superiority very soon.

"*The Soviet Union is . . . preparing for a Third World War as if it were unavoidable*," the panel declared emphatically. "The pace of the Soviet armament effort . . . is staggering; *it certainly exceeds any requirement for mutual deterrence.* The continuing buildup of the Warsaw Pact forces bears no visible relationship to any plausible NATO threat; it can better be interpreted in terms of intimidation or conquest."[15] Soviet leaders "probably believe that their ultimate objectives are closer to realization today than they have ever been before. *Within the ten year period of the National Estimate the Soviets may well expect to achieve a degree of military superiority which would permit a dramatically more aggressive pursuit of their hegemonial objectives.*"[16] It was a small step from strategic superiority to world dominance. The Cold War would be over. The West would have lost.

What was the basis for these claims? Not much. Little evidence was cited, and when the available evidence did not support their claims, they found a way to force it to. Consider this example. During the Cold War, submarines were a crucial part of the nuclear triad, so submarine detection was a crucial part of national defense. Most submarine surveillance was done with acoustics—we listened for the noise created by each other's submarines—but both sides had investigated other forms of detection, too. None of these worked very well. However, when the panel found evidence that the Soviets had spent large sums of money on nonacoustic antisubmarine warfare systems, but no evidence that they had ever deployed a nonacoustic system, they did not draw the obvious logical conclusion that those systems simply hadn't worked. Rather the panel concluded that they *had* worked, that the Soviets had deployed something, and covered it up. "The absence of a deployed system by this time is difficult to understand," they wrote. "The implication could be that the Soviets have, in fact, deployed some operational non-acoustic systems and will deploy more in the next few years."[17] The panel saw evidence that the Soviets had *not* achieved a particular capability as proof that it *had.* The writer C. S. Lewis once characterized this style of argument: "The very lack of evidence is thus treated as evidence; the absence of smoke proves that the fire is very carefully hidden."[18] Such arguments are effectively impossible to refute, as Lewis noted. "A belief in invisible cats cannot be logically disproved," although it does "tell us a good deal about those who hold it."[19]

The Team B panel also used the opportunity to push for new U.S. efforts in ballistic missile defense. The United States had tried twice before to develop and deploy ground-based antiballistic missile systems during the 1950s and 1960s, but found that they were very expensive and not very effective. These failures helped enable adoption of the Anti-Ballistic Missile Treaty—which limited each nation to a single installation of ABMs—since such systems likely wouldn't prove very valuable anyway. The Soviets built an installation to protect Moscow; we had built ours to protect an ICBM base near Grand Forks, North Dakota—and then shut it down less than a year later. Teller, and at least one member of the Strategic Objectives Panel, Lt. General Graham, wanted a new U.S. ABM program, so the panel report concluded—again without evidence—that the Soviets had "been conducting far more ambitious research in these areas," and it was "difficult to overestimate" the magnitude of their ABM efforts.[20]

While the tobacco industry had tried to exploit uncertainties where the science was firm, these men insisted on certainties where the evidence was thin or entirely absent. The "Soviet Union is" they repeatedly wrote, rather than "might be" or "appears to be." They understood the power of language: you could undermine your opponents' claims by insisting that theirs were uncertain, while presenting your own as if they were not.

THE TEAM B studies were written between October and December 1976, during the presidential campaign between President Gerald Ford and Democratic contender James Earl Carter. Just weeks before election day, one of the members of the Strategic Objectives Panel leaked the classified draft to the *Boston Globe*.[21]

The leak marked the beginning of an organized effort to assure that Team B's conclusions became public knowledge. Two days after Carter defeated Ford in the November election, a relic of the 1950s "red scare" was resurrected: the Committee on the Present Danger. Four of its members came from Team B. The committee spent the next four years currying media attention via press releases and opinion pieces, helping to push American foreign policy far to the right, often on the basis of "factual" claims with few facts behind them. Several Team B members—including Wolfowitz and Perle—became advisors to Ronald Reagan's 1980 presidential campaign; Reagan's victory made them the "A Team."[22] Their views became the basis for Reagan's confrontational foreign policy during his first term in office, and, most famously, his decision to pursue the Strategic Defense Initiative—better known as Star Wars.

Star Wars: The Strategic Defense Initiative

In March 1983, President Reagan called upon "the scientific community in this country, who gave us nuclear weapons, to turn their great talents to the cause of mankind and world peace; to give us the means of rendering these weapons impotent and obsolete."[23] The crux of the Strategic Defense Initiative (SDI) was to install weapons in space that could destroy incoming ballistic missiles. This would "shield" the United States from attack, making nuclear weapons obsolete.

The initiative was not just the result of Reagan's desire to achieve world peace; it was also a direct response to the nuclear freeze movement, which had crystallized in opposition to the Reagan administration's bellicose rhetoric. Movement leaders had asked both the United States and the Soviet Union to cease building, testing, and modifying nuclear weapons and their delivery systems. Since nuclear weapon cores decay over time and lose their explosive capacity, any agreement to stop building new ones was, in effect, an agreement to disarm.

The idea caught on quickly. By the end of 1981, twenty thousand nuclear freeze activists were working in forty-three U.S. states. The proposal was endorsed by major religious denominations and many local and state governments; by early 1982 it was being openly debated in the U.S. Congress. The movement's sudden growth was stunning, and it directly threatened the administration's foreign and military policies, as well as its reelection hopes for 1984.[24]

Reagan's answer was SDI, which many of his own advisors opposed. Some opposed it because they judged it technically infeasible, some because it would be seen as provocative by both the Soviets and domestic critics of the arms race, and still others because it would increase the risk of nuclear war. (If one side had an effective shield, it might be tempted to launch a first strike.) Reagan believed that SDI was technically possible and morally just. Like Team B, he thought the mutual assured destruction policy was repugnant—a suicide pact, in essence. No doubt he also saw political advantage in SDI, as it might undermine the nuclear freeze movement.

SDI was instantly controversial, creating a backlash among the very scientists Reagan would need to build it. While most physicists had long been accepting military R & D funds, they reacted differently to SDI, fomenting a coordinated effort to block the program. By May 1986, sixty-five hundred academic scientists had signed a pledge not to solicit or accept funds from the missile defense research program, a pledge that received abundant media coverage.[25] Historically, it was unprecedented. Scientists had never before refused to build a weapons system when the government had asked.

Why did scientists react so strongly to SDI? One reason was that they had a charismatic spokesman in the person of Cornell University astronomer Carl Sagan. Handsome and media savvy, Sagan had become famous during NASA's planetary missions in the 1960s and 1970s. Unlike most of his colleagues, Sagan thought that scientists should reach out to explain their work to the public. He had created a television series called *Cosmos*, broadcast in 1979, which presented the entire evolution of the universe, the solar system, the Earth, and human civilization in thirteen episodes. The final episode was the most controversial: Sagan used it to attack nuclear weapons as a threat to our survival. He also used *Cosmos* to promote environmental concern, a thread that had been woven through several episodes.

Sagan wasn't a weapons scientist, but he knew enough to know that what Reagan was proposing was as fantastical as the *Star Wars* films whose moniker they had borrowed. The reason was simple. No weapons system—indeed, no technological system—is ever perfect, and an imperfect defense against nuclear weapons is worse than worthless. It's a matter of arithmetic. If strategic defense is 90 percent effective, then 10 percent of the warheads still get through. The Soviets had an arsenal of about two thousand ballistic missiles capable of delivering over eight thousand warheads, 10 percent of which would more than suffice to destroy a nation.[26] But because the Soviets would never be certain how effective our defenses might be, SDI would provide incentive to build still more weapons, just to be sure, so SDI would fuel the arms race, not stop it. On the other hand, if the Soviets believed that SDI might actually work, then matters were even worse, because they might be tempted to "preempt"—to attack before the system was even built. SDI could provoke the Armageddon it was intended to prevent.

SDI was also untestable. The space missions that Sagan had been involved with had been tested on the ground meticulously to ensure they would work when actually launched; in the space business, you only get one chance. Nuclear war was the same—there'd be no second chance—but SDI couldn't be tested on the ground. Its satellites would have to be put in orbit, and then to test them we'd have to shoot large numbers of missiles at *ourselves*. The satellites, after all, were intended to destroy missiles launched from Europe and Asia at North America, not vice versa. And one or two missiles wouldn't do, because a system that's perfectly capable of shooting down one missile could very well fail in the face of ten, let alone thousands. To properly test SDI, we'd have to shoot a substantial fraction of our own missile inventory at ourselves.

Sagan was ill when SDI was announced, but he dictated a petition from his hospital bed and gave his wife a list of other scientists and heads of state to call for signatures.[27] Meanwhile, a student group at Cornell launched a campaign to get scientists to boycott SDI funding.[28] Many leading scientists were on board instantly, including Hans Bethe, head of the Theoretical Division at Los Alamos during World War II, and a major figure in the building of the H-bomb (despite considerable misgivings).[29] By the end of the year, the voices of opposition had swollen to a chorus, causing considerable consternation in the Reagan administration.[30]

The opposition infuriated Robert Jastrow. A longtime associate of Seitz, Jastrow was the founder of the Goddard Institute for Space Studies, NASA's theoretical arm in New York. With a background in astrophysics, Jastrow had been a prominent proponent of lunar exploration and had worked on various probes of the solar system, including *Pioneer*, *Voyager*, and *Galileo*. Like Sagan, he was a successful science popularizer, having written popular books on astronomy, space exploration, the origin of the universe, and the relation of science and religion. When he died in 2008, the *New York Times* described him as a man who brought "space down to earth for millions of Americans."[31]

Jastrow was not the scientist that Hans Bethe was, but he could give Sagan a run for his money on media savvy, having appeared on television more than a hundred times during the *Apollo* years. On the occasion of the *Apollo-Soyuz* flights, he appeared as a cohost on NBC with rocket engineer Werner von Braun, and on the occasion of the tenth anniversary of the first moon landing appeared on the *Today* show. At Columbia, some of Jastrow's students were struck more by his good looks and heavy smoking than by his science, dubbing him the "movie star."[32]

In 1981, Jastrow retired from NASA and moved to Dartmouth College as an adjunct professor of earth sciences, where he taught a popular summer course on the solar system. Two years later, he published a long article in the neoconservative magazine *Commentary* advocating strategic superiority. He had been motivated, he explained, by an article by Democratic senator Daniel Patrick Moynihan—a Team B supporter—published in the *New Yorker* in 1979. Moynihan had opposed the SALT II Treaty, moved by the Team B argument that the Soviets sought superiority, not stability. Moynihan imagined a counterfactual history in which the United States had engaged in an unlimited arms buildup during the 1970s, in order to spend the U.S.S.R. into ruin. That would have been a good thing, he thought, but we hadn't done it, and history would look back on that failure with regret. He feared that the 1980s would be remembered as the era in which "the peace of the world was irretrievably lost."[33]

Moynihan's article convinced Jastrow that something needed to be done. The Soviets had already achieved parity in ICBM accuracy, so we were now vulnerable to a Soviet first strike. "Within months," Moynihan wrote, "the Soviet Union would have the capacity to destroy the Minutemen, our land-based deterrent."[34] The United States needed to embark on a crash program to build an equivalent capability to destroy the Soviet missiles.[35] Ideally it would be a highly accurate mobile missile, but that would almost certainly be blocked by environmentalists, just as they had blocked nuclear power and almost stopped the Alaska oil pipeline.[36] Moynihan was referring to the MX missile, later called Peacekeeper, a large multiwarhead ICBM that was to be trundled around the desert states at night and hidden within various shelters in the daytime. This "multiple basing strategy" was eventually defeated, though not solely by environmentalists; many Americans didn't want nuclear missiles being trucked through their towns and cities. But the innuendo was clear: environmentalists served Soviet interests. Jastrow would pick up this idea and run with it.

Moynihan had not claimed that the Soviets were building a missile defense program, but Jastrow now did. This was one of the Team B claims, and in his article for *Commentary*, Jastrow emphasized that the strategy of Mutual Assured Destruction (MAD) depended upon an assumption that both sides accepted that there was no effective defense against nuclear attack. This "assumption has turned out to be false," Jastrow insisted. The Soviets "had implemented large programs for defending its citizens from nuclear attack, for shooting down American missiles, and for fighting and winning a nuclear war." MAD was now a "policy in ruins," and the United States faced the "greatest peril" in its history.[37]

According to Jastrow, the U.S.S.R. was in a position of strategic superiority from which it could dictate U.S. policy. It could, for example, invade Persian Gulf oil fields with impunity. It could absorb Western Europe without a fight. "A direct attack would not be necessary," Jastrow wrote. "Threat, accompanied by a general escalation of tension, would probably suffice to bring all of Western Europe under Soviet hegemony."[38] All this was possible because while the U.S. nuclear arsenal was sufficient to destroy the world twice over, the Soviet arsenal could destroy it three times.

· · ·

Team B's claims turned out to be more than a little exaggerated. Later analyses would show that the Soviet Union had not achieved strategic superiority, they had not implemented a missile defense system beyond their single Moscow installation, and they certainly never achieved the ability to dictate U.S. policy. One anecdote perhaps tells the whole story: A few years after the Soviet Union collapsed, one of Teller's protégés toured a site that the Team B panel had believed was a Soviet beam-weapon test facility; it turned out to be a rocket engine test facility. It had nothing at all to do with beam weapons.[39]

Nor did Soviet leaders ever believe that nuclear war was "winnable," even if they struck first. A series of interviews sponsored by the CIA in 1995 revealed that the Soviet leadership of the 1970s and 1980s believed quite the opposite: that nuclear war would be catastrophic, and that nuclear weapons use had to be avoided at all costs.[40]

Team B, Jastrow, and Moynihan had all overestimated Soviet capabilities, and greatly exaggerated the certainty of their claims. But their alarming arguments had the desired effect, providing "evidence" that the United States needed to act, and fast. It also demonstrated that you could get what you wanted if you argued with enough conviction, even if you didn't have the facts on your side. The Strategic Defense Initiative and its successor, the Ballistic Missile Defense Organization, were approved by Congress, at a cost of more than $60 billion.[41]

From Strategic Defense to Nuclear Winter

While the Team B arguments were being used to justify Ronald Reagan's massive military buildup, a new concern about nuclear weapons was developing in scientific circles. At NASA Ames Research Center, some of Sagan's colleagues had been using computer models to study the effects of atmospheric dust on surface temperature. Their goal was to understand the atmosphere on Mars, but they soon realized they could use the model to test a new hypothesis—being hotly debated in earth science circles—that a giant asteroid had struck the Earth at the end of the Cretaceous period (65 million years ago), wiping out the dinosaurs.[42] Geologists and biologists had generally supposed that the hapless dinosaurs, with their pea-sized brains, had been outcompeted by the newly evolving, smart, and agile mammals, but the asteroid hypothesis suggested something else. The dinosaurs had perished as huge dust clouds had been thrown into the atmosphere when the asteroid struck, which in turn blocked the Sun. The poor creatures had probably starved as the resulting deep freeze destroyed their food supplies.

The NASA-Ames scientists realized that their model could also be used to assess the climatic impact of a large-scale nuclear war. For what happened to the dinosaurs after the meteorite impact might happen to *us* after a nuclear war: death by deep freeze. If this were so, then nuclear war would have no winner. We would be the dinosaurs, and insects would inherit the Earth.

Using publicly available information on the effects of nuclear weapons and computer models of nuclear warfare, the NASA-Ames group investigated how nuclear exchanges of one hundred to five thousand megatons might affect global temperatures. (For comparison, the Mt. St. Helens eruption was equivalent to ten megatons.) Their model suggested that even the smallest nuclear exchange could send the Earth into a deep freeze: surface temperatures might fall below freezing even in summer. Larger exchanges could produce near-total darkness for many months.[43] The nuclear winter hypothesis had been born, but it could equally well have been called nuclear night. After even a modest nuclear exchange, we would indeed freeze in the dark.

The NASA-Ames scientists acknowledged many uncertainties in their model. It was not clear that data drawn from the Hiroshima and Nagasaki bombings and the above-ground testing programs of the 1950s adequately represented the blast and fire impacts of multiple detonations of modern weaponry. The destroyed Japanese cities were also not necessarily representative of how American and Soviet cities would burn. Forests and grasslands would burn with less intensity and have different impacts than urban conflagration. It was also not clear to what extent nuclear weapons bursts would cause fires outside the cities that they hit. Many details remained to be explained—what scientists call "second order" effects. Still, the big picture was clear: "The first-order effects are so large, and the implications so serious, that we hope the scientific issues raised here will be vigorously and critically examined."[44]

Word of "nuclear winter" theory spread quickly, and it provoked a rapid response. In June 1982, before the NASA-Ames group even had a chance to publish their results, Sagan was approached by executives of the Rockefeller Family Fund, the Henry P. Kendall Foundation, and the National Audubon Society about organizing a public conference on the long-term consequences of nuclear war. Sagan agreed, along with biologist Paul Ehrlich of Stanford University, author of the famous 1960s book *The Pop-*

ulation Bomb. Walter Orr Roberts, the founding director of the University Corporation for Atmospheric Research (UCAR), and George Woodwell, a prominent biologist at the Marine Biological Laboratory at Woods Hole, Massachusetts. These four men created a steering committee, deciding to base the conference around the NASA-Ames result.[45] First, they arranged a closed workshop, where a draft of the nuclear winter paper would be reviewed by scientific colleagues. If the nuclear winter concept held up to this scientific peer review, it would be analyzed for its biological implications by a group of prominent biologists. A public conference would only be scheduled if the paper survived this first effort.[46]

In April 1983, a month after Reagan's Star Wars speech, the workshop was held in Cambridge, Massachusetts. The nuclear winter paper survived peer review with only minor revisions. Then it was examined by the biologists, who found it sufficiently compelling to draft an article of their own on the biological consequences of nuclear winter and to schedule a "Conference on the Long-Term Worldwide Biological Consequences of Nuclear War" for October 31. Thirty-one scientific and environmental groups, including the Federation of American Scientists, the Union of Concerned Scientists, the Environmental Defense Fund, and the Sierra Club, contributed funds.[47] A satellite link to Moscow allowed Soviet scientists to participate, too.

This was all a bit out of the ordinary—particularly the use of a workshop to review a scientific paper (normally scientists get papers to review in the mail, or these days by e-mail)—but it didn't violate scientific protocols. Something else, however, did: Sagan jumped the gun on the public conference and on the formal publication of the paper in a manner that seemed designed to garner maximum public attention. The day before the conference, Sagan published a three-page summary of the nuclear winter hypothesis in *Parade*, the Sunday supplement magazine with a circulation of more than ten million. Sagan explained to *Parade* readers that in the model of a five-thousand-megaton nuclear exchange, "land temperatures, except for narrow strips of coastline, dropped to minus 25 degrees Celsius and stayed below freezing for months."[48] This would kill food crops and livestock, and lead to mass starvation among survivors who hadn't already perished in the blast. Sagan's words were accompanied by a set of scary drawings showing the dark nuclear clouds creeping inexorably across the Earth's surface. A text box—"Something you can do"—admonished readers to support nuclear arms reduction or the nuclear freeze, and to write to President Reagan and Soviet leader Yuri Andropov.

Sagan also used the nuclear winter hypothesis as the basis for a lengthy policy article published in the journal *Foreign Affairs*, which appeared around the same time as the public conference. He positioned nuclear winter as the realization of strategist Herman Kahn's "Doomsday Machine"—a concept lampooned in Stanley Kubrick's famous tragicomedy, *Dr. Strangelove or: How I Learned to Stop Worrying and Love the Bomb*. Kahn's idea was a device that would automatically and unstoppably destroy humanity in the event of a nuclear attack by launching the entire arsenal of nuclear weapons. It would be the ultimate deterrent, as no one in his right mind would risk setting it in motion. In the Kubrick film, however, one general is not in his right mind, an attack is launched, the Doomsday Machine is triggered, and the world is destroyed, as the soundtrack plays "We'll meet again . . ."

Nuclear winter *was* the Doomsday Machine, even if no one had planned it that way. Policy should be directed, Sagan argued, toward reducing the arsenals to levels below the threshold that would create climate catastrophe. This level was a total of five hundred to two thousand warheads—far less than the forty thousand or so *each* that the superpowers had.[49]

Finally the actual scientific paper came out. Its title was "Nuclear Winter: Global Consequences of Multiple Nuclear Explosions," but it came to be known as TTAPS for the last names of its authors: Richard Turco, O. Brian Toon, Thomas Ackerman, James Pollack, and Carl Sagan. Paired with the paper by Paul Ehrlich and colleagues on the biological consequences of nuclear war, it was published in the December 23 issue of *Science*, the most prestigious scientific journal in the United States. It was accompanied by an editorial by the magazine's publisher, William D. Carey, who congratulated the scientists for helping to bring to life the "conscience of science." Scientists, he argued, had the responsibility to "look squarely at the consequences of violence in the application of scientific knowledge."[50]

Carey's argument was hardly new; indeed, many of the atomic bomb's inventors had turned against their creation in the 1950s. This had created a schism within nuclear physics between defenders and opponents of nuclear weapons that never really healed, but the period of détente had made the old arguments seem irrelevant.[51] Reagan's revival of the Cold War reopened the old wounds, forcing the scientific community's members to take sides again, and they generally recapitulated the positions they had taken the first time around. Hans Bethe, who had opposed construction of the hydrogen bomb, now opposed Star Wars, while Teller, the hydrogen bomb's biggest advocate, personally pitched Star Wars to Reagan. And so on. Not much had changed in thirty years. But while the politics of nuclear weapons had perhaps stood still, the science of nuclear winter had not. The TTAPS paper quickly provoked scientific challenges, which led to new insights and a revised understanding.

Three scientists at the National Center for Atmospheric Research (NCAR) in Boulder, Colorado, quickly staked out a challenge to the TTAPS paper. Climate modeler Curt Covey used NCAR's three-dimensional Community Climate Model to reexamine the whole concept. The NCAR model included atmospheric circulation, so it would carry air warmed by the model's "ocean" over land, to give a more realistic view of the likely land cooling after a nuclear war. Their conclusion was qualitatively consistent with TTAPS: "for plausible scenarios, smoke generated by a nuclear war would lead to dramatic reductions in land surface temperature." But quantitatively it was less alarming: the model did not experience the 35°C drop that the TTAPS model had. Instead, it suggested drops of 10°C to 20°C— quite enough to cause crop failure in the growing season, but not really enough to be called "winter." One member of the group, Stephen Schneider, renamed the phenomenon "nuclear autumn."[52]

The NCAR team's 1984 paper addressed and overcame some of the TTAPS paper's weaknesses. Other papers followed. By mid-1988, John Maddox, the editor of *Nature* (the United Kingdom's premier scientific journal) concluded that nuclear winter had become "respectable academic work."[53] Whatever drama surrounded the initial work, the issue was now being handled appropriately in the mainstream scientific journals.

Two years later, the TTAPS team reviewed the now-substantial nuclear winter literature, and found that it supported the conclusion that an "average land cooling beneath the smoke clouds could reach 10 to 20 C and continental interiors could cool by up to 20 to 40 C with subzero temperatures possible even in summer."[54]

Yet, even here, as one reviewer of this book commented, the TTAPS team was "using a typical technique of Edward Teller's, looking to the top of the error bars to derive its claims."[55] (In other words, there was a range of physically plausible outcomes, and they were emphasizing the high end.) Of course, cooling of 10°C to 20°C was not the amount projected by their original 1983 paper, but the amount projected by the NCAR work the following year. Since 1984, further changes made to models had been relatively offsetting, so that the overall conclusion of the NCAR study hadn't changed. The physics of nuclear winter was now firmly established, and while the results weren't good for the promoters of winnable nuclear war, they also weren't as bad as the TTAPS group had originally thought.

On one level, then, the scientific process worked. Scientists took the nuclear winter hypothesis seriously, and worked through it, evaluating and improving the assumptions, data, and models supporting it. Along the way, they narrowed the range of potential cooling and the uncertainties involved, and came to a general consensus. Without actually experiencing nuclear war, there would always be quite a lot of "irreducible uncertainty" in the concept—no one denied that—but overall, the first-order effects were resolved. A major nuclear exchange *would* produce lasting atmospheric effects that would cool the Earth significantly for a period of weeks to months, and perhaps longer. It would not be a good thing.

But on another level, many scientists were unhappy with the way the whole thing had played out. Sagan's behavior—publishing in *Parade* and *Foreign Affairs* before the peer-reviewed TTAPS paper had appeared in *Science*—was a violation of scientific norms. Moreover, the *Parade* article presented the TTAPS worst-case scenarios and omitted most of the caveats, so to some scientists it didn't appear as an honest effort in public education. Some saw it as outright propaganda. Some decided it was appropriate to complain.

MIT professor Kerry Emanuel, a hurricane specialist, was particularly chagrined, and accepted an invitation from the editors of *Nature* to reply. In a letter entitled "Towards a scientific exercise," he attacked the nuclear winter movement for its "lack of scientific integrity." He criticized their work for failing "to quantify the large uncertainties associated with estimates of the war-initiated fires and their combustion products, [for] the highly approximate nature of the global circulation models used in the calculations, and [for] the appearance of the results in popular literature before being exposed to the rigors of peer review."[56]

Emanuel's first two points are a bit hard to credit. All models are simplifications, and accurate quantification of uncertainties about nuclear warfare is impossible without actually fighting the war—which nobody, not even Edward Teller, thought a particularly good idea. The point of a model is to explore domains that can't be explored otherwise; you build a model when you don't have access to the real thing—for reasons of time, space, practicality, cost, or morality. Emanuel was perhaps irritated because the global circulation models being used at that time ignored his own specialty:

mesoscale phenomenon, which he argued would matter. (Hurricanes, despite their gigantic size to us, are still too small to be explicitly modeled in global circulation models, even now.)

The third complaint was clearly legitimate, but just as Emanuel was irritated by Sagan's behavior, Covey's team was offended by Emanuel's, particularly his suggestion that their work was unscientific. For one thing, they had *not* neglected mesoscale effects. Rather, following an earlier National Research Council study, they had assumed that mesoscale processes would be responsible for half the smoke generated by nuclear fires being washed out of the air almost immediately. While they couldn't explicitly model those processes, they hadn't *ignored* them, and they had acknowledged that they were likely the largest uncertainty in the calculations. Covey and his colleagues were flummoxed by the suggestion that they had behaved inappropriately, suggesting that it was Emanuel, not they, who was out of line. "It is rare that scientists of high caliber characterize the work of their colleagues—even in controversial work—in terms as harsh as 'it has become notorious for its lack of scientific integrity.' "[57]

What was going on?

Clearly, the whole debate had started badly because of Sagan's decision to go public. On the other hand, Sagan's argument was based on scientific evidence—it was based on data—and he believed it was his duty as a citizen to explain the very real threat of catastrophe. Emanuel, however, thought he had gone too far, and thus violated his trust as a scientist by exaggerating what was really known about the risk.

While Covey was offended by Emanuel's questioning his own team's behavior, he accepted Emanuel's complaint about Sagan. In 1987, Covey raised the issue again, suggesting that the popular media had taken the TTAPS "baseline" result of 35°C cooling as "definitive truth," ignoring the later efforts that reduced it to 10°C to 20°C, and suggested that the TTAPS group should have done more to set the record straight. They were "guilty at least of lack of energy in combating distorted reporting of their model's results, as well as a tendency to imply that their original findings are as good as inscribed on stone tablets."[58] The popular media equated nuclear winter with a "deep freeze," but the current understanding of the phenomenon among active climate scientists was rather less dramatic. Covey also thought that the TTAPS team should have done more to credit the work of others, including (of course) himself. One was supposed to present the work of other scientists fairly, and Covey didn't think that had happened. Still, if the new estimates of 10°C to 20°C were about right, then the core of the hypothesis—that a nuclear exchange would have serious environmental consequences lasting long after the fires had gone out and the radiation diminished—was still intact.

Within the scientific community, then, the nuclear winter debate took place at two levels: over the details of the science and over the way it was being carried out in public. The latter created a fair bit of animosity, but the former led to resolution and closure. The TTAPS conclusions had been reexamined by others, and adjusted in the light of their research. Whether it was a freeze or a chill, scientists broadly agreed that nuclear war would lead to significant secondary climatic effects. Out of the claims and counterclaims, published and evaluated by relevant experts, a consensus had emerged. Despite the egos of individual scientists, the jealousies and the sour grapes, science had worked pretty much the way it was supposed to.

Robert Jastrow was not content.

The George C. Marshall Institute

Edward Teller, Robert Jastrow, and Fred Seitz had been particularly appalled at William Carey's decision to praise the TTAPS and Ehrlich work in *Science*. They especially disliked Paul Ehrlich, whose *Population Bomb* was one of the foundational works of the American environmental movement. Ehrlich had served as president of Zero Population Growth and of the Conservation Society, which linked him in their minds to the environmental left, which they viewed as largely Luddites, while Sagan's aggressive promotion of the nuclear winter thesis in *Foreign Affairs* and *Parade* further antagonized them.[59] But they didn't just complain among themselves, write a letter to the editor, or pen an op-ed piece. Like the tobacco industry before them, they decided to create an institute.[60]

Jastrow envisaged his institute serving as a counterweight to the Union of Concerned Scientists (UCS), which had been formed in 1969 by faculty and students at the Massachusetts Institute of Technology. A substantial body of MIT faculty members believed that the government nuclear weapons establishment—scientists working in national laboratories like Los Alamos and Lawrence Livermore—had grown far beyond any reasonable defensive purpose. This growth had been encouraged by a small group of "insider" physicists led by Edward Teller, while many academic scientists remained on the sidelines watching with concern. One of the five planks in the UCS policy platform had been "to express our determined opposition to ill-advised and hazardous projects such as the ABM system, the enlargement of our nuclear arsenal, and the development of chemical and biological weapons."[61] After Reagan's SDI speech, the union embarked on a new study of ballistic missile defense technologies.

The resulting report was largely written by two well-known physicists: Richard Garwin of IBM's Thomas J. Watson Research Center, and Hans Bethe. Both were longstanding opponents of antiballistic missile systems. During the Johnson and Nixon administrations, the United States had developed and started deployment of a defensive system, allegedly against Chinese ballistic missiles, although few in the defense business believed that claim (and in fact China did not obtain intercontinental ballistic missiles until 1981). This Sentinel system used two layers of ground-based interceptors, long-range Spartan missiles for area defense, and shorter-range Sprint missiles to destroy warheads that the Spartans missed. Both missiles used nuclear warheads of their own to destroy the incoming warheads. Garwin and Bethe had argued that Sentinel would be easy to outsmart with inexpensive dummy warheads, and the high-altitude nuclear explosions produced by the Spartan missiles would blind the radars guiding the shorter-range missiles, rendering them useless.[62]

The debate over Sentinel and its scaled-back descendant, Safeguard, split the American physics community.[63] Bethe had denounced it publicly at MIT on March 4, 1969, helping to spark the movement that spawned the Union of Concerned Scientists.[64] Other scientists, like Edward Teller and fellow physicist Eugene Wigner—Fred Seitz's mentor—supported Sentinel and its deployment in the fight against Communism. Garwin and Bethe's efforts were successful in first limiting its deployment to one site. By 1977, the United States had no ballistic missile defenses.

Six years later, the proponents of missile defense were trying again, but Garwin and Bethe's view hadn't changed. They argued that Reagan's vision would be colossally elaborate and expensive, the space-based "layer" requiring twenty-four hundred laser battle stations, weighing fifty to one hundred tons and costing around a billion dollars—*each*. Even then, it wasn't clear it would work. A very powerful computer would be necessary to control the network, and nobody knew how to test the system that would be required.

Jastrow didn't believe the UCS numbers. Knowing their history of opposing ballistic missile defenses, he assumed their work was slanted. He also claimed to have heard rumors from defense insiders that the numbers were far off. Certainly existing estimates varied widely. An earlier analysis produced at the Los Alamos National Laboratory had argued that "only" ninety battle stations would be necessary; an Office of Technology Assessment consultant had said merely hundreds. Jastrow convinced himself that they'd all made serious errors—or had deliberately fudged the numbers to make SDI look "impractical, costly and ineffective"—and attacked Garwin and Bethe for exaggerating the number of satellites by a "factor of about twenty-five."[65]

Jastrow decided that it was time for a more organized response to the UCS, and by September 1984 he had decided to create it: a union—or at least a coterie—of scientists who shared *his* concerns about national security and had faith in the capacity of science-based technologies to address them. Next to Edward Teller, Frederick Seitz, who was chairman of the official SDI Advisory Board, was perhaps the most famous physicist in America who shared Jastrow's views; Jastrow invited Seitz to be the founding chairman of the board. He also invited physicist William Nierenberg, the recently retired director of the Scripps Institution of Oceanography in La Jolla, California, to join them. Nierenberg had known Seitz for decades, and they had served on Reagan's transition team together. He had also attended the same high school in the Bronx as Jastrow, and both Nierenberg and Jastrow had received their Ph.D.s in physics at Columbia in the 1940s. All three men had been associated through a variety of high-level advisory committees on which they had served. In short, they had much in common: all physicists, all retired or semiretired, all political hawks, all sons of the Cold War.

Named for General George C. Marshall—the American architect of European reconstruction of Europe after World War II, which had been designed in part to head off the spread of Communism—the Institute was "intended to raise the level of scientific literacy of the American people in fields of science with an impact on national security and other areas of public concern." Jastrow raised initial funds for the Institute from the Sarah Scaife and John M. Olin foundations, well-known funders of conservative causes (until the mid-1990s, he avoided taking corporate money).[66]

The Institute would promote its message through the distribution of "readable reports, books, films, etc." They would also hold "training seminars" for journalists, on the fundamental technologies of Strategic Defense, starting with one in December 1984, and also for congressional staffers. In a letter to Nierenberg, Jastrow explained how he'd also been busy writing articles and op-ed pieces to get their views on the radar screen and provoke debate. His latest *Commentary* article, he boasted, "seemed to have been effective. *Commentary* and the *Wall Street Journal* have been getting calls and letters from Sagan, Bethe, Carter, et al."[67] The debate was now on.[68]

The initial training seminar in December 1984 was not widely reported, but the Marshall Institute's next move—a critical review of the Office of Technology Assessment's full report on SDI—was, and Jastrow followed up with an entire book illustrating how the Union of Concerned Scientists and OTA had distorted the "facts" of strategic defense.[69] Replaying the tobacco strategy, they began urging journalists to "balance" their reports on SDI by giving equal time to the Marshall Institute's views. When they didn't, Jastrow threatened them, invoking the Fairness Doctrine. In 1986, public television stations across the nation were preparing to air a program on SDI, which Jastrow considered "one-sided." Jastrow and the Marshall Institute board of directors sent letters across the country warning "the managers of these stations that by airing the UCS program they could incur obligations, under the Fairness Doctrine, to provide air time for presentation of contrasting viewpoints."[70]

The Fairness Doctrine had been established in the late 1940s, when radio and television licenses were scarce and tightly controlled by the U.S. government.[71] A Federal Communications Commission license was thought to come with an obligation to serve public purposes, one of which was "fairness." But does fairness require equal time for unequal views? After all, sixty-five hundred scientists had signed the petition against SDI, and the Marshall Institute—at least at this early stage—consisted of Robert Jastrow and two colleagues.[72]

Whether or not it was fair, Jastrow's approach worked. Jastrow reported with satisfaction: "Very few public TV stations aired the program."[73]

Jastrow believed that if the American people understood SDI, they would support it. Over the next two years, the Institute built up its program activities in the manner that Jastrow had hoped. By this time, it had clarified its goal and was moving toward getting its message directly where it counted: namely, to Congress, through press briefings, reports, and seminars directly aimed at members and their staff. By 1987, he'd moved the Institute from New York to Washington and hired a full-time executive director.

Jastrow's approach was underlined by his strongly anti-Communist orientation. He believed that the opponents of SDI were playing into Soviet hands.[74] As evidence, he cited a letter written by Soviet secretary general Mikhail Gorbachev to MIT professor and UCS founder Henry Kendall to congratulate him on the union's "noble activities in the cause of peace."[75]

Jastrow found it alarming that Gorbachev approved of Kendall's work, suggesting that Kendall and the union were stooges of the Soviet Union, noting "the intensification—one could say almost, the ferocity—of the efforts by the UCS and Soviet leaders to undermine domestic support for SDI."[76]

A major debating point was whether SDI violated the Anti-Ballistic Missile Treaty. The Institute insisted that it did not, an argument used in England by Conservative Member of Parliament Ian Lloyd in a House of Commons debate. Quoting directly from Marshall Institute materials, Lloyd insisted that SDI did not violate the ABM treaty, because the treaty did not prohibit research, which was all that was being proposed (as yet). He closed with a familiar Cold War argument: that the goal of the arms race was not to maintain a balance of terror, but to free the Soviet people. SDI was linked to that goal:

> A fundamental Western interest is the survival of the Russian people as a whole long enough for them to understand, evaluate, and eventually escape from the yoke of their self-imposed tyranny. That is in the interests of the civilised world. The perspective of this decision on SDI on both sides is one that extends well into the next century and clearly embraces that possibility. Our purpose is not merely the survival, but ultimately the legitimate enlargement, of the free world by the voluntary actions of convinced peoples.[77]

Of course, no one could say whether SDI would eventually aid the cause of freeing the Soviet people from Communism. In this sense, there was a crucial difference between the debate over SDI and nuclear winter and the earlier debates over tobacco: while there was enormous evidence of tobacco links to cancer and other health problems—people had been smoking cigarettes for decades—there were no *facts* to be had over strategic defense or nuclear winter. Strategic defense and nuclear winter were hypotheticals—logical constructs from theory. No one had ever built a full-scale, functioning, orbital strategic defense, and no one had ever fought a two-sided nuclear war. The claims and counterclaims were just projections—even useful fictions.

JASTROW HAD SUCCEEDED in creating a debate about Stars Wars, but he wanted to go further, and take a stand against fraudulent science—at least if that science was being used against nuclear weapons. In a 1986 fundraising letter to the Coors Foundation—a group committed to supporting "self-sufficiency" and education, especially in the area of free enterprise[78]—Jastrow insisted that the change from nuclear winter to nuclear autumn demonstrated that the TTAPS authors had been willfully deceptive and that the climate effects of nuclear winter would be "minor to negligible."[79] The antinuclear scientists were playing into Soviet hands, since one of the "prime objectives of Soviet leaders is to convince the people of the western democracies that nuclear weapons in any numbers cannot be used without risking destruction of humanity. The Nuclear Winter scenario could not serve the needs of Soviet leaders better if it had been designed for that purpose."[80] TTAPS were at best dupes, at worst, accomplices. Jastrow concluded the worst, accusing the authors of deliberately ignoring the effects of the oceans and the fact that smoke would rain out.

TTAPS *had* in fact mentioned both mitigating circumstances in their paper; Jastrow was misrepresenting their work to suggest that they had intentionally downplayed elements that would lessen the impact, and played up the worst-case scenario. Having planted the suggestion of scientific fraud to his potential donors, he then hired a spokesman to push the claim in public.[81]

A Wholesale Attack on Science

A cousin of Frederick Seitz, Russell Seitz was affiliated with the Harvard Center for International Affairs, and later with the John M. Olin Institute for Strategic Studies—a policy center funded by the conservative Olin Foundation.[82] (The President of the Olin Foundation was William Simon, Secretary of the Treasury in the Nixon Administration. Deeply committed to laissez-faire capitalism, Simon believed in the sovereignty of the individual, and considered capitalism the only "social system that reflects this sovereignty . . .")[83] In 1984, he had written a letter to *Foreign Affairs* attacking the nuclear winter concept; he now developed a full-fledged attack, published in the fall 1986 issue of the *National Interest*.[84] The theme of "In from the Cold: 'Nuclear Winter' Melts Down" was that scientists were not to be trusted. Russell Seitz declared nuclear winter dead: "Cause of death: notorious lack of scientific integrity."[85] He recapitulated the history of nuclear winter theory, focusing his readers' attention on the network of foundations that had played some role in publishing or advertising the nuclear winter research: the Audubon Society, the Henry P. Kendall Foundation, the Union of Concerned Scientists, Physicians for Social Responsibility, and the Federation of American Scientists—in short, a gaggle of liberal environmental groups.

Seitz then summarily dismissed models as bad science. "The TTAPS model," he said, "postulated a featureless bone-dry billiard ball [instead of a realistic Earth] . . . [and] instead of realistic smoke emissions, it simply dumped a ten-mile thick soot cloud into the atmosphere instantly. The model dealt with such complications as east, west, winds, sunrise, sunset and patchy clouds in a stunningly elegant manner—they were ignored."[86]

"One way to see the TTAPS model is as a long series of conjectures," he continued, "if this smoke goes up, if it is this dense, if it moves likes this, and so on. This series of coin tosses was represented to laymen and scientists alike as a 'sophisticated one-dimensional model'—a usage that is oxymoronic, unless applied to Twiggy."[87] Of course the models were simplified, no one denied that. Every model is, in a sense, a conjecture, just as every scientific theory is. But just as theories are tested by observation, models are built on established theory and observation. The models Seitz was attacking were state-of-the-art: the most sophisticated approach available. If they weren't perfect, Seitz wasn't offering anything better. And neither was anyone else.

Having dismissed the TTAPS model as unscientific and casting doubt on the objectivity of its authors by linking them to liberal and environmentalist organizations, Seitz completed the picture for his readers by alleging ulterior motives. "Political considerations subliminally skewed the model away from natural history; in retrospect, the politics in question can be seen as those of the nuclear freeze movement."[88] Of course, the nuclear freeze movement *was* part of the larger social and political story surrounding nuclear winter, but so was Ronald Reagan's desire to build strategic defense. All science exists in a social context, but that doesn't prove that the relevant scientific work is skewed by that context in any *particular* direction. After all, a conservative scientist might have wanted to downplay nuclear winter just as much as a liberal one might have sought to highlight it. Scientists are well aware of these issues, which is why they have mechanisms like peer review to flag failures of objectivity, and scientists of various political persuasions had long wanted to avoid Armageddon. They still do.

Seitz was interested in none of these subtleties. He insisted that nuclear winter was not science at all: it was left/liberal/environmental politics dressed up as computer code. "No one who is familiar with the malleability of computer projections can be surprised at the result."[89]

Seitz made sure that his readers understood that nuclear winter had been diminished to "a barely autumnal inclemency."[90] He summarized Sagan's alleged exaggerations, pairing them with dismissive comments on the science by famous physicists, such as Caltech theoretical physicist Richard Feynman and Princeton theoretical physicist Freeman Dyson. "It's an absolutely atrocious piece of science but I quite despair of setting the public record straight," Seitz quoted Dyson.[91] (Dyson later gave himself two weeks to disprove the TTAPS physics and was "disturbed" when he couldn't do it.[92] Dyson, a self-proclaimed "heretic," would also later dismiss the climate models that demonstrated global warming, even though he had been one of the early scientists to express concerns about warming in the 1970s.)[93]

In a section of the article entitled "Physics Meets Advertising," Seitz reflected on the construction of the iconography of nuclear winter ("fill your airbrush with acrylic flat black and obliterate the northern hemisphere") and its linkage to the apocalyptic rhetoric adopted by Sagan. He drew on his readers' recognition that he meant *the* Apocalypse simply by capitalizing it midsentence. "Activists asked scientists for a consciousness-raising tool and were given a secular Apocalypse with which to preach for our deliverance from nuclear folly." The *Parade* vision of a black, cancerous cloud spreading across the world was "in many ways . . . more important than the research it illustrated."[94]

Finally, Seitz expanded his attack to encompass *all* science, and the scientific establishment itself. Perhaps reflecting his older cousin's turn against his own, Seitz the younger insisted that scientists had betrayed the public trust. Citizens regard "the scientific profession as a bulwark of objectivity and credibility in an otherwise untrustworthy world," he noted, but they shouldn't. Drawing on a popular book by *New York Times* science writers William Broad and Nicholas Wade, *Betrayers of the Truth*, which chronicled failures of honesty and objectivity in the history of science, Seitz insisted that "science bears little resemblance to its conventional portrait."[95] Instead, scientists are guided by such "non-rational factors as rhetoric, propaganda, and personal prejudice." Moreover, for two generations, scientists had been pressured to take responsibility for their work, a legacy of having "known sin" in the Manhattan Project. These "politically motivated scientists," he concluded, "have achieved an easy dominance in matters of science and public policy."[96]

Broad and Wade had argued something that was quite noncontroversial among historians of science: that the portrait of science as the sum of the work of rational individuals just wasn't true, that scientists were fallible, and that more than a few of them had committed fraud. Historians and sociologists in the 1960s and '70s had stressed that scientists work in communities where they are buffeted by the same social forces that prevail in all human communities, plus a few distinctive ones. One of these distinctive pressures was the pressure to innovate, which at times encouraged individuals to cut corners. No academic scholar would have considered this a novel claim; indeed, Broad and Wade had built their arguments on mainstream academic work and acknowledged the assistance of several professors of the history of science. Moreover, in their conclusion, Broad and Wade allowed that "most scientists, no doubt, do not allow the thirst for personal glory to distort their pursuit of the truth."[97] But the right wing had seized on the book, viewing it as a means to undercut science that contradicted their views. Indeed, tobacco industry executives wondered if Wade might be recruited to their cause.[98]

Russell Seitz wasn't interested, however, in advancements in the history of science. He was interested in challenging the concept of nuclear winter, and so he continued to insist that the scientific community was corrupted by left-wing politics. Following on the antiwar movement of the 1960s and the environmental movement of the 1970s, Seitz suggested, left-wing activists had taken over the mainstream of American science. Who were they? The "Federation of American Scientists and Union of Concerned Scientists exercise an almost unopposed and largely invisible role as a coherent force for political action and editorial direction in a broad coalition of organizations and foundations—educational, scientific, and journalistic."[99] The list of activist groups also included the American Association for the Advancement of Science, which publishes *Science*. The president of the AAAS and the president of the American Physical Society "serve on interlocking boards of the FAS, the UCS, the Arms Control Association and the Pugwash movement" (a scientists' organization established after World War II to promote international cooperation, arms control, and disarmament).[100] It would be a surprise if their obvious political biases were not found in their journals, Seitz implied.

This network of bias extended even into the National Academy of Sciences. The NAS had been "politically transformed," according to Seitz, by the election of officers associated with the science advisers to the Kennedy, Johnson, and Carter administrations.[101] These new National Academy officers had then recruited a staff linked back to the UCS and to the FAS, forming what seemed to be a permanent liberal political network at the heart of American science, corrupting the entire apparatus of American science. "The tendency away from objectivity has reached alarming and notorious dimensions in the overselling and subsequent stonewalling that have characterized the 'Nuclear Winter' episode," Seitz concluded.[102]

"Does all this matter?" he asked rhetorically. Indeed it did. Seitz was painting a canvas of politically motivated exclusion—conservative victimhood, as it were. If all this were true—or even if *any* of it were true—it meant that science, even mainstream science, was just politics by other means. Therefore if you disagreed with it politically, you could dismiss it as political.

Was Seitz's presentation objective? Hardly. One would never have known from his discussion that political conservatives had played major roles in the SDI and nuclear winter debate and had been quite able to publish, even in supposedly biased journals. Edward Teller, for example, published an article in *Science* in January 1984 promoting the development of ABMs and his view of nuclear winter in *Nature*—the leading scientific journal in the world.[103] (The *Nature* article accepted that nuclear "winter" might be severe enough to cause widespread crop failure; Teller concluded that the correct solution was to increase food storage.)[104] S. Fred Singer published an attack on nuclear winter in *Science*, and *Science* published a letter from Russell Seitz, as well.[105] Kerry Emanuel, the hurricane specialist who attacked the TTAPS group, was by his own account at that time a conservative.[106]

While Seitz insisted that the scientific establishment was controlled by a liberal agenda, he neglected to mention that the conservative minority, led by Edward Teller, were deeply influential in the Reagan White House. Edward Teller had access to the Reagan White House, and so did William Nierenberg. Both Nierenberg and Jastrow were at one point considered as science advisor to the Reagan administration.[107] Teller and Nierenberg served on the official SDI advisory committee; Fred Seitz was the committee chairman.[108] And many advisors who had served Presidents Kennedy and Johnson continued to serve President Nixon; they had been chosen for their scientific bona fides, not their party credentials. It was scarcely true that conservative scientists were excluded from power.

The National Academy of Sciences might not have been as conservative as the Seitz cousins *wanted* it to be, but it was still viewed by most scientists as a deeply conservative organization. Most historians of science would say that the Academy has an intrinsic conservatism stemming from its dependence on the executive branch. Most of its studies are funded by executive-branch agencies—NASA, the EPA, the Department of the Interior, sometimes the White House—and the Academy has little interest in offending its sponsors, so it tends to step lightly around scientific controversies. Moreover, Academy reports are normally consensus reports, which have to be approved by all of the committee members, as well as by independent reviewers chosen by a Committee on Reviews. The result is often a "least common denominator" conclusion, with a text innocuous enough that everyone involved can agree. Radical claims rarely pass through this process intact—even ones that later turn out to be true.

"Most of the intellectual tools and computation power necessary to demolish TTAPS's bleak vision were already around in 1983; the will, and perhaps the courage, to utilize them was lacking," Seitz insisted, but this statement was plainly false. Other modeling groups, especially at the National Center for Atmospheric Research, took up the subject immediately.[109] And why blame the entire scientific community for the misdeeds of Carl Sagan? Many years later, the right wing continued to lambast Sagan well after the man was dead, while Seitz's attack on nuclear winter was reprised by Rush Limbaugh in the 1990s and by novelist Michael Crichton in the 2000s.[110] What was going on?

The answer is that the right-wing turn against science had begun.

Since the 1970s, scientists had generally supported the goals of arms control and even disarmament, which Teller, Nierenberg, Jastrow, and both Seitzes rejected. Teller and his followers believed that America could achieve permanent military supremacy through weapons engineering (so long as sufficient funds were available), while most other scientists—certainly Bethe, Sagan, and Garwin—thought the arms race could only be managed (and never won), and this would be done primarily through diplomacy. This was disturbing enough to these men, as it threatened to undermine their whole understanding of the role of science and technology in national defense, and thus the role that they, personally, had played in the Cold War. But science was also threatening to undermine these men's views in another, ultimately more important, way.

One of the great heroes of the American right of the late twentieth century was neoliberal economist Milton Friedman.[111] In his most famous work, *Capitalism and Freedom,* Friedman argued (as its title suggests) that capitalism and freedom go hand in hand—that there can be no freedom without capitalism and no capitalism without freedom. So defense of one was the defense of the other. It was as simple—and as fundamental—as that.[112] These men, committed as they were to freedom—liberty as they understood it, and viewing themselves as the guardians of it—were therefore also committed capitalists. But their scientific colleagues were increasingly finding evidence that capitalism was failing in a crucial respect: it was failing to protect the natural environment upon which all life—free or not—ultimately depends.

Working scientists were finding more and more evidence that industrial emissions were causing widespread damage to human and ecosystem health. The free market was causing problems—unintended consequences—that the free market did not know how to solve. The government had a potential remedy—regulation—but that flew in the face of the capitalist ideal. It's not surprising, then, that Russell Seitz's broadsides against science were promoted in business-oriented journals, or that Jastrow's early defense of SDI was published in *Commentary* (a principal voice of neoconservatism) and in the *Wall Street Journal*. Indeed, in 1986, the *Wall Street Journal* published a twenty-four-hundred-word version of Seitz's attack on science—*on page 1*.[113] If science took the side of regulation—or even gave evidence to support the idea that regulation might be needed to protect the life on Earth—then science, the very thing Jastrow, Nierenberg, Teller, and Frederick Seitz had spent their working careers trying to build up, would now have to be torn down.

The attack on nuclear winter was a dress rehearsal for bigger fights yet to come. Barry Goldwater famously argued that extremism in the defense of liberty was no vice. Our story will show that it is.[114]

CHAPTER 3

Sowing the Seeds of Doubt:
Acid Rain

WHILE THE DEBATE OVER STRATEGIC defense and nuclear winter was playing out, another rather different issue had come to the fore: acid rain. While the science of nuclear winter was entirely different from that of acid rain, some of the same people would be involved in both debates. And as in the debate over tobacco, opponents of regulating the pollution that caused acid rain would argue that the science was too uncertain to justify action.

The story begins in 1955, when the U.S. Department of Agriculture established the Hubbard Brook Experimental Forest in central New Hampshire. Experimental forest might seem like an oxymoron—forests are natural; experiments are man-made—but the idea was the same as what scientists do in laboratories: take an object or question and investigate it intensively. In this case the object was the "watershed ecosystem"—the forest, the diverse plants and animals associated with it, and the water flowing through it.

Hydrological studies at Hubbard Brook had been pioneered by a U.S. Forest Service scientist named Robert S. Pierce, who teamed up with F. Herbert Bormann, a biology professor at Dartmouth College, and two bright young assistant professors, biologist Gene E. Likens and geologist Noye M. Johnson. In 1963, Bormann, Likens, Johnson, and Pierce established the Hubbard Brook Ecosystem Study. That same year they discovered acid rain in North America.[1]

"Discovered" is perhaps too strong a word, because naturally acidic rain— caused by volcanoes or other natural phenomena—had been known since the Renaissance, and man-made acid rain had been recognized since the nineteenth century in areas close to industrial pollution in the British Midlands and central Germany.[2] But Hubbard Brook was located in the White Mountains of New Hampshire, a refuge where New Yorkers and Bostonians sought shelter from the haste and waste of urban centers, far from any major cities or factories. Yet its rain had a measured pH of 4 or less (neutral pH is 6, ordinary rain is around 5); one sample measured 2.85—about the same as lemon juice, acidic enough to burn a cut. Acid rain in this remote a setting was new, and worrisome.

The Hubbard Brook work came at a crucial time, coinciding with a shift in American thinking about environmentalism. In the first half of the twentieth century, conservationists such as Theodore Roosevelt, John D. Rockefeller, John Muir, and Gifford Pinchot sought to preserve and protect America's beautiful and wild places, in part by creating special areas—like Yosemite, Yellowstone, and the Grand Tetons—set aside from daily use and development. "Preservationist" environmentalism was broadly popular and bipartisan; Roosevelt was a progressive Republican, Rockefeller a captain of industry. Preservationism was mostly driven by aesthetics and moral values, and by the desire for restorative recreation. It did not depend on science. Preservationists were often interested in science—particularly the natural historical kind, like geology, zoology, and botany—but they did not need science to make their case.

For decades, preservationist environmentalism remained bipartisan. When the Wilderness Act of 1964 designated over nine million acres of American lands as "areas where man himself is a visitor and does not remain," it passed the U.S. Senate by a vote of 73–12, and the House of Representatives by a vote of 373–1.[3] Richard Nixon, a president not generally recalled as a visionary environmentalist, created the Environmental Protection Agency and signed into law several signature pieces of environmental legislation: the Clean Air Act Extension, the Clean Water Act, the Endangered Species Act, and the National Environmental Policy Act. Things were changing, though, and within a few years, Ronald Reagan would begin to shift the Republican Party away from both environmental preservation and environmental regulation, a position that would separate the party from its historic environmentalism, and put it on a collision course with science.

Bills like the Clean Air Act reflected a shift in focus from land preservation to pollution prevention through science-based government regulation, and from local to global. These were profound shifts. *Silent*

Spring—Rachel Carson's alarm bell over the impacts of the pesticide DDT—led Americans to realize that local pollution could have global impacts. Private actions that seemed reasonable—like a farmer spraying his crops to control pests—could have unreasonable public impacts. Pollution was not simply a matter of evil industries dumping toxic sludge in the night: people with good intentions might unintentionally do harm. Economic activity yielded collateral damage. Recognizing this meant acknowledging that the role of the government might need to change in ways that would inevitably affect economic activity.

Collateral damage was what acid rain was all about. Sulfur and nitrogen emissions from electrical utilities, cars, and factories could mix with rain, snow, and clouds in the atmosphere, travel long distances, and affect lakes, rivers, soils, and wildlife far from the source of the pollution. At least, this is what the Hubbard Brook work seemed to show. Throughout the mid to late 1960s and into the 1970s, the Hubbard Brook scientists studied the phenomenon in great detail, writing numerous scientific articles and reports. Then, in 1974, Gene Likens took the lead on a paper submitted to *Science*, declaring unequivocally: "Acid rain or snow is falling on most of the northeastern United States."[4] The phenomenon appeared to have reached Hubbard Brook about twenty years before, they explained, and was associated with the introduction of tall smokestacks in the Midwest.[5] The government would have to take acid rain into account when it set rules and regulations for air pollution.

Chemical analysis showed that most of the acidity was due to dissolved sulfate and the rest mostly to dissolved nitrate, by-products of burning coal and oil. Yet fossil fuels had been burned enthusiastically since the mid-nineteenth century, so why had this problem only arisen of late? The answer was the unintended consequence of the introduction of devices to remove particles from smoke and to reduce local air pollution.

In industrial England, particulate pollution was so bad it killed people—famously in the great smog of London in 1952—and dramatic steps had been taken to reduce it by using taller smokestacks to disperse the pollution more widely, and by installing particle removers, or "scrubbers," at power plants. However, scientific work subsequently showed that the offending particles also neutralize acid, so that removing them inadvertently increased the acidity of the remaining pollution. Particles also tend to settle back to Earth fairly quickly, so while tall smokestacks had successfully decreased local pollution, they had increased regional pollution, transforming local soot into regional acid rain.[6]

But was acid rain a problem? As we will see in later chapters, studies of global warming and the ozone hole involved predicting damage before it was detected. It was the prediction that motivated people to check for damage; research was intended in part to test the prediction, and in part to stimulate action before it was too late to stop—so too, here. It was too soon to tell whether or not widespread and serious ecological damage was occurring, but the potential effects were troubling. They included leaching of nutrients from soils and plant foliage, acidification of lakes and rivers, damage to wildlife, and corrosion of buildings and other structures. Still, if the point were to prevent damage before it happened, then such arguments were necessarily speculative. A careful scientist would be in a bit of a bind: wanting to prevent damage, but not being able to *prove* that damage was coming.

So scientists looked for early warning signs, and they found them. Studies in Sweden suggested that acid precipitation was reducing forest growth. Studies in the United States and elsewhere documented the damaging effects of acidity on plant growth, leaf tissue development, and pollen germination. In Sweden, Canada, and Norway, acidification of lakes and rivers was correlated with increased fish mortality.

Many of the details had been published in very specialized journals (which few journalists or congressional staffers routinely read) or in government reports. The Swedish results were, not unreasonably, mostly published in Swedish.[7] This technical difficulty had also been true for the damage from DDT, much of which had been documented in government reports, which Rachel Carson gathered together in *Silent Spring*, and for the risks of taking the birth control pill, which were first documented in specialized ophthalmology journals when otherwise healthy young women developed mysterious blood clots.[8] This is a characteristic pattern in science: first there is scattered evidence of a phenomenon, published in specialist journals or reports, and then someone begins to connect the dots.

Likens and his colleagues were connecting the dots, and so was Swedish meteorologist Bert Bolin—who would later help to create the Intergovernmental Panel on Climate Change. In 1971, Bolin led a panel on behalf of the Swedish government in preparation for a United Nations Conference on the Human Environment, cochaired by Svante Odén, one of the first Europeans to document the impacts of acid rain on soils.[9] Their report, *Air pollution across national boundaries: The impact on the environment of sulfur in air and precipitation, Sweden's case study for the United Nations conference on the human environment,* laid out the essentials. It explained the evidence of acid rain, the chemistry of how it formed, the physics of how it dispersed, and the effects it had or was likely to have on human health, plant life, soils, lakes and rivers (and the fish in them), and buildings and other structures. (Among other things, the report illustrated corrosion from acid rain of a set of nickel door handles.)[10]

Although the exact magnitude of the acid rain effects was uncertain, their existence and gravity was not, and the Swedes warned against discounting the effects just because they weren't immediate, or fully documented. Although occurring gradually, the effects were serious, and potentially irreversible. However, the situation was not all bleak, because the cause was known, and so was the remedy. "A reduction in the total emissions both in Sweden and in adjacent countries is required."[11]

In science, this sort of clear demonstration of a phenomenon should inspire fellow scientists to learn more. It did. Over the next ten years, scientists around the globe worked to document acid rain, understand its dimensions, and communicate its significance. In 1975, the U.S. Department of Agriculture sponsored the first International Symposium on Acid Precipitation and the Forest Ecosystem.[12] In 1976, the International Association for Great Lakes Research held a symposium, cosponsored by the U.S. Environmental Protection Agency and Environment Canada, on the effects of acid rain on lakes.[13] That same year, Canadian scientists documented the extinction of fish species in acidified lakes in the nickel-mining district of Sudbury, Ontario.[14]

As acid precipitation came to be seen as a global problem, scientists working on it were increasingly able to get their papers published in high-profile journals. In 1976, two Norwegian scientists reported in *Nature* on massive fish kills associated with pH shock, caused by a sudden influx of spring meltwaters from acidic snow and ice.[15] Gene Likens summarized these results in *Chemical and Engineering News*—the official magazine of the American Chemical Society—explaining that acid rain and snow were having "a far-reaching environmental impact." This included sharp declines of fish in lakes and streams, damage to trees and other plants, and corrosion of buildings, and maybe damage to human health.[16]

A few years later, skeptics would argue that the science was not yet really firm, but Likens's summary shows otherwise. However, the way *Chemical and Engineering News* framed it also shows that resistance to the scientific evidence was already beginning to emerge. Likens's argument was clear—acid rain was happening, it was caused by pollution, and it was killing fish and trees and possibly harming people—but in a caption that sat above the article's title, the editors wrote, "The acidity of rain and snow falling on parts of the U.S. and Europe has been rising—for reasons that are still not entirely clear and with consequences that have yet to be well evaluated."[17]

Were the reasons not entirely clear? It depended on what you meant by *entirely*. Science is hard—why so many kids hate it in school—and nothing is ever *entirely* clear. There are always more questions to be asked, which is why expert consensus is so significant—a point we will return to later in this book. For acid rain, the consensus of experts was that anthropogenic sulfur was implicated, but exactly how that sulfur moved through the atmosphere and exactly how much damage it could do was still being worked out. On the other hand, negative effects on fish and forests were clear, so why did *Chemical and Engineering News* suggest otherwise?

Herbert Bormann, at this point teaching at Yale, thought that ambiguity arose from confusing different types of uncertainty. There was no question that acid rain was real. Rainfall in the northeastern United States was many times more acidic than it used to be. The uncertainty was about the *precise* nature of its cause: tall smokestacks—dispersing sulfur higher in the atmosphere—or just increased use of fossil fuels overall?[18] Moreover, while the broad picture was emerging, many details were still to be sorted out, some of them quite important. Chief among these was the question: did we know *for sure* that the sulfur was anthropogenic—made by man—rather than natural? This question would recur in debates over ozone and global warming, so it's worth understanding how it was answered here.

Bolin and his Swedish colleagues had made "mass balance arguments": they considered how much sulfur could be supplied by the three largest known sources—pollution, volcanoes, and sea spray—and compared this with how much sulfur was falling as acid rain. Since there are no active volcanoes in northern Europe, and sea spray doesn't travel very far, they deduced that most of the acid rain in northern Europe had to come from air pollution. Still, this was an indirect argument. To really prove the point, you'd want to show that the actual sulfur in actual acid rain came from a known pollution source. Fortunately there was a way to do this—using isotopes.

Scientists love isotopes—atoms of the same element with different atomic weights, like carbon-12 and carbon-14—because they are exceptionally useful. If they are radioactive and decay over time—like carbon-14—they can be used to determine the age of objects, like fossils and archeological relics. If they are stable, like carbon-13—or sulfur-34—they can be used to figure out where the carbon or sulfur has come from.[19] Different sources of sulfur have different amounts of sulfur-34, so you can use the sulfur isotope content as a "fingerprint" or "signature" of a particular source, either natural or man-made. In 1978, Canadian scientists showed that the isotopic signature of sulfur in acid rain in Sudbury was identical to the sulfur in the nickel minerals being mined there. In later years, some skeptics would argue that the acid in acid rain came from volcanoes (they would say the same about fluorine and ozone depletion, and about CO_2 and global warming), but the isotope analysis showed that *couldn't* be true.[20] In any case there are no active volcanoes in Ontario.

Meanwhile, Noye Johnson—the geologist in the original Hubbard Brook team—and his colleagues had made a crucial discovery. The acid rain story contained an anomaly: rain at Hubbard Brook was acidic, but the pH of the local streams was mostly normal. Why didn't the acidity affect the local streams? Johnson and his colleagues now explained why: the acidic rain was neutralized as it moved through soils. Acid precipitation fell onto the forest floor, where it reacted with minerals in the soils. These reactions stripped the soils of essential nutrients—particularly calcium—and simultaneously buffered the acidity of the water. The buffered water then percolated into local streams. This explained why the pH of the streams was largely unaffected even while the soils were being damaged and overall stream chemistry being changed. The results were reported first in *Science*; then Johnson took the lead on a more detailed paper that would become the third most cited scientific paper ever written on acid rain—published in the elite journal *Geochimica and Cosmochimica Acta*.[21]

The basic science of acid rain was now understood. Scientists had been working steadily on the question for nearly twenty-five years, demonstrating the existence of acid rain, its causes, and its effects on soils, streams, and forests. Major articles had been published in the world's most prominent scientific journals, as well as in many specialist journals and government reports. In 1979, when Likens and his colleagues summarized the arguments for the general scientific reader in *Scientific American*, the magazine's editors did not cast doubt or raise uncertainties. In a summary below the article's title, the editors encapsulated: "In recent decades, the acidity of rain and snow has increased sharply over wide areas. The principal cause is the release of sulfur and nitrogen by the burning of fossil fuels."[22] Not a maybe, possibly, or probably in sight.

Scientific American is often viewed as the place where well-established science is explained to the general public. If so, then we can say that 1979 was the year in which the American people were told about acid rain. As if to seal the case, an eight-year Norwegian study designed to integrate all the evidence related to acid precipitation was reviewed in *Nature* in the summer of 1981. The message? "It has now been established beyond doubt that the precipitation in southern Scandinavia has become more acidic as a result of long-distance transport of air pollution."[23] If this were a court of law, the jury would now have ruled the defendant guilty beyond a reasonable doubt. But science is not a courtroom, and environmental problems involve far more than science. Acid rain had become the first global environmental problem, and with that came global challenges.

Political Action and the U.S.-Canadian Rift

In 1979, the United Nations Economic Commission for Europe passed the Convention on Long-range Transboundary Pollution. Based on the Declaration of the U.N. Conference on the Human Environment—the one for which Bert Bolin's report had been prepared—the convention insisted that all nations have responsibility to "ensure that activities within their jurisdiction or control do not cause damage to the environment of other states or of areas beyond the limits of national jurisdiction."[24] Henceforth, it would be illegal to dump your pollution on someone else, whether you did it with trucks or with smokestacks.

The 1979 convention committed its signatories to control any emissions into the air that could harm human health, property, or the natural environment. Article 7 specifically focused on sulfur, with its impacts on agriculture, forestry, materials, aquatic and other natural ecosystems, and visibility. When the signatories met again in 1985, they set firm limits on sulfur emissions, mandating 30 percent reductions.[25]

Meanwhile, the United States and Canada had started their own bilateral negotiations, and in July 1979, the two countries issued a Joint Statement of Intent to move toward a formal agreement. The statement outlined eight general principles, including prevention and reduction of transboundary air pollution, and development of strategies to limit emissions. The overall goal was "a meaningful agreement that will make a real contribution to reducing air pollution and acid rain."[26]

While negotiations were proceeding behind the scenes, a confluence of scientists, environmentalists, and political leaders convened in Canada in November for an "Action Seminar on Acid Precipitation." The U.S.

government was represented by Gus Speth, chairman of President Carter's Council on Environmental Quality. Speth thought the way forward was clear. Some years before, he noted, industry leaders had objected to emissions reductions, arguing that tall smokestacks could remedy the problem by dispersing the pollutants high in the atmosphere where they would "finally come down in harmless traces." One electric utility, he recalled, had been particularly shrill, taking out newspaper and magazine ads blasting "irresponsible environmentalists" who insisted on absurdly strict emission standards at the expense of jobs and the economy.[27]

Those "irresponsible" environmentalists had been right: the emissions had *not* come down in harmless traces, but as acid rain. This could have been avoided had the power companies done the right thing and controlled pollution at the source, rather than attempting to get around air quality standards by building taller smokestacks and attacking environmentalists. Still, Speth was optimistic, because "both at home and internationally, we are beginning to address the acid rain problem with the seriousness it deserves."[28]

The Carter administration tried to. As Environment Canada concluded that more than half the acid rain falling in Canada was coming from U.S. sources, President Carter signed the Acid Precipitation Act of 1980, which established the National Acid Precipitation Assessment Program (NAPAP), a comprehensive ten-year research, monitoring, and assessment program to determine the effects of sulfur and nitrogen oxides on the environment and human health.[29]

Carter also created the federal Acid Rain Coordinating Committee and began negotiations with the Canadian federal government for scientific and political cooperation on acid rain. Canada and the United States signed a Memorandum of Intent concerning transboundary air pollution, committing both nations to enforcing air pollution control laws and establishing a series of technical working groups to evaluate the scientific basis for a new, stronger treaty to stop acid rain.

Then the political winds in America changed.

Skepticism in the Reagan White House

In 1980, Ronald Reagan came to power in the United States on a platform of reducing regulation, decreasing the reach of the federal government, and unleashing the power of private enterprise. Government, the new president insisted, was not the solution but the problem. Reagan was charismatic, his demeanor relaxed and genial, and his worldview put his administration on a collision course with the scientists working on acid rain.

The new administration did not oppose NAPAP.[30] Diverse groups and constituencies agreed that it made sense to reduce the scientific uncertainties, particularly if the cost of mitigation would be high. But as events unfolded, the administration's position began to diverge from the scientific community, and strongly.

In 1983, the technical working groups established under the 1980 Memorandum of Intent affirmed that acid rain caused by sulfur emissions was real and causing serious damage. The solution was to reduce these emissions—the necessary technology already existed—and if reductions were not made, damage would increase.[31] At the last minute, however, the U.S. representatives seemingly backpedaled. When the working group results were summarized, the U.S. versions were much weaker than the Canadians expected.

The Canadian government asked the Royal Society of Canada to review the documents compiled by the working groups. Chaired by F. Kenneth Hare, a distinguished meteorologist and provost at the University of Toronto, the review panel included two scientists from the United States, one from Sweden, and one from Denmark. They also consulted with other several other experts, including Bert Bolin.

The panel began by noting a common problem among scientists: the tendency to emphasize uncertainties rather than settled knowledge. Scientists do this because it's necessary for inquiry—the research frontier can't be identified by focusing on what you already know—but it's not very helpful when trying to create public policy. The panel wished that the working group scientists had begun with a "clear statement of what is known." This, in their view, included three crucial facts: one, that detrimental acidification of large areas of the continent had been occurring for decades; two, that acid deposition could be quantitatively related to anthropogenic emissions through long-range atmospheric transport; and three, that emissions and pollutants were crossing the U.S.-Canadian border in both directions, so both countries had a stake in preventing them.[32] "The evidence supporting these conclusions is persuasive, and, in the opinion of most Panel members, overwhelms residual uncertainties in our knowledge . . . The existence of a severe problem of environmental acidification . . . is not in doubt."[33] But the U.S. summaries seemed to suggest otherwise.

The reports of the technical working groups revealed overall broad agreement, especially on big picture issues: "The facts about acid deposition are actually much clearer than in other environmental *causes célèbres*," Hare's panel concluded, but when these facts had been gathered, something peculiar had happened.[34] There had been numerous "changes in scientific content" as the report went through successive drafts, changes that made the summaries more ambiguous than the reports themselves.[35] Moreover, while most of the report was "agreed text," the U.S. and Canadian groups had submitted different versions of the conclusions. The U.S. version saw far greater uncertainty than the Canadian one. It did *not* accept that cause and effect had been established, on the grounds that the relative importance of different contributing factors had not been quantified, and potentially off-setting processes had not been fully investigated.

This was like saying that we know that both cigarettes and asbestos cause lung cancer, but we can't say either is proven, because we don't know exactly how much cancer is caused by one and how much by the other, and we don't know whether eating vegetables might prevent those cancers. The Canadian group fell short of accusing the United States of tampering with the evidence, but they certainly implied it. In the panel's words: "The U.S. version of the text cannot be reconciled with the evidence as presented in the *agreed* text."[36] The following year, Environment Canada put it this way: "In each country independent peer review experts have indicated the need for action based on what we now know."[37] But that was not how the U.S. government saw it, and in January 1984 Congress rejected a joint pollution control program. What had happened?

SCIENCE IS NEVER FINISHED, so the relevant policy question is always whether the available evidence is *persuasive*, and whether the established facts outweigh the residual uncertainties. This is a judgment call. Chris Bernabo, who worked at the White House Council on Environmental Quality at the time and served as research director for the Interagency Task Force on Acid Precipitation, suggests that because so much more was at stake for Canada—70 percent of their economy at the time came from forests and fish or tourism related to them—it was only natural that they would interpret the evidence as more dire than their U.S. counterparts would.[38] Pollution went across the border in both directions, but by far the larger share came from the United States, which would therefore bear most of the burden of cleanup. As Bernabo puts it, for any problem, the degree of scientific certainty demanded is proportional to the cost of doing something about it. So the United States was more resistant to accepting the evidence and demanded a high level of certainty.[39]

No doubt this is true, but it doesn't quite explain the gap between the science and the summaries. Scientists are supposed to summarize science, and let the chips fall where they may. However, the summaries were not written by the scientists who had done the research. They were written, at least in part, by interagency panels—groups of scientists from U.S. government agencies, including the Department of Energy and the EPA, with relevant (or roughly relevant) expertise.[40] Government scientists are usually conscientious individuals who strive to be objective, but sometimes they come under political pressure. Even when they don't, they often can't help but be mindful of the positions of their bosses. And the position of the U.S. boss was clear. Gene Likens recalls that both agencies were very reticent "to do anything that would jeopardize their positions in the Reagan White House."[41] Richard Ayres, chairman of the National Clean Air Coalition, who worked to ensure passage of the acid rain control amendments to the Clean Air Act, recently put it more bluntly: "This was during the Reagan years, when acid rain was almost as verboten [to acknowledge] as global warming under George W. Bush."[42]

Getting a Third Opinion

In 1982, while the technical working groups were at their task, the White House Office of Science and Technology Policy (OSTP), under the direction of physicist George Keyworth, commissioned its own panel to review the evidence on acid rain. The National Academy of Sciences had already reviewed the available evidence the previous year—so some wondered why the OSTP needed yet another report.[43] The *New York Times* reported that most observers assumed that the review of the joint U.S.-Canadian work would be done by a joint panel of the National Academy and the Canadian Royal Society, and called it "unusual" that the administration would bypass the Academy and use "an outside group," picked by the White House.[44] The *Washington Post* noted that the president was "certainly entitled to appoint his own panel of experts," but he had done it in a manner that was "far from reassuring."[45]

The *Post* was right about presidential prerogative—the president may of course ask anyone he likes for information—and there are plenty of occasions where scientists agree that more information is needed.[46] But that wasn't the case here. In 1981, the Academy had stated unequivocally that there was "clear evidence of serious hazard to human health and the biosphere," and that continuing business as usual would be "extremely risky from a long-term economic standpoint as well as from the standpoint of biosphere protection." And they concluded that the situation was "disturbing enough to merit prompt tightening" of emissions standards, perhaps by as much as 50 percent.[47] A major EPA report the following year agreed. The *Wall Street Journal* reported on the EPA study under the headline ACID RAIN IS CAUSED MOSTLY BY POLLUTION AT COAL-FIRED MIDWEST PLANTS, STUDY SAYS, and quoted an EPA spokesman explaining how the twelve-hundred-page report had been compiled over two years by forty-six industry, government, and university scientists to produce a "scientifically unimpeachable assessment."[48]

The administration's outright rejection of the conclusions of the nation's most distinguished and qualified experts caused considerable consternation in scientific and regulatory circles. But what is particularly striking to our story is that the man they asked to assemble and chair the panel was someone we have already met—a man who had never worked on acid rain, but was well-known to the Reagan White House—Marshall Institute cofounder and SDI defender William A. Nierenberg.

Nierenberg already had ties to the Reagan White House. When Reagan was elected in the autumn of 1980, Nierenberg had been approached as a candidate for the position of president's science advisor. It was a position any scientist would covet and Nierenberg did, soliciting supporting letters from numerous colleagues.[49] Nierenberg was also interviewed by National Security Advisor Henry Kissinger for a special position as a liaison between his office and the science advisor.[50]

Ultimately, the nod for science advisor went to Keyworth, and the special position didn't materialize. Nierenberg was offered a job as the head of the National Science Foundation, which he turned down. However, he served the administration in several other ways. He was invited to be a member of Reagan's Transition Advisory Group on Science and Technology (to make suggestions for scientists to serve in high-level positions), and he served as a member of the Townes Commission to select a launching platform for the MX mobile ballistic missile. In March 1982, Nierenberg received a personal note from "Ron" thanking him for this work, and that November, a nomination to the National Science Board—a prestigious position that Nierenberg had asked numerous friends and acquaintances to suggest him for, including the Republican mayor of San Diego (and later governor of California), Pete Wilson.[51]

When the United States began to run into conflict with Canada over acid rain, Nierenberg was putting the finishing touches on a major report of the National Academy of Sciences on the impact of carbon dioxide on climate, arguably the first comprehensive scientific assessment on the subject. Its conclusions were fully in line with the position of the administration—that no action was needed other than more scientific research—and the administration used it publicly to counter work being done at the time by the EPA with a graver outlook.[52] So it is perhaps unsurprising that when the administration needed someone to grapple with acid rain, they turned to Bill Nierenberg.

Like his fellow physicists Frederick Seitz and Robert Jastrow, Nierenberg was a child of the atomic age, a man for whom the global anxieties and national challenges of the Cold War had offered remarkable personal opportunities. Raised in the Bronx by immigrant parents, Nierenberg had attended the prestigious Townsend-Harris High School (as did Robert Jastrow) and the City College of New York, where he studied physics, won a coveted fellowship to spend a year in Paris, and returned to New York in 1939 fluent in French and fearful of fascism.

In September 1942 he entered Columbia University for his Ph.D. He soon found himself working on isotope separation: how to isolate fissionable uranium for the atomic bomb. After graduating, he taught nuclear physics at the University of California, Berkeley, and in 1953 he became director of Columbia University's Hudson Laboratory, created to continue scientific projects begun on behalf of the U.S. Navy during World War II, particularly underwater acoustic surveillance of submarines. He subsequently held a series of positions at the interface between science and politics, including succeeding Seitz as NATO's assistant secretary general for scientific affairs. In 1965, Nierenberg became director of the Scripps Institution of Oceanography in La Jolla, California, an institution busy at the time applying scientific knowledge to national security problems, particularly in research linked to underwater surveillance of Soviet submarines and targeting submarine-launched ballistic missiles.[53]

Like Seitz and Teller, Nierenberg hated environmentalists, whom he viewed as Luddites (particularly for their opposition to nuclear power), and like Seitz and Teller he was an unapologetic hawk. He had been a fierce defender of the Vietnam War. Three decades later he still harbored bitterness toward academic colleagues who had failed to defend military-sponsored work, as well as residual anger over the disruption and violence that left-wing students had brought to campuses in the 1960s. Recalling an incident in which students at the University of California, San Diego, had threatened to march onto the nearby Scripps campus in protest of military-sponsored work, Nierenberg became visibly upset. Moreover, he insisted that the students were mistaken—because there was no classified work being done at Scripps. But this was untrue; many Scripps scientists—including Nierenberg—had security clearances to work on secret military projects and had done so for years, even decades.[54]

Nierenberg was a man of strong will and even stronger opinions—a good talker but not always a good listener. Some colleagues said that the old adage about famous physicists definitely applied to him: he was sometimes in error but never in doubt. And he was fiercely competitive, often debating until his adversaries simply gave up. Still, Nierenberg was a highly respected scientist and administrator, and if at times he was overconfident, it wasn't without justification: even his detractors thought he was brilliant. He had a way of keeping a conversation going, because he knew so much. He was an authority, but an accessible one. He pushed you around, but somehow you didn't mind. He was interesting to be around. He could even be fun. Perhaps in part for these reasons, when he asked you to serve on a committee, you'd most likely say yes. One of the people who said yes to Nierenberg was Sherwood Rowland.

In 1982, Rowland was already pretty famous. In the early 1970s he had realized that certain common chemicals—the so-called chlorinated fluorocarbons, or CFCs, used in hairsprays and refrigerants—could damage the Earth's protective ozone layer. In the mid-1980s, a giant ozone hole was discovered, and in the 1990s, Rowland, together with colleagues Mario Molina and Paul Crutzen, would be awarded the chemistry Nobel Prize for this work. After that, he would never lack for people eager to hear what he had to say, and usually to agree with it.

But in the early 1980s, when Nierenberg asked him to serve on the acid rain panel, Rowland worried that he would be lonely. He was fairly certain acid rain was a real problem, but he wondered if the rest of the panel felt the same way. Things got off to an inauspicious start at the first meeting, where Nierenberg had arranged for a briefing by Dr. Lester Machta, an expert on radioactive fallout. Rowland had encountered Machta in the 1950s, when radioactive strontium had been detected in the baby teeth of children in St. Louis. Scientific work showed that it came from the U.S.

weapons testing site in Nevada, but for a long time the official position was to blame *Soviet* fallout. Machta had been a spokesman for that view. The prospects for an unbiased acid rain panel didn't look good.

But Rowland soon found that he was not alone. Nierenberg's panel also included Gene Likens, and after the Machta presentation, Rowland, Likens, and a few others discovered over dinner that they were in general agreement about acid rain. As Likens recalled, the food was extremely good, too.[55] Rowland felt that things were going to be all right. Events turned out to be more complicated.

The Nierenberg Acid Rain Peer Review Panel

Nierenberg's panel was charged with reviewing the output of the technical working groups that had been impaneled under the U.S.-Canada bilateral agreement. They concluded that it was "basically sound and thorough," and they affirmed that acid rain was serious and sufficiently documented to warrant policy action now.

Nierenberg's panel summarized:

> Large portions of eastern North America are currently being stressed by wet deposition of acids, by dry-deposition of acid-forming substances, and by other air pollutants . . . The principal agent altering the biosphere acidity is traceable to man-made sulfur dioxide (SO_2) emission . . . The panel recommends that cost effective steps to reduce emissions begin now even though the resulting ecological benefits cannot yet be quantified.[56]

Of course, there were still details to be worked out, but these might take "ten, twenty, or fifty years" to resolve, and that was too long to wait.[57] There was no need to wait to dot every scientific *i* and cross every technical *t*, because you had enough information to begin to act now. This was a pretty strong conclusion. It would have been even stronger, but for political interference.

Bill Nierenberg had boasted about how six of the nine of the members of his acid rain panel were members either of the National Academy of Sciences or the National Academy of Engineering. He had also boasted that he had handpicked all the members—that is, all but one.[58] That one was S. Fred Singer, who had been suggested to Nierenberg by the White House

Office of Science and Technology, and would contribute an appendix suggesting that, despite the conclusions of the Executive Summary, we really *didn't* know enough to move forward with emissions controls.

Why was Singer on this committee?

Like Jastrow, Seitz, and Nierenberg, Fred Singer was a physicist who owed his career to the Cold War. While a graduate student at Princeton during World War II, he had worked for the navy on underwater mine design; after the war, he moved to the Applied Physics Laboratory at Johns Hopkins University, where he pursued upper atmosphere rocketry research. And also like Jastrow, Seitz, and Nierenberg, Singer rapidly moved into administrative positions at the interface between science, government, and the military. In the early 1950s he served as a scientific liaison officer for the naval attaché in London, and later as the first director of the U.S. National Weather Satellite Center, an organization that drew on military rocketry and expertise to develop civilian weather prediction. However, despite his bona fides, Singer's relations with his colleagues were sometimes testy. Some colleagues think Singer's attitude problems began in the mid-1950s, when scientists were making plans for what would become the International Geophysical Year (IGY)—an international collaborative effort to collect synoptic geophysical data around the globe.

An illustration of a satellite orbiting the globe would later become the official symbol of IGY, but in the mid-1950s it was unclear whether satellites were even feasible, and whether scientists might have to make do with rockets that penetrated the upper atmosphere without going into orbit. Singer, who had been using rockets to study cosmic rays and the Earth's magnetic field, became a strong advocate for a satellite. As NASA historian Homer Newell recounts it, Singer's outspokenness generated friction in part because of his aggressive demeanor, and in part because he acted as if the idea of using satellites for scientific research was his alone. Scientists working with the navy and air force had been trying to determine if a satellite was feasible, but because of security restrictions they couldn't discuss it openly. Their calculations suggested that Singer's proposal was overly optimistic; it could be done, but not as readily as Singer said.[59] In the end, the International Geophysical Year did include geophysical instrumentation of satellites, but Singer felt he'd been insufficiently credited, and continued to antagonize colleagues by implying that he had invented the satellite concept.[60]

Shortly after the IGY incident, Singer moved to the National Weather Satellite Center. This center had been organized as part of the Weather Bureau, rather than as part of the space program, setting up further conflict between Singer and scientific colleagues at NASA who thought all satellites should be overseen by the space agency.[61] In the years that followed, Singer moved away from science and into government and policy.[62] In the 1970s, he served in the Nixon administration as deputy assistant secretary in the Department of the Interior under Walter J. Hickel, and then as deputy assistant administrator at the EPA. So Singer and Nierenberg had much in common—both physicists, both conservative politically, both with a history of working at the interface between science and government. Indeed, the commonalities went perhaps even deeper. Born in Vienna in 1924—the s stood for Siegfried—Singer had personally witnessed the threat of looming fascism, just as Nierenberg had during his year in France in 1939. However, there was one interesting difference. Throughout the 1960s, Singer had been an environmentalist.

In a book published in 1970 (and reprinted in 1975), based on a symposium held by the American Association for the Advancement of Science (AAAS) on "Global Effects of Environmental Pollution," Singer made clear that he shared the view later famously credited to Roger Revelle: that human activities had reached a tipping point. Our actions were no longer trivial; we were capable of changing fundamental processes on a planetary scale. Numerous emerging problems—acid rain, global warming, the effects of DDT—made this clear.

Like most of his colleagues, Singer believed there was a need for more science, but in 1970 he argued that one cannot always wait to act until matters are proven beyond a shadow of a doubt. Singer cited the famous essay "The Tragedy of the Commons," in which biologist Garrett Hardin argued that individuals acting in their rational self-interest may undermine the common good, and warned against assuming that technology would save us from ourselves. "If we ignore the present warning signs and wait for an ecological disaster to strike, it will probably be too late," Singer noted. He imagined what it must have been like to be Noah, surrounded by "complacent compatriots," saying, " 'Don't worry about the rising waters, Noah; our advanced technology will surely discover a substitute for breathing.' If it was wisdom that enabled Noah to believe in the 'never-yet-happened,' we could use some of that wisdom now," Singer concluded.[63]

Singer made a similar argument in a book on population control published in 1971, in which he framed the debate about population as a clash between neo-Malthusians, who focused on the limits of resources, and Cornucopians, who believed that resources are created by human ingenuity and are therefore unlimited. In 1971, Singer did not take sides, but stressed that the Cornucopian view hinged on the availability of energy: if population increases and one has to work harder to obtain available resources, then "per capita energy consumption must necessarily increase."[64] Energy was key; the other crucial issue was protecting the quality of life. "Environmental quality is not a luxury; it is an absolute necessity of life,"[65] Singer wrote, and so it was "incumbent upon us . . . to learn how to reduce the environmental impact of population growth: by conservation of resources; by re-use and re-cycling; by a better distribution of people which reduces the extreme concentrations in metropolitan centers; but above all by choosing life styles which permit 'growth' of a type that makes a minimum impact on the ecology of the earth's biosphere."[66]

Somewhere between 1970 and 1980, however, Singer's views changed. He began to worry more about the cost of environmental protection, and to feel that it might not be worth the gain. He also adopted the position he previously attributed to Noah's detractors: that something *would* happen to save us. That something would be technological innovation fostered in a free market. Singer would come down on the Cornucopian side.[67]

In 1978 Singer developed an argument for cost-benefit analysis as a way to think about environmental problems in a report for the Mitre Corporation—a private group that did extensive consulting to the government on energy and security issues. "In the next decade," he wrote, ". . . the nation will spend at least 428 billion dollars to reach and maintain certain legal air and water standards. To know whether these costs are in any sense justified, one must carry out a cost-benefit analysis. This has not been done."[68]

In the years to come economists would grapple with how to value species conservation, clean air and water, beautiful views, pristine landscapes. The problem then, as it largely remains today, is that it is easier to calculate the cost of a pollution control device than the value of the environment it is intended to protect: who can calculate the benefit of a blue sky? Meanwhile Singer did his own analysis, focusing on the fairly well-known costs of emissions control, and glossing over the admittedly harder-to-quantify benefits of clean air and water. In doing so, he radically changed his views. "The public policy conclusion from our analysis is that where a choice exists, one should always choose a lower national cost, i.e. a conservative approach to air pollution control, which will not inflict as much economic damage on the poorer segment of the population."[69] Singer had emphasized the potential cost to those who could afford it least—a point with which many liberals would concur—but if you left off his final phrase, you had a view that many free market conservatives, as well as polluting industries, found very attractive.[70]

When Nierenberg had been finding scientists to serve the Reagan administration, Singer had sent Nierenberg his CV.[71] He stressed that he was a longtime Republican and member of the Republican National Committee, with close ties to George H. W. Bush and Virginia Republican senator John Warner. Above all, he had "the right political-economic philosophy to mesh with the Reagan administration."[72]

Singer also sent Nierenberg two articles he had written on oil markets, which showed how he had moved away from his earlier environmentalism to embrace a market-based approach. The gist of Singer's argument was supply and demand: if the price of oil went up, supply would increase—either directly, due to more exploration or more efficient refining, or indirectly, as the price of other fuels, such as nuclear, became competitive—so there was no need for government intervention. The "oil industry is making . . . major adjustments in response to market forces, without specific government help or advice," Singer wrote. To increase supply, one simply needed to deregulate the natural gas industry, license nuclear power plants more quickly, and expand oil drilling in Alaska and offshore. In other words, just unleash the power of the marketplace by decreasing government regulation and restriction of economic activity.[73] In an article published in the *Wall Street Journal* in February 1981, Singer predicted that by the 1990s, the world would be using "less than half of the oil it uses today," and by 2000 the U.S. "oil dependence on the Middle East" would "become a thing of the past."[74] Too bad he wasn't right.

Singer had high ambitions, suggesting himself to run either NASA or NOAA—the National Oceanic and Atmospheric Administration. He was also interested in the Department of Energy, the Department of the Interior, and the EPA, where he thought he could serve in the number two position, or having an impact at the State or Treasury Department, or even "a greater impact on government operations . . . from OMB."[75] Singer was offered the number two spot at NOAA, which he turned down on the grounds that it would not permit him to accomplish "any substantial policy initiatives." However, if the administration had a future opening where he could exert some policy influence, such as on a presidential commission, he remained interested.[76] In 1982, the opportunity to influence policy arose.

When the White House asked Nierenberg to chair the Acid Rain Peer

Review Panel, Nierenberg sent a detailed list of proposed members, with various options including "a foreigner, if wanted." (The foreigner was Svante Odén—one of the original discoverers of acid rain—but he was not wanted.) The White House accepted most of the people on Nierenberg's list, but rejected Gordon MacDonald, a geophysicist and former advisor to Richard Nixon who had warned about global warming in 1964, and who Nierenberg had labeled "A must!"[77] They also rejected biologist George Woodwell, the ecologist we met in chapter 2 considering the biological impacts of nuclear winter, who Nierenberg described as "deeply concerned about environmental degradation and active in environmental protection issues."[78] And despite plenty of names still left over, they added one of their own: Fred Singer.[79]

Besides being the only member proposed by the White House, Singer was also the only member without a regular, full-time academic appointment. He was affiliated with the conservative Heritage Foundation in Washington, D.C., which advocated unrestricted offshore oil development, transfer of federal lands to private hands, reductions in air-quality standards, and faster licensing of nuclear power plants.[80] (Heritage continues to oppose environmental regulation: in 2009, their Web site featured the article "Five Reasons Why the EPA Should Not Attempt to Deal with Global Warming.")[81]

Nierenberg did not propose Singer, but he did know Singer's views on acid rain. In January 1982, Gordon MacDonald had made a presentation to the State Department on acid rain, and in a three-page letter to Nierenberg two weeks later, Singer raised numerous doubts about it. While most studies focused on sulfur, MacDonald had called attention to NO_x—oxides of nitrogen, mostly from automobiles, that can also contribute to atmospheric acidity—suggesting that tighter emissions standards for cars might be needed. Without exactly saying that MacDonald was wrong (and later research would show that he wasn't), Singer insisted that the problem was very complex, it was premature to suggest remedies, and in any case technological solutions might obviate the need for emissions controls.[82] This was pretty much the same tack he took on the acid rain panel.

When Nierenberg's panel convened in January 1983, they began by discussing what their procedure would be.[83] The panel agreed that any conflicting or dissenting views would be included in the report; there was no discussion of any appendices.[84] In June, the White House Office of Science and Technology Policy asked the panel for an interim report and summary of research recommendations. The OSTP then prepared a press release.[85] The word was out on the street that the report would be a strong one, and the *Wall Street Journal* reported on June 28: REAGAN-APPOINTED PANEL URGES BIG CUT IN SULFUR EMISSIONS TO CONTROL ACID RAIN.[86] They were right.

The draft version of the press release, which was admittedly long at nearly five full single-spaced pages, pulled no punches. It began by noting that the United States and Canada together emitted more than 25 million tons of sulfur dioxide per year, and then stated: "The incomplete present scientific knowledge sometimes prevents the kinds of certainty which scientists would prefer, but there are many indicators which, taken collectively, lead us to our finding that the phenomena of acid deposition are real and constitute a problem for which solutions should be sought."[87] It was a little verbose, but the point was clear. Lakes were acidifying, fish were dying, forests were being damaged, and the time had come to act.[88] "Steps should be taken now which will result in meaningful reductions in the emission of sulfur compounds."[89]

The strongest part of the press release was perhaps the two paragraphs on the fourth page that dealt with long-term damage. The first noted that the damage being discussed might not be irreversible in an absolute sense, but that it was legitimate to use that term when discussing damage that could take more than a few decades to repair. The second paragraph dealt with the most worrisome issue: that soil damage might set off a cascade of effects at the base of the food chain. "The prospect of such an occurrence is grave."[90]

However, when the draft came back to Bill Nierenberg from the White House, these two paragraphs had been struck out, and someone at the OSTP—probably senior policy analyst Tom Pestorius, the committee's official liaison with the OSTP—had placed a set of numbers in the margins suggesting that the remaining paragraphs be presented in a different order. Rather than start with the fact of the 25 million tons of SO_2 emissions per year, the White House wanted to start with a statement that earlier actions taken under the Clean Air Act were a "prudent first step," and then proceed to the discussion about incomplete scientific knowledge. In other words, the White House version would *not* begin by stressing the problem—massive sulfur emissions that caused acid rain—but by stressing that pollution was already partially controlled, and then moving straight on to the uncertainties that might be taken to suggest that further controls were not justified.

A second document, "Overall Recommendation of the Acid Rain Review

Panel," also came back to Nierenberg with suggested revisions. "Enclosed is a draft substitute first paragraph written by Fred Singer," with Singer's initials on the document. Singer's version again began differently from the panel's: "Acid Deposition (A.D.) is a serious problem, but not a life-threatening one. It is at once a scientific problem, a technological problem, as well as an institutional problem." The summary then made three enumerated points. The first sentences of each read as follows:

1) <u>Scientifically</u> we are not certain of all the causes of A.D. . . .

2) <u>Control technologies</u> are still costly and unreliable . . .

3) <u>Institutionally,</u> the Clean Air Act, and successive amendments, have [*sic*] wrestled with the problem of setting air standards to protect human health and property.[91]

Singer suggested that he was proposing a reasonable middle ground. "We would recommend a middle course: Removing a meaningful percentage of pollutants by a least-cost approach and observing the results, before proceeding with a more costly program."[92] This might have been a reasonable recommendation. It might even have been correct. But it was not what the peer review panel had said.

So now there were two different versions of the problem. One, written by the panel, acknowledged the uncertainties but insisted that the weight of evidence justified significant action. The other, written by Singer (perhaps with help from the White House), suggested that the problem was not so grave, and that the best thing was to make only small adjustments and see if they helped before considering anything more serious. These were not the same view at all. Which one would prevail?

Throughout the panel deliberations, Singer highlighted uncertainties in the science and emphasized the costs of emissions controls. On more than one occasion, he presented views that echoed those promoted or circulated by the electric power industry. One of these was the suggestion that forests in Germany were not actually in decline—or if they were, it wasn't because of acid rain—a view promoted by Chauncey Starr, a nuclear physicist at the Electric Power Research Institute (EPRI). In a letter to Keyworth in August, copied to Nierenberg, Starr had insisted that the panel review should contain a "comprehensive societal benefit/risk/cost analysis," because "public anxiety" was being unnecessarily inflamed.[93] What was really needed was more research.[94] Starr continued the argument in additional letters to

Nierenberg; in November, Singer presented a set of arguments that largely paralleled Starr's points. He also circulated a paper produced by the so-called National Council of the Paper Industry for Air and Stream Improvement arguing that acid rain had not been shown to affect tree growth, and a set of papers arguing for market-based, rather than regulatory, approaches to clean air (even though it was well outside the charge of the committee to consider, much less propose, solutions).[95] Perhaps to suggest that other forms of pollution were more serious than acid rain, he circulated a paper outlining crop damage from ground-level ozone.[96]

When Nierenberg circulated a draft of research recommendations in August 1983, Singer added several comments consistent with the idea that the problem might be overstated and the cost of fixing it too high. Where the report said that a pressing need was to understand the ecological consequences, Singer changed this to "ecological and economic consequences." In a discussion of emissions data, Singer added, "A better characterization of natural sources is required."[97]

That the science was uncertain, that more research was needed, that the economic consequences of controlling acid rain would be too great, and that acid rain might be caused by natural sources: these claims were all part of the position taken by the electrical utility industry. As *Time* magazine put it, the utility industry was "vociferously opposed to any emission control program without further research into the causes of acid rain," and insisted that "installing scrubbers could break the economic backbone of the Midwest."[98]

But the cause of acid rain *was* known, and it was *not* natural. Singer found himself out on a limb among his scientific colleagues. Rowland and Likens's memory is that no one supported Singer's views, which were in any case seen as irrelevant to the panel's charge to summarize the *science*. No one, that is, except Tom Pestorius from the White House Office of Science and Technology Policy. In April 1983, Pestorius had forwarded to the committee some "unsolicited" material from a representative of the Edison Electric Institute—a utility group—which Gene Likens dismissed as "uncritical propaganda" from a man with a "track record for obfuscating the obvious and for generating 'red herrings' . . . pleasing to his employer."[99]

Someone on the panel also circulated a document produced by a private consulting firm criticizing earlier National Academy work on acid rain. The consultants' report asserted that the scientific arguments for adverse effects from acid rain were "speculative" and "oversimplified," the conclusions "premature" and "unbalanced," and also added that some crops might benefit from acid rain.[100] While the record doesn't say who circulated this report to the panel, its complaint that "relative costs and benefits of available options are not considered" certainly resonated with Fred Singer's views. But economic analysis was neither within the charge nor the expertise of the Academy scientists, so they were being criticized for not doing something they had not been asked to do.

A few weeks later, Singer sent a set of materials to John Robertson, a major at the West Point who was serving as the committee's executive secretary. Writing on Heritage Foundation letterhead, Singer asked Robertson to distribute to the panel a long document that "set forth the Administration's *general* perspective and policy on global issues."[101] These included the claims that "although important 'global' problems do exist, recent . . . projections . . . are less alarming than most previous studies." Moreover, these problems "all seem amenable to solution . . . and promising new approaches and technologies are emerging." Above all, the administration wished to stress the "importance of the market place for achieving environmental quality goals." A primary goal of U.S. policy in the 1980s would be to "improve the functioning of the market place by removing trade barriers and . . . in particular to expand food, minerals and energy availability over the long term."[102]

Whether or not these claims were true and the policy goals reasonable was irrelevant—or should have been irrelevant—to the panel. Their job was to summarize and critique the science of the U.S.-Canada technical working groups. That is what it means to do a scientific peer review. The White House's perspectives were irrelevant to that task, but Fred Singer didn't see it that way.

Gene Likens recalls one particularly frustrating moment, when he blurted out, "Fred, you're saying that lakes aren't valuable. They *are* economically valuable. Let me give you an example. Let's say every bacterium is worth $1. There are 10^4–10^6 bacteria [ten thousand to a million] in every milliliter of water. You do the math." Singer replied, "Well, I just don't believe a bacterium is worth a dollar," and Likens retorted, "Well, prove that it isn't." Twenty-six years later, Likens recalled, "It was the only time I ever shut him up."[103]

Singer was effectively insisting that if the scientists couldn't *prove* the value of things (like bacteria), then they had no value. It was a foolish argument, and no one on the committee accepted it, not even Bill Nierenberg.[104] "If we went by absolute science," Nierenberg noted at another juncture, "there would be nothing to do."[105] When the panel's report came out in the summer of 1984, Nierenberg summarized its gist: "Even in the absence of precise scientific knowledge, you just know in your heart that you can't throw 25 million tons a year of sulfates into the Northeast and not expect some . . . consequences."[106]

Having failed to sway his fellow panelists, Singer tried another tack. In September 1983, civil engineer William Ackermann, the panel's vice-chair, had presented the committee's interim conclusions to the House of Representatives Committee on Science and Technology.[107] Singer wrote a six-page letter to the committee chair taking issue with Ackermann's testimony, which he claimed was unsupported by sufficient data. He argued that evidence of damage was lacking, or limited, that a good deal of soil acidification is natural, that only certain kinds of soils were susceptible to acid damage, and that acidification might in some cases be beneficial. Some of Singer's claims—for example, that some soils are naturally acidic—were true, but irrelevant. Others were misleading, insofar as he was the only member of the committee who held the opinion that the evidence of potential soil damage was "insufficient."[108] Whether or not the House Committee chairman believed Singer's claims, his letter certainly would have had at least one effect: to make it appear that the committee was divided and there was real and serious scientific disagreement. The committee *was* divided, but it was divided 8–1, with the dissenter appointed by the Reagan White House.

Singer was supposed to be writing the final chapter of the report, on the feasibility of estimating the economic benefits of controlling acid pollution. It was to be an investigation of how you *might* try to place a dollar value on nature—and what would be lost if you didn't.[109] Somehow, along the way, it turned into the claim that if you did nothing, it cost you nothing. Singer was continuing to equate the value of nature to zero. This was not something the others would accept, so the panel had three choices: keep working until they came to agreement, delete the chapter altogether, or relegate it to an appendix.

As the panel neared completion of their report, this issue remained unresolved. When the report finally appeared, the third solution had been chosen. While the rest of the report was jointly authored—the norm for National Academy and other peer review panel reports—Singer's appendix was all his own. It began with a strange claim: that the benefits as well as the costs of doing nothing were zero. This was patently at odds with the rest of the report, which stressed repeatedly the ecological costs of acid deposition.

If the panelists were correct, then the cost-benefit question at stake was how much money should be spent on pollution abatement to avoid or minimize these ecological costs. Singer ignored this, considering cost only in terms of the cost of pollution control—ignoring the cost of ecological damage. Moreover, one *could* calculate the cost of ecological damage and the value of avoiding it: in 1979, the White House Council on Environmental Quality had done just that, and placed the value of air quality improvements since the passage of the Clean Air Act at $21.4 billion—*a year*.[110]

Singer also presumed that the costs were mostly accrued in the present, but the benefits in the future, and therefore the latter had to be discounted in order to make them commensurate with the former. (That is to say, a dollar in the future is not worth as much to you as a dollar now, so you "discount" its value in your planning and decision making. How much you discount it depends in part on inflation, but also in part on how much you value the future.) Discounting would later become a huge issue in assessing the costs and benefits of stopping global warming, as long-term risks can be quickly written off with a sufficiently high discount rate.[111]

Was Singer doing that here? Not quite. He acknowledged that the choice of discount rate was "important," but then changed the subject to argue that because there are many sources of pollution you could spend a great deal of money addressing one source without any immediate benefit.[112] In principle this was true, but it was not what the scientists had said about acid rain. They had concluded that there was one dominant cause—sulfur dioxide—and that cutting it by 25 percent would yield rapid benefits. Singer also asserted that because pollution control often was applied only to new sources—think of automobiles—this also made it very hard to achieve quick results. True again, but the analogy to cars was faulty, because while it was very hard to put new pollution control devices on old cars, the available technology to control sulfur at power plants could be easily applied to old plants as well as new ones. Singer himself acknowledged that there was a strong argument in favor of applying new regulations to both old and new sources, lest you create a perverse incentive to stick to obsolete technologies. How regulations worked depended upon how policy makers designed them, and that was a matter of political power and will, rather than a law of nature.

Singer's appendix did not actually include the analysis he insisted was needed. When he reached the point of actually making the analysis, he demurred, arguing that both the costs and the benefits were extremely difficult to quantify, and simply jumped to his preferred conclusion: that the most practical approach would be a market-based one. Using transferable emissions rights, the government would determine the maximum allowable pollution, and then grant or sell the right to pollute to parties who could then use, sell, or trade those rights.[113]

In later years, emissions trading *would* be used to reduce acid pollution—and today many people are looking to such a system to control the greenhouse gases that cause global warming. Yet economists (and ordinary people) know that markets do not always work.[114] Indeed, many economists would say that pollution is a prime example of market failure: its collateral damage is a hidden cost not reflected in the price of a given good or service. Milton Friedman—the modern guru of free market capitalism—had a name for such costs (albeit an innocuous one): he called them "neighborhood effects."[115]

Friedman tended to dismiss the significance of neighborhood effects, suggesting that the evils of expanded government power to prevent them generally outweigh any plausible benefit. "It is hard to know when neighborhood effects are sufficiently large to justify particular costs in overcoming them and even harder to distribute the costs in an appropriate fashion," he wrote in his classic work, *Capitalism and Freedom*.[116] So in the vast majority of cases it would be better to let the market sort things out—and this is pretty much what Singer concluded about acid rain. Without any analysis of the details or an example of a successful market-based pollution control scheme, he simply asserted that a system of transferable emissions rights "would guarantee that the market will work in such a way as to achieve the lowest-cost methods of removing pollution."[117] For a man who worried enormously about scientific uncertainties, he was remarkably untroubled by economic ones.

Singer's final sentence was a question: "Will a reduction in emissions produce proportionate reductions in deposition and in the environmental impacts believed to be associated with acid rain?"[118] In posing the question, he left the reader with the impression that the answer, perhaps, was no. So a report that was otherwise clear on the reality and severity of acid rain now ended with doubt.

Singer's appendix left the reader with an impression very different from what the rest of the text had said. Yet it was very similar to what Reagan officials had been saying for some time. In 1980, David A. Stockman, director of the powerful Office of Management and Budget, asked in a speech to the National Association of Manufacturers, "How much are the fish worth in these 170 lakes that account for 4 percent of the lake area of New

York? And does it make sense to spend billions of dollars controlling emissions from sources in Ohio?" On another occasion, Stockman put the cost of eliminating acid rain at $6,000 for every fish saved.[119] The acid rain panel report was supposed to be a scientific peer review, but Singer had placed within it a policy view consistent with that of the Reagan administration, but seemingly at odds with the science that had been reviewed.

The full report was sent to the White House in early April, just as a key House of Representatives subcommittee was considering legislation to control acid rain. Secretary of State George Shultz had reassured the Canadians that he and EPA director William Ruckelshaus held acid rain to be a high priority, but the Canadians were worried.[120] Canadian government spokesman Allan MacEachen noted that they believed there was enough evidence to justify abatement measures, but the U.S. view was "that the scientific conclusions are not clear."[121] The Canadians were right on both counts. In May the House subcommittee voted 10–9 against the legislation, effectively killing congressional action on the issue. The panel report was finally released to the public on the last day of August.[122]

Press coverage was extensive and critical. "Prove it," was how *Newsweek* later characterized the Reagan administration position, neglecting to point out that scientists had, in fact, proved it.[123] "Who'll stop acid rain? Not Ronald Reagan," said the *New Republic*.[124] *Nature* concluded that "Canada must act alone."[125]

The business press, however, began to pick up on Singer's theme. *Fortune* ran an article by a researcher at the Hudson Institute, a progrowth think tank founded by Cold Warrior Herman Kahn. "Maybe acid rain isn't the villain" asserted that it "could eventually cost Americans about $100 billion . . . to achieve a major reduction in sulfur dioxide emissions. Before committing to any program of this magnitude, we should want to be more certain that acid rain is in fact a major threat."[126] The article didn't just misrepresent the state of the science, it misrepresented its history, too. "It's not surprising that there should be sharp disagreements about acid rain. The rain has been studied only for about six years." (You'd think think tank researchers could do arithmetic: the elapsed time between 1963 and 1984 did not come to six years.) The *Wall Street Journal* ran a piece on its editorial page by a consultant for Edison Electric named Alan Katzenstein entitled "Acidity is not a major factor," questioning the scientific evidence and suggesting that the real "villain in the acid-rain story" might be aluminum.[127] One forest ecologist responded in a letter to the editor: "Katzenstein made several assertions about the research findings [and] all of them are incorrect!"[128] Who was Katzenstein? An ecologist? A chemist? A biologist? No, he was a business consultant who previously had worked for the tobacco industry.[129]

Many of these pieces were published before the panel's final report was actually released; some of them were based on the interim findings published the previous summer. So maybe it didn't really matter whether or not the report had been delayed. Why *had* it been delayed? If the report was sent to the White House in April, why was it not released until August?

Manipulating Peer Review

On August 18, Maine senator George Mitchell and New Hampshire congressman Norman D'Amours issued statements saying that the report had been suppressed by the White House. Both the *New York Times* and the *Los Angeles Times* covered the story. "The 78-page study . . . directly rebuts the Reagan Administration position that pollution controls should not be ordered until further studies are conducted," the *Los Angeles Times* concluded. The administration shelved the report, D'Amours was quoted as saying, to avoid giving legislators "the ammunition we needed to push acid rain controls through Congress." The *New York Times*, however, quoted an OSTP spokesman explaining that the final report had not been received until mid-July, and quoted Nierenberg explaining, "We were making changes right up to mid-July."[130] That was true. But they were not changes that the panel had authorized.

A few weeks later, *Science* magazine suggested that the congressional vote might have been different had the Nierenberg report been released beforehand. *Science* quoted one panel member saying that "paragraphs were re-ordered and material added . . . that changed the tone of the original summary. The net effect [was] that the new summary weakens the panel's message that the federal government should take action now."[131] A Canadian paper repeated the charge: "The U.S. Administration suppressed a report that told it to cut acidic air pollution during a crucial congressional vote."[132]

The historical record shows that something irregular had indeed occurred. Changes *were* made after the report was finished, at least some of them were made without the agreement of the full panel, and they did weaken the message.

The report had been more or less completed by March, when Nierenberg sent a draft to the panelists, asking for final comments; in April, the final version was ready. Somehow proceedings were delayed, and a plan developed to present it to EPA administrator William Ruckelshaus on June 27.[133] After the article in *Science*, Nierenberg wrote to the journal to protest that, contrary to the article, the panel's report had *not* been changed since June. Who was right? After all, it was Nierenberg's panel; surely he knew what had gone on.

The historical record supports *Science* magazine. Documents show that the panel report *was* forwarded to the White House in April, it was ready to be released in June, and it was not actually released until August (albeit with a July date). The record also shows that changes had been made to the text. In fact, it shows that *two* sets of changes were made—one set in the spring, and a second set in the summer. Fred Singer had played a role in these changes—and so had Bill Nierenberg.

On May 21, Tom Pestorius sent a telecopy of the Executive Summary to Nierenberg. The first paragraph was completely different: a strong statement about the reality of acid rain was replaced with a historical introduction as to how and why the panel came to be. The original opening paragraph, which began, "Large portions of eastern North America are currently being stressed by . . . deposition of acids . . . [and t]he principal agent altering the biosphere is acidity traceable to sulfur dioxide," had been buried as the penultimate paragraph.[134]

The changes made in the summer were even more serious, and when the panel realized what had happened, they protested loudly. A red flag was raised in September by panel member Kenneth Rahn, an atmospheric chemist who had studied pollution dispersion. Rahn thought the claim that northeastern acid rain definitely came from midwestern pollution *was* a bit premature—that it might be best to do more research before implementing policy solutions, and he had testified in Congress to this effect— so no one would have considered him an alarmist.[135] But he now sent a very alarmed, three-page, single-spaced letter to the panel members, detailing what he had learned.

The penultimate draft of the report had been compiled in February, and this "was the last that most of us saw," Rahn recalled. "We read it over for a last time and sent any remarks back, and from this a final version was constructed. The principal change in the final version was Fred Singer's Chapter VIII, which was made into the signed appendix 5."[136] This account was consistent with other documents; a letter sent from John Robertson to the panel in February referred to the report as "the 'almost final' draft."[137]

John Robertson had been handling all the compiling of the various sections and was responsible for making the editorial changes suggested by the panelists. In a memo on February 24, 1984, he had summarized for the panel the major changes he had made, based on their input. There were only five, and most just dealt with organization and style. Only one was substantive: adding a recommendation to include control of nitrogen oxide emissions. However, Robertson did remind the panel of one "unresolved" issue: "the form, placement, and content of [Singer's] chapter VIII." Two panelists did not want it to have the status of a chapter, but would accept it as a signed appendix; four would accept it as a chapter but only if "the conclusions . . . are removed."[138]

In fact, the report had been almost ready nearly a year before. In March 1983—eleven months earlier—Robertson had written to the panel asking for their comments on a draft of what was expected to be the final report. "I have compared the draft issued to you in November with the enclosed 'Final Report,'" Robertson wrote. "All changes are attached. Rahn, Rowland and Ruderman should be in a position to finish all writing within the next two weeks . . . I would like to have all input by 21 March."[139] This version included Singer's contribution as a chapter. But the March 1983 version did not prove final. In July 1983, a draft of the report, with revisions, was sent to the panel, but somehow this too failed to become the final version. As noted above, in August, Singer sent Nierenberg a round of suggested corrections on the draft; Singer also sent panel members various memos and materials suggesting that acid rain might not be as serious as they believed. Another half year elapsed before the committee agreed on a final report. Among other things, Singer's chapter had been moved to the appendix, after all.

According to Robertson, only three copies of the final report were made: one each for him, Nierenberg, and the OSTP. That report was finished in March 1984, and sent to the OSTP in the first week of April. "We all know that the March report was regarded by us all as the final one," Rahn reminded his colleagues.[140]

Why was Singer's chapter converted into an appendix? It would appear that the OSTP was hoping to avoid the obligation of getting the committee to sign off on what Singer had done—something they had already refused to do. That was one of two major changes. Rahn explained the other. "By sometime in May, OSTP had decided to request that the Executive Summary be changed; they proposed this to the Chairman and followed up with a draft version of a revised Executive Summary which contained the types of changes they wanted the panel to consider. There followed several exchanges of new versions between the Chairman and OSTP, and at one point, Dr. Keyworth became personally involved." The OSTP had told Nierenberg what changes it wanted, and Nierenberg had made them.

So two parts—the Executive Summary, and Singer's appendix—had *not* been approved by all panel members. Most of the members didn't even know that the summary had been changed. But Rahn had now read them side-by-side, and he sent them to the rest of the panel to compare.

In Rahn's opinion, nothing important had been added or deleted, but changes in order, in adjectives, and in tone had changed the tenor of the report, so the reader was left with a very different impression. "The new message carries a softer message than the old one did. All parties who have carefully read both versions agree on this point, and in fact OSTP freely admits that their goal was to soften the tone." The structure of the summary had been changed, too. Whereas the original closely followed the report, beginning with the policy recommendation to act to control SO_2 emissions, the new one left that recommendation to the end. In so doing, "the remarks on policy which OSTP found most unjustified are thereby diminished in stature."[141]

It's a frequently asked question in scientific circles whether scientists should make policy recommendations about complex issues. The OSTP is the Office of Science and Technology *Policy*, so it was perhaps reasonable for them to suggest adjustments to the report's policy recommendations. Or was it? No, because this was a peer review panel. Their charge was to review, summarize, and critique what the technical working groups had done, and this included summarizing *their* policy recommendations. Peer review is a crucial part of science. For the OSTP to alter those recommendations was to interfere with scientific process. The report released by the OSTP on August 31 was simply not the report the panel had authorized.

"In short, our report has been altered since we last saw it," Rahn concluded. " 'Tampered with' may not be too strong . . . In light of the changes made, which I judge to be substantial, I suspect that we would not have approved it if we had been given the chance."[142]

Other panelists drew the same conclusion. "I am very distressed to learn that the Executive Summary for our Report from the Acid Rain Peer Review Panel has been rewritten and changed from the version our Panel prepared and authorized last spring," Gene Likens wrote to Nierenberg. "These revisions were done without informing the members of our Panel and without gaining their approval . . . My understanding is that these un-approved changes in the Executive Summary originated within the White House/ OSTP. Frankly, I find such meddling to be less than honest and extremely distasteful." Likens was clearly angry, but he held his temper, ending with a straightforward question: "Is there some explanation for what happened?"[143]

Panel member Mal Ruderman was also deeply disturbed, particularly because the *Science* article gave the impression that he had participated in the tampering. In a letter to Nierenberg, Ruderman wrote, "I am extremely upset by the description in *Science* of what happened to our Executive Summary between April and June." Ruderman *had* seen a version in June with certain proposed changes, but Nierenberg had not explained to him that this version had *already* been altered in May, with the text rearranged, and that the proposed changes in June were *additional* ones. Moreover, Ruderman now reminded Nierenberg that he (Ruderman) had *rejected* the proposed changes. "Some of the suggested changes altered the meaning of certain sentences and I did my best to change them back to conform to what our Committee had agreed on . . . I feel strongly that my role in all of this was defending against substantial changes in the Executive Summary given to me [and] I am counting on you to set the record straight on all of this to our Committee and to *Science*. It is a matter of great importance to me."[144]

Rahn made a similar point. The press coverage made it seem as if the whole committee had participated in the alterations, noting an article in the *New York Times* in August quoting "an OSTP spokesman as saying that the authors 'were making changes right up to mid-July,' which is very mis-leading."[145]

Nierenberg responded by suggesting that he too had been misled, or at least confused. "I received communications from [Kenneth] Rahn and Gene L[ikens] which partially confused me," he replied to Ruderman. "Your letter cleared up what was my major confusion . . . I did not realize till just now (any more than you did) that there had been prior rearrange-ment of the text . . . I am not sure what should be done. We could ask *Science* to publish the original summary. I am reasonably sure they would. We could also ask them to publish both at the same time and let the readers judge." He added a P.S. "I also have [now] compared the various versions and agree that there was considerable rearrangement."[146]

It is common practice for the head of a panel to meet, at the start of a re-view, with the office or agency that is commissioning the report to receive the charge. It is also accepted practice for the committee to meet with government officials to present the finished report, but it's highly irregular for a government official to *change* a report without the committee's knowledge and permission. If the White House had done that, it's hard to believe that Nierenberg would have been so calm about it. He should have been outraged.

Moreover, the record does not support the idea that Nierenberg had no idea what the White House was up to. When word got back to the OSTP that Ruderman was asking for the record to be set straight, Pestorius laid the blame squarely on Nierenberg's shoulders: "Bill told me that 'Mal is out here with me working on the Executive Summary.' "[147]

Ruderman was not content with Nierenberg's explanation, and in November he wrote to Nierenberg again. "I think that there still exists a need to explain to the Acid Rain committee members just what did happen between the original submission of our report to the OSTP (April?) and the receipt of an amended telecopied Executive Summary by you from Tom Pestorius in late July."[148] While Nierenberg had offered to set the record straight, there is no evidence in the published record or in his own files that he did.

When asked recently about Nierenberg's role, Gene Likens said simply, "He was the one talking to politicians in power. He pushed it through . . . Nierenberg was definitely responsible for the changes." Some of the panelists sought advice from colleagues at the National Academy about what to do, but to no avail. Likens recalls again, "We went to our sources, but ours weren't as powerful as Nierenberg's."[149]

Historical documents confirm Likens's account. In Bill Nierenberg's files, there is a second copy of the telecopied Executive Summary from May 21, but this time dated, by hand, 7/10/84—and the note next to the date reads: "Changes wanted by Keyworth." Nierenberg *had* changed the Executive Summary, and it was the science advisor to the president who had asked him to do so.[150]

Republicans in general were pleased with Nierenberg's work. In July, he received a letter from the Republican congressman from Lansing, Michigan. "I am delighted that you were chosen for this task," the congressman wrote.[151] In September, Nierenberg received an autographed photograph of President Reagan.[152] In 1984, Nierenberg sent Attorney General Ed Meese a copy of a crossword puzzle he had completed in which one of the answers was "Meese" (the clue was "Reagan aide"). In 1985, Nierenberg was considered once again for the position of science advisor to the president. One referee described him as "a strong, loyal, and vocal supporter of the Administration's policies . . . a [real] team player."[153]

THERE WOULD BE no legislation addressing acid rain during the remainder of the Reagan years. The administration would continue to insist that the problem was too expensive to fix—a billion-dollar solution to a million-dollar problem. There would, however, be plenty of further scientific research. William Ruckelshaus, the EPA administrator who had banned DDT in the Nixon years and was viewed by most people as an honest broker, appeared on ABC news in August 1984 to explain the administration's position. He was asked by conservative commentator George Will, "Isn't the evidence now in on acid rain?" Ruckelshaus replied, "Well, no it's not . . . We don't know what's causing it."[154]

"We don't know what's causing it" became the official position of the Reagan administration, despite twenty-one years of scientific work that demonstrated otherwise. "We don't know" was the mantra of the tobacco industry in staving off regulation of tobacco long after scientists had proven its harms, too. But no one seemed to notice this similarity, and the doubt message was picked up by the media, which increasingly covered acid rain as an unsettled question. We've already noted how *Fortune* ran an article insisting that the "standard scientific view of acid rain's effects may be simply wrong." (At least they acknowledged that there was a standard scientific view.) Echoing Fred Singer, the author, William Brown, associated with the Cato Institute, asserted that it "could eventually cost Americans about $100 billion in today's dollars to achieve a major reduction in sulfur dioxide." Given this enormous cost, "we should want to be more certain that acid rain is in fact a major threat to the country's environment."[155] The fact that all the relevant scientific panels *had* concluded that it was a major threat was ignored.

Likens tried to set the record straight with an article in *Environmental Science and Technology* entitled "Red Herrings in Acid Rain Research."[156] But in a pattern that was becoming familiar, the scientific facts were published in a place where few ordinary people would see them, whereas the unscientific claims—that acid rain was not a problem, that it would cost hundreds of billions to fix—were published in mass circulation outlets. It was not a level playing field.

And it wasn't just *Fortune* that misrepresented the science and the situation. *BusinessWeek* attacked the EPA as "activist" for trying to take action on acid rain—in effect, for doing its job.[157] *Consumers' Research Magazine* (which despite its name was a journal that consistently took probusiness positions) demanded to know: "Acid Rain: How Great a Threat?"[158] William Brown reprised his earlier article in *Fortune* with a new piece in 1986, "Hysteria about Acid Rain."[159] A few months later *Fortune* repeated his claim yet again, insisting that "delay makes sense because we still have a lot to learn about acid rain."[160] The *Futurist* joined the chorus, insisting that "the jury is still out on acid rain."[161]

Conflict, it is sometimes said, makes good copy, and when a lonely scientist took up the right-wing charge that acid rain might not be a serious problem, the press were quick to pick up his claims. Edward Krug was a soil scientist at the Connecticut Agricultural Research Station who began to argue that a good deal of soil acidification in northeastern forests was natural or associated with land use changes.[162] Krug called his argument a "a new perspective," but it wasn't new at all; natural acidification had been considered and found to be inadequate to explain the observations.[163] Still, Krug's argument was presented in *Policy Review*, published by the Hoover Institution,[164] and taken up by *Reason* magazine, which insisted that new evidence showed that "acid rain was not a problem."[165] He even appeared on *60 Minutes*, where he claimed that NAPAP had shown that acid rain simply wasn't a serious problem—a claim that almost no one else associated with NAPAP agreed with.[166]

As the World Wide Web developed in the 1990s, many sites began to quote Krug as having demonstrated that acid rain was not the crisis that environmentalists made it out to be. Many of these sites are still live today.[167] One complains that Krug was cited in the mainstream media only nine times between 1980 and 1993, while Gene Likens was cited thirty-nine times. (Given their relative standing in the scientific community and the depth and breadth of their acid rain research, this figure suggests that the mainstream media were biasing their coverage in *favor* of Krug.)[168] Print media kept up the drumbeat, as *Fortune* continued in the 1990s to claim that acid rain was "a relatively minor problem on which it would be absurd to spend billions of tax dollars."[169] Fred Singer, citing his own contributions to the 1983 Nierenberg report, claimed in *Regulation*—the journal of the Cato Institute—that avoiding premature action on acid rain had saved from $5 billion to $10 billion per year.[170]

Many people became confused, thinking that the acid rain issue was unsettled, that scientists had no consensus. When a group of NAPAP scientists met in 1990 at Hilton Head, South Carolina, National Public Radio reported that there was a "general consensus among the scientific community that acid rain is . . . complicated."[171] And while we are embarrassed to admit it, in the early 1990s one of us (N.O.) used Krug's arguments in an introductory earth science class at Dartmouth College to teach "both sides" of the acid rain "debate."

Meanwhile the Reagan administration, having gotten some but perhaps not all of what they wanted from Bill Nierenberg, commissioned yet another report. This one was led by Columbia University geochemist Laurence Kulp, who was well-known for his conservative religious views; colleagues at Columbia referred to him as a "theochemist" for his efforts to reconcile geological evidence with Christian belief."[172] Kulp's report concluded that acid rain was not as great a threat as many believed, a conclusion that most scientists described, in the words of the *New York Times*, as "inaccurate and misleading." With echoes of the 1984 Nierenberg report, the *Times* reported that several scientists "suggested that the [Executive] summary had been tailored . . . in the belief that policy-makers and journalists would read it [the summary] and not the report itself."[173]

It would take six years and a new administration to pass legislation to control acidic emissions. In 1990, under the administration of George H. W. Bush, amendments to the Clean Air Act established an emissions trading—or "cap and trade"—system to control acid rain. The system resulted in a 54 percent decline in sulfur dioxide levels between 1990 and 2007, while the inflation-adjusted price of electricity *declined* during the same period.[174] In 2003, the EPA reported to Congress that the overall cost of air pollution control during the previous ten years was between $8 billion and $9 billion, while the benefits were estimated from $101 billion to $119 billion—more than ten times as great.[175] Singer's "billion-dollar solution to a million-dollar problem" was just plain wrong.

The energy industry had often accused environmentalists of scaremongering, yet this is just what they had done with their claims of economic devastation. Protecting the environment didn't produce economic devastation. It didn't lead to massive job losses. It didn't cost hundreds of billions of dollars. It didn't even cause the price of electricity to rise. And the science was correct all along. As Mohamed El-Ashry of the World Resources Institute was quoted in *Newsweek*, "When we waited for more research on acid rain, we ended up realizing that everything we knew 10 years earlier was true."[176]

But even if the scientists got the science right, perhaps Republican policy *was* right to focus on market mechanisms to control pollution. Cap and trade to control sulfate emissions was widely considered a success and is now the leading model for controlling the greenhouse gases that cause global warming. Perhaps Singer was right to push for a market-based solution to acid rain. Perhaps, except that scientists close to the issue have reservations as to whether cap and trade has really worked.

Well after acid rain was off the headlines, Gene Likens and his colleagues continued to work at Hubbard Brook. By 1999, they had concluded that the problem had not been solved. "Acid rain still exists," Likens wrote in the *Proceedings of the American Philosophical Society*, "and its ecological effects have not gone away." Indeed, matters had gotten worse, as additional stresses such as global warming were making the forests "even more vulnerable to these anthropogenic inputs of strong acids from the atmosphere."[177] The net result was that "the forest has stopped growing."[178]

Over the next ten years, Likens and his colleagues pursued the question of net forest health. In 2009 they spoke out frankly. "Since 1982, the forest has not accumulated biomass. In fact, since 1997, the accumulation . . . has been significantly negative."[179] The forest was shrinking, "under siege" from multiple onslaughts of climate change, alien species invasion, disease, mercury and salt pollution, landscape fragmentation, and continued acid rain. The sugar maple—beloved by both Canadians and New Englanders—"is dying . . . [and s]cientific research suggests that by 2076, the 300th birthday of the United States, sugar maples will be extinct in large areas of the northern forest."[180] First on their list of threats to forest sustainability is acid rain, which "remains a major problem . . . as emissions were not fully controlled" by the Clean Air Act Amendments.[181] The cap and trade system simply did not do enough. Not only did it not eliminate acid rain, it did not even reduce it sufficiently to stabilize the situation. Forest decline has continued.

Likens and his colleagues do not rule out the continued use of market-based mechanisms to help save the forests, but they also note that some issues "require national and even global regulation."[182] But the real issue in either a cap and trade system or its alternative—setting pollution limits through command and control—is *where* you set the cap, and whether or not you have a mechanism to adjust it (either up or down) if future information suggests you should. The ongoing scientific work shows that, among other things, the Clean Air Amendments set the caps too high, perhaps in part because the arguments made by Fred Singer and his allies—and then taken up by the Reagan administration and much of the media—suggested that since we weren't entirely sure about the problem and its severity, it would be foolish to take excessively dramatic action. And so we didn't. We took modest steps, and then did nothing to strengthen them as time went on, even as the science increasingly indicated that we needed to. We went on faith that the market would do its "magic."

Magical thinking still informs the position of many who oppose environmental regulation. As recently as 2007, the George Marshall Institute continued to insist that the damages associated with acid rain were always "largely hypothetical," and that "further scientific investigation revealed that most of them were not in fact occurring."[183] The Institute cited no studies to support this extraordinary claim.

Moreover, there is reason to believe that a straight-out command and control approach might have better results than cap and trade in one important respect: research shows that regulation is an effective means to stimulate technological innovation. That is to say, if you want the market to do its magic—if you want businesses to provide the goods and services that people need—the best way to do that, at least in terms of pollution prevention, appears, paradoxically, to be to mandate it.

David Hounshell is one of America's leading historians of technology. Recently he and his colleagues at Carnegie Mellon University have turned their attention to the question of regulation and technological innovation. In an article published in 2005, "Regulation as the Mother of Innovation," based on the Ph.D. research of Hounshell's student, Margaret Taylor, they examined the question of what drives innovation in environmental control technology. It is well established that the lack of immediate financial benefits leads companies to underinvest in R & D, and this general problem is particularly severe when it comes to pollution control. Because pollution prevention is a public good—not well reflected in the market price of goods and services—the incentives for private investment are weak. Competitive forces just don't provide enough justification for the long-term investment required; there is a lack of driving demand. However, when government establishes a regulation, it creates demand. If companies know they have to meet a firm regulation with a definite deadline, they respond—and innovate. The net result may even be cost savings for the companies, as obsolete technologies are replaced with state-of-the art ones, yet the companies would not have bothered to make the change had they not been forced to.

Of course, regulation is not the only possible government action. Governments can invest directly in R & D, provide tax credits and subsidies, or facilitate knowledge transfer. Many economists prefer these alternatives to

straight-out regulation, thinking they provide companies greater flexibil-
ity, increasing the likelihood that resources will be allocated in appropriate
ways and the desired goals actually met. But Hounshell and his colleagues
show that this presumption may be wrong. The empirical evidence shows
that regulation may be the most effective means, because clear and stringent regulation provides a strong and continuous stimulus for invention.[184]
Necessity *is* the mother of invention, and regulatory compliance is a powerful form of necessity.

If the U.S. government had established a strong regulatory regime on
acid emissions, then the industry might have done more to innovate. And if
technological advancement had made it easier and cheaper to control emissions, then industrial resistance to tightening the caps as time went on
would have lessened, and it might well have been easier to tighten the regulations over time, giving the forests the protection that science showed they
really needed.

This is admittedly speculative. We will never know what would have happened had a different approach been taken. However, one thing we do know
for sure is that doubt-mongering about acid rain—like doubt-mongering
about tobacco—led to delay, and that was a lesson that many people took to
heart. In the years that followed, the same strategy would be applied again,
and again, and again—and in several cases by the same people. Only next
time around, they would not merely deny the gravity of the problem; they
would deny that there was any problem at all. In the future, they wouldn't
just tamper with the peer review process; they would reject the science itself.

CHAPTER 4

Constructing a Counternarrative: The Fight over the Ozone Hole

A T THE SAME TIME AS ACID RAIN was being politicized, another, possibly even more worrisome problem had come to light: the ozone hole. The idea that human activities might be damaging the Earth's protective ozone layer first entered the public mind in 1970. Awareness began with the American attempt to develop a commercial airliner that could fly faster than the speed of sound. The "supersonic transport," or SST, would fly inside the stratospheric ozone layer, and scientists worried that its emissions might do damage. While the SST did not turn out to be a serious threat, concern over it led to the realization that chemicals called chlorofluorocarbons were.

In 1969, MIT commissioned a major study of human environmental impact. "Man's Impact on the Global Environment: Report of the Study of Critical Environmental Problems" (mercifully abbreviated SCEP) was released a year later, and contained the first major statement on the state of the stratosphere and the probable impact of the SST.[1] A panel chaired by William Kellogg of the National Center for Atmospheric Research (NCAR) in Boulder, Colorado, took on the question. Water vapor is the second-largest combustion product of jet engines, after carbon dioxide, and like carbon dioxide, it's a greenhouse gas, so the scientists worried that water vapor from engine exhaust could cause climate change. Water vapor also makes clouds, which in turn affect weather. The scientists concluded that although stratospheric water vapor concentration would increase—by as much as 60 percent with a large SST fleet—it probably would not appreciably change surface temperatures. Just to be sure, however, they recommended the development of a stratospheric monitoring program.[2]

An article in *Science* by a scientist at Boeing Laboratories—the research arm of the SST's builder—accidentally undermined the SCEP argument. Using a different model, he calculated that the water vapor produced by a fleet of 850 SSTs *would* deplete the ozone column by 2 to 4 percent. Most of this reduction would occur in the northern hemisphere due to the high concentration of air routes there, producing a temperature rise on the Earth's surface of about 0.04°C.[3]

This tiny amount of warming was indistinguishable from natural variation, so one could say it was no different from the SCEP result, but it unexpectedly opened a new and more controversial issue. James E. McDonald of the University of Arizona, a member of the National Academy of Sciences' Panel on Weather and Climate Modification, found the Boeing findings startling. McDonald had been a panelist on an earlier study of stratospheric ozone depletion that had concluded depletion was very unlikely to happen, but the new argument caused him to reconsider. Moreover, by 1970, medical scientists believed that ultraviolet radiation caused certain kinds of skin cancer.[4] The ozone layer protects us from that UV radiation. If ozone depletion did occur, then skin cancer incidence would increase. Indeed, McDonald believed there was a sixfold magnification factor: each 1 percent reduction in ozone concentration would produce a 6 percent increase in skin cancer occurrence. McDonald testified to this effect before Congress in March 1970.[5]

Meanwhile, the issue had come to the attention of a University of California atmospheric chemist named Harold Johnston, who started to think about another by-product of jet engines: oxides of nitrogen—or NO_x. Johnston was the author of the leading textbook on ozone chemistry, and his expertise got him invited to a Department of Transportation–sponsored study of stratospheric flight held in Boulder, in March 1971. Johnston rapidly became annoyed at the proceedings because the conferees seemed to accept the conclusions of the SCEP study that nitrogen oxides would not be a significant cause of ozone depletion. Johnston's knowledge of ozone chemistry suggested to him that this was wrong.

Johnston spent much of that night working out calculations showing that oxides of nitrogen would be far *more* potent stratospheric ozone scavengers, and in the morning he handed out a handwritten paper that estimated NO_x-derived depletion of 10 to 90 percent. This was a huge range, but there was a good reason for it: the depletions would be much worse over the North Atlantic, where air routes are concentrated. The range was caused in part by geographic diversity, not just scientific uncertainty.

Still, no one was convinced. At an impromptu "workshop" organized in the men's washroom and held in a small conference room later that day, the discussion stayed focused on water vapor. Participant Harold Schiff, a chemist from York University in Toronto, recalls that Johnston lashed out at the group for ignoring the NO_x reactions, finally prompting the other chemists to grapple with his idea—if only in self-defense.

No one knew what the stratosphere's "natural" concentration of NO_x was, because no one had ever measured it. But if you didn't know how much NO_x there was, then you couldn't calculate its effect. If the stratosphere already had plenty of these chemicals, then small amounts added by a fleet of SSTs wouldn't make a difference. If, on the other hand, the stratosphere had very little or none at all, then the SST effect might be devastating. The conference ended with a recommendation that more research was needed.[6]

Back in Berkeley, Johnston turned his calculations into a formal paper, which he sent to several colleagues in early April. On April 14, he sent a substantially revised version to *Science*. The journal's peer reviewers deemed the paper unsatisfactory for two reasons, and recommended that Johnston rewrite it. First, Johnston had failed to cite a key paper that suggested the stratosphere's high sensitivity to nitrogen oxides.[7] Second, Johnston's tone was unacceptable. Scientists were supposed to be dispassionate, in order to at least *appear* to be unbiased and objective. Johnston didn't do that. He bluntly contended that an SST fleet could cut ozone concentration over the Atlantic corridor in half and allow enough radiation to reach the Earth's surface to cause widespread blindness.

So Johnston's paper was delayed as he made changes (including removing the blindness claim), and it was finally published in August. But his preliminary draft—the one he had sent to colleagues—had been leaked to a small California newspaper, the *Newhall Signal*, leading the University of California's Public Relations Office to release it officially. Reaction was swift. Sensational summaries of Johnston's draft sped east on the wire services and on May 17 the story made the *New York Times*. Two weeks later, a lengthy follow-up article by famed science editor Walter Sullivan put the issue solidly before the public.[8] According to Johnston, Sullivan reported, SSTs could have devastating consequences for humanity.

Johnston's paper had no impact on the SST program, though, because it had already been canceled. The House of Representatives had refused to finance the project back on March 17 for economic reasons, not environmental ones.[9]

But Johnston's paper did have an impact on stratospheric science. The Department of Transportation still wanted to restart SST development, and the Anglo-French Concorde SST was being developed, so the questions Johnston had raised still had to be answered. Congress financed $21 million for a Climate Impact Assessment Program (CIAP, pronounced sy-ap), to answer them. This three-year effort involved nearly a thousand scientists across many agencies, universities, and several other countries, and was one of the first efforts to assess the potential impact of a technology not yet in widespread use.

CIAP was controversial, because after enormous amounts of work by scientists around the world, the Department of Transportation tried to whitewash the findings. The program's scientists had found that a fleet of five hundred Boeing-type SSTs was likely to deplete the ozone layer by 10 to 20 percent. More important, there would be vastly worse depletions over the highly traveled North Atlantic routes. Harold Johnston might have been right.[10]

But the report's Executive Summary, a twenty-seven-page summation of a seventy-two-hundred-page study, didn't say that. Instead, it claimed that an improved SST, to be developed in the future with a sixfold reduction in emissions, *wouldn't* deplete the ozone layer. At the press conference held on its release, in January 1975, the study's director emphasized this "techno-fix." The resulting newspaper headlines said things like SST CLEARED ON THE OZONE. But the report hadn't cleared the Boeing SST, or the Concorde; it had only cleared an imaginary technology that didn't exist.

The scientists had said there *was* a problem, and they now worried "they would probably be . . . branded alarmists and fellow travelers of the disaster lobby."[11] They were right. Columns in the *San Francisco Chronicle*, the *Christian Science Monitor*, and the *Pittsburgh Press* promptly attacked them. The *Pittsburgh Press* announced that CIAP had shown scientists concerned about ozone depletion had spoken "unscientific nonsense. The phony ozone argument has no place in rational scientific discourse and no place in the SST debate."[12]

The CIAP scientists were furious about the misleading presentation of their work. After an ugly meeting in Boston, shortly after the Executive Summary's release, both Johnston and Thomas M. Donahue of the University of Michigan tried to publish corrective letters in several newspapers, but without success. Newspapers simply declined to publish their letters.

Donahue had chaired a committee that reviewed the larger CIAP study for accuracy; he finally got *Science* to publish a letter laying out the correct interpretation of the study. This forced the Department of Transportation's official to respond in *Science* as well, acknowledging the misleading nature of the summary.[13] But that admission would only be read by the scientists who read *Science*; once again, scientific claims were being published in scientific journals, where only scientists would read them, but unscientific claims were being published in the mass media. The public was left with the impression that the ozone layer was fine, and the "alarmists" had got it wrong.

MEANWHILE SCIENTIFIC ATTENTION had started to stray from the SST's impact on ozone. The Anglo-French Concorde failed to sell, and production ceased after twenty were built—a low enough number to pose no threat. But thinking about ozone had led a handful of NASA scientists to consider yet another issue: the potential impact of the agency's new space shuttle, the boosters of which used a chlorine-based propellant. And that was a different story.

The 1970 National Environmental Policy Act required the shuttle program office to prepare an Environmental Impact Statement. For the atmospheric portion they had contracted with the University of Michigan, where scientists Ralph Cicerone and Richard Stolarski found some troubling results: the exhaust from the shuttle's solid rocket boosters would release chlorine, a highly reactive element known to destroy ozone, directly into the stratosphere.[14] Stolarski says that their report was initially "buried" by the program office in Houston, but NASA headquarters reversed that decision and scheduled a workshop on the problem, for January 1974.[15] In the meantime, at a conference in Kyoto, Japan, Cicerone and Stolarski presented a paper on *volcanic* chlorine as a potential ozone scavenger— not because they were really worried about volcanoes, but because NASA had asked them not to say anything about the shuttle—or even to mention the fact that NASA was paying for their work. Since volcanoes do emit chlorine, the two realized they could present their work without having to say that the source of chlorine they were actually worried about was the shuttle.

Volcanoes or no, other people were thinking about chlorine, too. Paul Crutzen, a brilliant Dutch atmospheric scientist who would later dub the current period of Earth's history the "Anthropocene" because of humans' extensive environmental impact, also presented a paper on chlorine at the Kyoto meeting.[16] A few months later, the British journal *Nature* published a paper by F. Sherwood Rowland and Mario Molina, who argued that the decomposition of a set of common industrial chemicals—chlorofluorocarbons—would release large quantities of chlorine monoxide into the stratosphere.[17] In 1970, British scientist James Lovelock had documented the widespread presence of chlorofluorocarbons in the Earth's troposphere (the lower portion of the atmosphere). Lovelock had calculated that given the known concentration of CFCs in the atmosphere virtually all of the billions of pounds that had been manufactured were still in the atmosphere. If Lovelock was right—there were no chemical processes, or "sinks," that could remove CFCs from the lower atmosphere—then eventually the atmosphere's circulation would move them into the stratosphere. There, Rowland and Molina argued, they would finally decompose under the impact of ultraviolet radiation, where they'd be converted into fluorine and chlorine compounds. And some of these compounds were known from laboratory studies to be ozone scavengers.

Billions of pounds of CFCs were produced every year for use in spray cans, air-conditioners, and refrigerators. In comparison, the four shuttles' exhaust would be utterly trivial.

The revelation that mundane items like hair spray could destroy the Earth's ozone and increase cancer rates produced a media firestorm. The U.S. National Academy of Sciences, unusually, moved quickly to understand the stratospheric implications of this research. The chairman of the National Research Council's Division of Chemistry and Chemical Technology arranged for a one-day, ad hoc panel to decide whether a full committee-level study was in order. Donald Hunten of the Kitt Peak National Observatory convened the ad hoc study in October; his group recommended to Academy president Philip Handler that a panel be established as an adjunct to the Climatic Impact Committee.[18] Congress also moved with unaccustomed speed. In December 1974, a month before the ill-fated CIAP press conference, the House held its first hearings on CFCs and ozone depletion.[19]

The Ozone War

The aerosol industry responded almost immediately to Rowland and Molina's work. They already had two trade associations, the Chemical Specialties Manufacturers Association and the Manufacturing Chemists'

Association, which responded with research on chlorofluorocarbon effects of their own. The Manufacturing Chemists' Association established a sub-panel to dispense $3 million to $5 million in research grants, which largely went to university scientists.[20] The industry then established two more organizations for public relations purposes: the Aerosol Education Bureau and the Council on Atmospheric Sciences.[21] A little later, a group of aerosol can fillers formed the Western Aerosol Information Bureau. Their job was "defense of the product" in the public sphere. If you had followed the tobacco story, it would have been déjà vu all over again.

Meanwhile, the Ford administration established an interagency task force on so-called Inadvertent Modification of the Stratosphere (IMOS) in January 1975. The IMOS panel held a contentious public hearing in February, then worked quietly on a report for several more months. Finally, the panel announced that unless new evidence was found absolving CFCs, "it would seem necessary to restrict [their] uses . . . to . . . closed recycled systems or other uses not involving release to the atmosphere."[22] Absent new evidence, CFC emissions to the atmosphere should be banned entirely. It was a stunning conclusion.

IMOS placed the burden on determining whether CFC regulations were necessary not on a government agency such as the EPA, but on the National Academy of Sciences. This was unusual, because the Academy isn't a regulatory body. Its usual role is to generate a summary of the current state of a particular scientific field, and to suggest what further research would be needed to advance it. The Academy leadership was not pleased to be stuck with the job of deciding whether or not to extinguish the $1 billion per year CFC spray industry. Nonetheless, Academy president Handler accepted the task, appointing a Panel on Atmospheric Chemistry to examine the state of the art, and a second panel, the Committee on Impacts of Stratospheric Change, to evaluate the science and its policy implications. In the 1970s, it was common to exclude scientists with known positions from the panels, so Hal Johnston, Sherry Rowland, Mario Molina, and Tom Donahue, already strong advocates of CFC control, were not invited. The panel called on their expertise to help them understand the issues, but not to participate in the writing.[23]

Handler chose John Tukey, a distinguished statistician at Princeton University and Bell Labs, as the chairman of the Committee on Impacts of Stratospheric Change, and Herbert Gutowsky, director of the School of Chemical Sciences at the University of Illinois, as chairman of the Panel on Atmospheric Chemistry.[24] Their report deadline was April 1, 1976. They had a difficult task. The scientific research front was advancing quickly; the panel would be analyzing a moving target, as it were. They also had to do their work under a media magnifying glass, enduring efforts by reporters to get them to discuss the panel deliberations, attacks by the industry's PR machinery, and pressure from both advocates and opponents of immediate regulation.[25]

While the two panels struggled with understanding and assessing the science, the industry's Committee on Atmospheric Science began their resistance campaign. Their initial star witness against Rowland and Molina was a British professor of theoretical mechanics at Imperial College, Richard Scorer. The Chemical Specialties Manufacturer's Association arranged a U.S. tour so that Scorer could publicly denounce the ozone depletion work. Scorer called the CIAP study "pompous claptrap," and seems to have held the same opinion of the ongoing CFC-related research.[26]

Scorer's main point during his tour was one that would become a common refrain among anti-environmentalists in the years to come. He insisted that human activities were too small to have any impact on the atmosphere, which he called "the most robust and dynamic element in the environment."[27] He dismissed the idea of ozone destruction as a "scare story" based on little scientific evidence. Even in Los Angeles, struggling with a tremendous smog problem that created widespread respiratory distress during the summer, he insisted that humans were incapable of harming the environment.[28]

Scorer's tour was intended to generate proindustry press, not to contribute to the scientific effort, and after a *Los Angeles Times* reporter exposed his connection to the industry lobby, calling him a "scientific hired gun," he lost whatever PR effectiveness he had.[29] But while Scorer would go away, his arguments wouldn't.

The industry's Committee on Atmospheric Sciences had an idea, perhaps generated by Stolarski and Cicerone's work-around for their chlorine paper in Kyoto: to blame volcanoes. Magmas contain dissolved chlorine, and when volcanoes erupt, this chlorine can be released into the atmosphere. When volcanoes erupt catastrophically they send ash, dust, and gases into the stratosphere. If there were a lot of volcanic chlorine floating around the stratosphere, then a small amount of additional CFCs might not make much difference. If volcanoes supplied most of the chlorine, and the ozone layer hadn't been destroyed yet, then chlorine couldn't be a big deal. Or so the industry argument went.

But volcanoes also erupt a lot of water vapor, and soot-and-dust-laden rain (often black) falls during or just after eruptions, as the water vapor condenses. Chlorine is easily dissolved in water and some of it therefore rains out. This phenomenon was understood qualitatively in the mid-1970s but not quantitatively, so the industry Council on Atmospheric Sciences decided to make a big show of proving that most of the chlorine would reach the stratosphere. They held a press conference in October 1975 to announce their "research" program on an Alaska volcano expected to erupt soon. The volcano erupted at the end of January 1976, but evidently it did not do what they were hoping, as the industry group never announced results, beyond stating they were "inconclusive."[30] Yet the claim that volcanoes were the source of most stratospheric chlorine was repeated well into the 1990s.

The volcano idea was just one part of the industry effort to impugn CFC science. As Harold Schiff put it, "They [the CFC industry] challenged the theory every step of the way. They said there was no proof that fluorocarbons even got into the stratosphere, no proof that they split apart to produce chlorine, no proof that, even if they did, the chlorine was destroying ozone."[31] Each of these claims was defeated by evidence during 1975 and 1976. A scientist from the University of Denver showed that CFC concentrations in the stratosphere over New Mexico had doubled between 1968 and 1975, proving that they *were* reaching the stratosphere in increasing amounts. Only a few months later, National Center Atmospheric Research scientists measured CFCs at the key altitude that Rowland and Molina had predicted they would break up, and showed that the concentration declined as the altitude increased—exactly as expected.[32] Experiments conducted at the National Bureau of Standards then showed that CFCs did, in fact, release chlorine atoms when exposed to ultraviolet light, just as Rowland and Molina's chemical model had said they would.[33]

Rowland and Molina had proposed that ozone depletion occurred through a complex set of interlocking chemical reactions, a complexity that helped industry cast doubt on the issue. But there was one key molecule predicted by their theory whose existence in the stratosphere would prove that chlorine was reacting with ozone to destroy it: chlorine monoxide, or ClO. Theory predicted that ClO would form from the reaction of a free chlorine atom with ozone. The free chlorine would strip one of the three oxygen atoms from the ozone molecule, leaving a normal oxygen molecule (O_2) and chlorine monoxide (ClO) as the by-products. But chlorine monoxide is very hard to measure, because it rapidly reacts with the surface of most instruments and vanishes. Scientists measured the chemistry of the stratosphere by sending up refrigerated flasks, opening them, collecting some air, and having the flask parachuted back to Earth. Then the air would be tested in ground-based instruments. This didn't work for ClO, because it reacted with the walls of the flasks.

The problem was solved by a young scientist named James Anderson, who had worked with Tom Donahue at Michigan. He developed an instrument that drew a continuous flow of air through itself, and used a laser to detect ClO. As long as the instrument was able to maintain a smooth airflow through itself, the air in the middle of its air passage wouldn't touch the walls, and if the ClO were there, the laser would detect it.

The chlorine monoxide *was* there. It was the smoking gun they'd been waiting for.[34] ClO had been detected in the stratosphere, and there was no explanation for its presence except that it was caused by CFCs reacting with ozone to produce it. In effect, it was a fingerprint—a telltale sign that CFCs had been there.

As Anderson was making his measurements, Sherry Rowland realized that the computer models being used by the National Academy panels had incorrect data for another important compound, chlorine nitrate. Chlorine nitrate is a "sink"—a species that keeps chlorine (and nitrogen, another ozone destroyer) from destroying yet more ozone—but it is also a photochemical, meaning that it breaks down in sunlight, where it releases chlorine and nitrogen. German data from the 1950s suggested that it broke down in just minutes, but these data were pretty old; Rowland had his group remeasure the chlorine nitrate destruction rate. He found that it survived much longer than expected and the new figures reduced their computer model's estimate of ozone depletion by 20 to 30 percent. Rowland announced the result immediately, even though it undercut his own position favoring an immediate CFC ban.

The Academy panels were thrown into turmoil, as they tried to make sense of these new data. The Academy decided to delay both reports, annoying the federal agencies that were funding the study (NASA, NOAA, NSF, and EPA), who didn't like the additional cost involved or the delay in their schedules. But Academy president Handler was adamant about getting the science right: he would cancel the reports entirely rather than publish them without having digested the new data properly. The agencies backed off (and NASA scraped up more money). The chemical industry trumpeted the delay as a sign that the Academy might be preparing to announce that CFCs weren't a threat after all, calling a press conference in

May to announce that the new data brought depletion estimates "nearly to zero."[35]

That was simply not true.

When reports were finally released on September 15, 1976, the scientists had concluded that steady state depletion was likely to result from continuing releases of CFCs at 1973 levels. The depletion was pretty much proportional to the amount of CFC released, so doubling emissions would roughly double ozone loss.[36] This was less than earlier predictions, but still substantial.

It wasn't the job of the Academy panels to propose regulations, but the scientific results clearly showed that CFCs did serious damage. Tukey's committee tried to walk the fine line between scientific deduction and policy prescription, suggesting that while more research was needed to reduce uncertainties in the predictions, they were not proposing a long delay. "No more than two years need be allowed," they wrote, before beginning selective regulation of CFCs.[37]

The political process now moved rather rapidly. The chairman of the president's Council on Environmental Quality argued that "the criminal defendant's presumption of innocence was not appropriate for regulatory decisions under uncertainty," and said that he would prod the federal regulatory apparatus toward immediate rule making.[38] The federal interagency committee on the stratosphere, IMOS, also called for immediate rule making. On October 12, the EPA and FDA stated together that they would get to work, suggesting that European countries follow suit.[39]

The rapid move toward regulation caught the aerosol industry by surprise. Their PR machinery tried to spin Tukey's judicious prose—and his suggestion that at most scientists needed two more years—into well-publicized claims that regulation shouldn't be undertaken for *at least* two years. The Western Aerosol Information Bureau argued that what the studies "really" said was that "we don't know what is going on, we don't know if what we are measuring has anything to do with what the problem may be, but we're sure that nothing is going to happen one way or the other for the next couple of years."[40]

Meanwhile something very interesting had happened: American people had already started to change their habits. By the time Food and Drug Administration head Donald Kennedy announced regulations, in 1977, CFC propellant use had already dropped by three quarters. The public had realized that that there were many (often less expensive) substitutes for

CFCs, such as roll-on antiperspirants and pump sprays for kitchen cleansers. The ban on propellant use, which took effect in 1979, was merely the coup de grace.

When the CIAP program ended, NASA took over leadership of the nation's stratospheric research program. The agency's leaders were still concerned about the potential threat that regulations posed to the shuttle, and also wanted to improve the quality and relevance of its research programs. The NASA earth science manager supported the development of a number of new balloon and space-borne instruments for measurement of various trace species associated with ozone depletion. The agency had also received a mandate from Congress that required a quadrennial report on the state of the science. This forced NASA to engage scientists around the nation in a regular process of assessment. Over the next several years, that assessment process forged an active research community around the question of ozone depletion.

Holes in the Ozone Layer

In 1985, the British Antarctic Survey announced the existence of an area of severe ozone depletion over Antarctica, initiating a new phase of the ozone conflict. The British scientists had actually detected it four years earlier, but had disbelieved their own results. No known version of the ozone depletion hypothesis could explain such extremely low ozone levels, so they returned each year to gather more data. No one could accuse them of having rushed to judgment.

The Survey's paper came out when many of the leading stratospheric scientists were meeting in Les Diablerets, Switzerland, to review the draft of the 1985 NASA assessment. This assessment differed from its predecessors in that its manager, Robert T. Watson of NASA, had approached the World Meteorological Organization about making the assessment international, reducing the possibility of competing interpretations.[41] Watson had been a chemical kineticist at the Jet Propulsion Laboratory (JPL) prior to becoming the Upper Atmosphere Research Program manager in 1980; he had also been a postdoctoral researcher in Hal Johnston's lab at Berkeley.

The 1985 assessment document updated the status of laboratory-based efforts to improve understanding of the chlorine and nitrogen reactions, and to incorporate this understanding into chemical models. It also summarized the recent history of stratospheric measurements, filling more than a thousand pages. All this work had left the research community thinking they understood things pretty well, especially when a 1985 Shuttle *Challenger* flight demonstrated that all the predicted trace chemicals really were there.[42] It seemed that the scientific community had sorted it all out. So the Antarctic Survey's announcement was a shock.

Some of the conferees at Les Diablerets were already aware of the paper, as *Nature*'s editor had circulated it to referees in December 1984.[43] The authors had raised the question of a link to chlorine and nitrogen oxides, in keeping with the prior hypotheses regarding depletion mechanisms, but did not provide a convincing chemical mechanism. The known chemistry did not appear to permit such large ozone losses. Although the published paper had come too late to incorporate into the 1985 assessment, it was much discussed in the informal hallway and dinner conversations.

NASA also had a satellite in orbit that should have detected the ozone hole—if it was there—and some of the informal attention at the conference focused on the fact that it hadn't. No data are ever perfect, and the ground-based "Dobson" network—named for G. M. B. Dobson, the British meteorologist who pioneered systematic ozone measurement in the 1930s—was known to have problems. Moreover, the Survey's announcement had been based on the data from a single instrument, so most of the participants were willing to dismiss it in favor of the satellite data. They were soon proved to be wrong.

Richard Stolarski, who had moved to Goddard Space Flight Center in Maryland, decided to take another look at the satellite data. It turned out the satellite *had* detected the depletion. What happened is a lesson in both the limits and the strength of scientific research, so it's worth explaining.

Satellites don't just "collect" data in the way that nineteenth-century geologists collected rocks or biologists collected butterflies; they detect signals and process them. The electronics and computer software involved are very complex and sometimes things go awry, so procedures are included for screening and rejecting "bad" data. This was the case here. The satellite processing software contained computer code designed to flag ozone concentrations below a certain level—180 Dobson units—as unrealistically low, and therefore probably bad data.[44] Concentrations that low had never been detected in the stratosphere and could not be generated by any existing theoretical model, so it seemed like a reasonable choice. When some of the Antarctic ozone retrievals had come in well below 180, they were catalogued as errors. The instrument's science team had a map that showed the errors concentrated over the Antarctic in October, but they had ignored it, assuming the instrument was faulty. A healthy skepticism about their machinery led them to dismiss crucial data.

When Stolarski double-checked, he found that the depleted region covered *all* of Antarctica—and the "ozone hole" was born. It wasn't an instrument error. It was a real phenomenon. It *had* been detected by the satellites. And it defied expectation.

AT A MEETING in Austria that August, the principal investigator of the satellite group showed images generated from the data for 1979–1983 depicting a continent-sized region in which ozone levels dropped to 150 Dobson units.[45] The images offered visual confirmation that the Antarctic Survey data were not in error. They also demonstrated that the ozone hole was not a localized phenomenon. The ground-based measurements were point measurements—that is, taken at specific points in Antarctica—while the satellite data covered the whole Earth. Scientists were right to think that other things being equal the satellite data might be more reliable, because it was so much more comprehensive, but other things had not been equal. Now they were.

Like the famous images of the Earth from space, these ozone depletion maps were viscerally powerful. While almost no one lived within the boundaries of the depleted region, if it grew very much, it would reach populated landmasses in Australia and South America. And since no one knew the mechanism that produced the hole, no one could be certain that it would not grow.

NASA and NOAA sponsored two expeditions to Antarctica to learn more. The first, the National Ozone Experiment, led by atmospheric chemist Susan Solomon, was mounted at the U.S. research station at McMurdo Bay at the so-called winter fly-in in October 1986. Instrument teams from NOAA, SUNY–Stony Brook, and Jet Propulsion Laboratory examined various aspects of the atmosphere's chemistry. Before the team left Antarctica, they held a press conference to explain their results, and raised what was probably an inevitable controversy. Solomon, who was too young to have participated in the earlier round of ozone debates, gave an honest answer: their evidence supported the CFC depletion hypothesis.[46] This was the view that most atmospheric chemists endorsed.

But some meteorologists had a different view. They believed that upwelling atmospheric currents could carry ozone-poor tropospheric air into the stratosphere, creating the appearance of a hole, without any stratospheric ozone being destroyed at all. Widely quoted in the mainstream press, Solomon's statement outraged proponents of this meteorological hypothesis.[47] In reality, Solomon's team felt that meteorology probably played some role in creating the hole, but their quickly assembled crew didn't have the equipment or expertise to examine all possibilities.

Meanwhile, while Solomon's team was still in Antarctica, NASA and NOAA officials had been planning a second expedition whose goals would include closer attention to meteorological effects.[48] The Airborne Antarctic Ozone Experiment (AAOE) flew two planes out of Punta Arenas, Chile, in October and November 1987, with four hundred scientists on-site doing daily data reductions. The expedition involved nearly everyone in the small community of stratospheric research.

The 1987 Antarctic expedition had posed the question: what explains the extremely low ozone levels over Antarctica? The new evidence provided the answer: the combined effects of very high levels of chlorine from CFC breakdown and the peculiar meteorology of the Antarctic. The extreme Antarctic cold produced distinctive clouds—polar stratospheric clouds (PSCs)—made of ice crystals. Meanwhile, the polar vortex—a powerful wind that blows around the pole—confined the extremely cold air in the region. The ice crystals dramatically accelerated the chemical reactions that released chlorine, while the vortex prevented undepleted air from the midlatitudes from mixing in. The net result: when the Sun rose in the Austral spring, chlorine concentrations became far higher than any model had predicted, and ozone levels fell far lower. It was complex, but it made sense. The expedition provided the "smoking gun" by showing a strong correlation between high chlorine levels and low ozone levels.

Creating an Adaptive Regulatory Regime

Meanwhile international negotiations over CFC emissions were also taking place. Despite the U.S. ban, CFC emissions were still rising in Europe and the Soviet Union. The general conclusion of the ozone assessment—even prior to the ozone hole discovery—was that current emissions levels were tolerable, but higher levels would not be.[49] Some controls would be necessary, but how strict? And how would they be enforced?

The 1985 Vienna Convention for the Protection of the Ozone Layer—sponsored by the United Nations Environment Programme (UNEP)—imposed no restrictions on CFCs at all.[50] It was simply a procedural framework for future negotiations on a protocol—a set of amendments to the convention—which might include actual production cuts.

It took two more years of negotiations before UNEP gained international agreement for CFC production cuts. The resulting Montreal Protocol on Substances that Deplete the Ozone Layer specified cuts of 50 percent by the CFC-producing nations, to be implemented over several years.[51] An important innovation in the protocol was that it required participating nations to meet every few years to revisit the treaty in the light of new evidence, such as that being produced by the in-progress Airborne Antarctic Ozone Experiment. It was the opposite of hysterical: it allowed for scientific uncertainty, and included a mechanism to respond to new evidence, whether that evidence suggested the need for tighter restrictions, or looser ones.

Over the next few years, evidence accumulated that tighter restrictions were needed. In addition to the results of the AAOE, announced after the conclusion of the Montreal negotiations but before it was submitted to the U.S. Senate for ratification, other scientific initiatives began to produce additional relevant data. Most alarmingly, these data showed that Antarctica wasn't the only place where ozone was being depleted. The northern midlatitudes, where most people live, seemed to be experiencing ozone depletion, too.

A key initiative was the creation of the Ozone Trends Panel, established by Robert Watson to resolve a new conflict between the ground-based Dobson network data and data obtained by a satellite-based instrument called Total Ozone Mapping Spectrometer—or TOMS. The scientists working with the TOMS data had circulated a preprint—a copy of a paper that they planned to submit to a journal—claiming very large midlatitude ozone depletion. But it was well-known among satellite specialists that their instruments had a tendency to decay in space, causing unpredictable errors. So which data were correct, the Dobson data or the TOMS data?

The Ozone Trends Panel had twenty-one primary panel members, and was also subdivided into many subpanels that drew in still more members of the active research community. Over the next year and a half, the panel revisited the two data sets, and concluded that there *was* winter ozone depletion in the midlatitudes. They also concluded that the satellite was reporting more depletion than there actually was—but this overreporting could be corrected mathematically.[52] The downward trend found in the reanalysis was still twice that predicted by the theoretical models.[53]

The panel's findings were released on May 15, 1988, one day after the U.S. Senate voted to ratify the Montreal Protocol, but its general conclusions were already well-known. Policy makers and politicians had already been briefed, and, inevitably, the gist of the briefings had been leaked. Still, their formal press conference drew attention. Bob Watson and Sherry Rowland stated unequivocally that human activity was causing rapid increases in CFCs in the stratosphere, and that these gases controlled ozone levels.

The Ozone Trends Panel included a chemist from the DuPont Corporation, which had also provided financial support for the Antarctic field expeditions. After the panel's announcement, he convinced his own management that the results had to be taken seriously; they, in turn, approached the corporation's executives. After three days of intense discussion, DuPont's executives decided that the panel had demonstrated an appropriate level of harm. On March 18, they decided that DuPont would cease production of CFCs within about ten years.[54]

An Arctic Ozone Hole?

Scientific research did not stop with Senate ratification of the Montreal Protocol. While the evidence was strong enough to warrant regulatory action, from the scientists' perspective there were enough remaining uncertainties to justify more research. While the overall picture was clear, scientists still didn't fully understand the *precise* reason for unexpectedly high levels of chlorine in the Antarctic, and the exact meteorological conditions that had led to the "hole." Since the protocol had been negotiated on the basis of the gas-only chemistry, it might well need to be altered to reflect new insights about the role of ice crystals in the relevant reactions.

And what about the Arctic? Would an ozone hole develop there, too? Far more people (including those paying for the research) lived in the northern hemisphere than did in the southern, so the human and political implications were obvious.

Two indications already suggested trouble in the Arctic. In February 1988, an Ames aircraft had carried its Antarctic payload on a flight from its home at Moffett Field north to Great Slave Lake, Canada. This was south of the Arctic polar vortex, but the flight data nevertheless showed highly elevated levels of chlorine monoxide. A second indication came from Aeronomy Lab scientists, who had carried the spectrometer they had used in the Antarctic to Thule, Greenland, inside the Arctic vortex, in the last week of January 1988. They found elevated levels of chlorine dioxide and very depressed levels of nitrogen dioxide, suggesting that at least some of the relevant reactions were taking place there, too.

Bob Watson pondered what to do: consider another Arctic expedition that year, or wait until the following winter? Nearly every qualified scientist had been involved in either expeditions or postexpedition conferences of one sort or another since January 1987; it was asking a lot to send them back into the field right away. But they decided to do it anyway. The Airborne Arctic Stratospheric Expedition (AASE), to be flown out of Stavanger, Norway, was scheduled for January 1989.

The AASE made thirty-one flights into the northern polar vortex in January and February 1989. The expedition found that the chemistry of the Arctic polar vortex was highly disturbed. The low levels of nitrogen species and high levels of chlorine mirrored those in the Antarctic, and the final flight found *higher* chlorine monoxide levels than had ever been measured in the Antarctic. It certainly looked like the same reactions that happened in the Antarctic were happening in the Arctic, too.[55]

If all the chemical conditions necessary to reproduce the Antarctic ozone hole existed in the Arctic, then why was there no hole? This question they *could* answer: the Arctic was simply not as cold and its polar vortex not as strong. For a hole to form over the Arctic, very cold temperatures would have to prevail into March, and the atmospheric waves that normally roiled the Arctic stratosphere would have to be quiescent. Such conditions were not impossible, but they were much less likely than in the Antarctic. The observed patterns now made sense.

The combined results of the Ozone Trends Panel and the field expeditions caused the Montreal Protocol to be renegotiated. The results also convinced the industry that their products really were doing harm, and opposition began to fade. CFCs would now be regulated based on what had already happened, not on what might happen in the future. Because the chemicals had lifetimes measured in decades, there was no longer any doubt that more damage would happen. In a series of meetings culminating in London in June 1990, the protocol was revised to include a complete ban on the manufacture of chlorofluorocarbons, as well as other chemicals that introduced chlorine into the stratosphere. CFC production was scheduled to cease in 2000; the other chemicals had deadlines ranging from 2005 to 2040.[56] Step-by-step, the science had been worked out, and regulations were based on them.

Constructing a Counternarrative

If environmental regulation should be based on science, then ozone is a success story. It took time to work out the complex science, but scientists, with support from the U.S. government and international scientific organizations, did it. Regulations were put in place based on the science, and adjusted in response to advances in it. But running in parallel to this were persistent efforts to challenge the science. Industry representatives and other skeptics doubted that ozone depletion was real, or argued that if it was real, it was inconsequential, or caused by volcanoes.

Probably the most notorious dismissal of ozone depletion came from President Reagan's secretary of the interior, Donald Hodel, who proposed a "personal protection plan" in 1987 against ozone depletion: wearing hats and long-sleeved shirts.[57] He gave environmentalists an easy target and didn't last much longer in the administration.

Unfortunately, Hodel was not alone. During the early 1980s, anti-environmentalism had taken root in a network of conservative and Libertarian think tanks in Washington. These think tanks—which included the Cato Institute, the American Enterprise Institute, the Heritage Foundation, the Competitive Enterprise Institute, and the Marshall Institute, variously promoted business interests and "free market" economic policies, and the rollback of environmental, health, safety, and labor protections. They were supported by donations from businessmen, corporations, and conservative foundations.[58]

One of these groups, the Heritage Foundation, grew directly out of the SST debate of 1971.[59] Two days *after* the crucial congressional vote killing the SST, the American Enterprise Institute had provided a briefing supporting the project. This wasted briefing infuriated a pair of Republican congressional aides, who started a new foundation intended to provide a "quick response capability" in support of conservative, "probusiness" policy objectives. The two gained their start-up funding from Joseph Coors and Richard Mellon Scaife. By the mid-1980s, the Heritage Foundation was supported by a wide range of corporations and banks, including General Motors, Chase Manhattan, and Mobil Oil.[60]

One aspect of the effort to cast doubt on ozone depletion was the construction of a counternarrative that depicted ozone depletion as a natural variation that was being cynically exploited by a corrupt, self-interested, and extremist scientific community to get more money for their research. One of the first people to make this argument was a man who had been a fellow at the Heritage Foundation in the early 1980s: Fred Singer.[61]

Now CHIEF SCIENTIST for the U.S. Department of Transportation, Singer first protested what he called the "ozone scare" in an article that the *Wall Street Journal* ran on page one.[62] In this article, Singer admitted that ozone depletions had been observed, but he dismissed them as "localized and temporary" and insisted there was no proof that CFCs were responsible. "Some scientists believe that ozone is not lost at all but simply moves about as atmospheric motions bring in ozone-depleted air for a few weeks," he wrote. This was in April 1987—so it was true that definitive proof of CFCs' role wasn't in yet—but it was still an absurd claim, because the satellite data were global; if the ozone had just "moved about," then the satellite would have detected ozone increases somewhere else.

Singer also recycled the old tobacco tactic of refutation by distraction, noting that there are many causes of skin cancer, including "viruses, genetic predisposition, environmental carcinogens, population shifts to the Sun Belt, changes in life style, earlier detection of melanomas, and even diet."[63] All true, but beside the point: the point was that if ozone depletion continued, it would lead to additional skin cancers, on top of those already generated by other causes.

Finally Singer created a straw man that scientists would have to contest for the next decade. Just as he had alleged, contrary to the consensus of the scientific community, that fixing acid rain was a billion-dollar solution to a million-dollar problem, he now claimed that scientists had wrongly worried that water vapor from the SST would destroy ozone. "According to then-prevailing scientific wisdom," he wrote, "water vapor from the SST exhaust was supposed to destroy ozone, admitting more ultraviolent [*sic*] radiation to the earth's surface."[64] In truth, the notion of significant water-vapor-induced depletion had been rejected by the organized assessments back then. It wasn't "prevailing scientific wisdom" at all, but a hypothesis that had been rapidly discredited. But Singer went on. It was the beginning of a counternarrative that scientists had overreacted before, were overreacting now, and therefore couldn't be trusted.

In 1988, Singer laid out his own idiosyncratic interpretation of the ozone hole. He found it conspicuous that the hole appeared suddenly in 1975, at essentially the same time as a global surface warming trend on Earth had started. Accepting that high levels of chlorine played a role (though not necessarily from CFCs), Singer argued that the real cause of the hole was the stratospheric cooling, and this cooling was just part of the Earth's natural climate variability.[65]

If this were true, then there was no need to regulate CFCs—they were irrelevant. Since the Montreal Protocol had been "spurred by the belief that the [ozone hole] may just be the precursor of a general global decline in stratospheric ozone," it was clearly unnecessary.[66] The implication was that natural warming would in due course bring everything back to normal.

Singer's views were not preposterous, and they didn't violate the laws of nature. They just went against the accumulated work of hundreds of experts over the previous decade, and they just happened to lead to the conclusion that no regulation was needed.

Singer's article is also interesting for what it doesn't say. His source for the argument that stratospheric cooling was linked to surface warming was a recent article by V. Ramanathan, a leading atmospheric chemist at the University of Chicago, entitled "The Greenhouse Theory of Climate Change: A Test by an Inadvertent Global Experiment." The experiment Ramanathan referred to was the human emission of greenhouse gases that were changing the composition of the atmosphere. By this time, it was well understood among scientific researchers that humans had been increasing the atmospheric levels of greenhouse gases, and Ramanathan had summarized their likely effects. As greenhouse gas levels increased, they would trap heat in the lower portion of the Earth's atmosphere—the troposphere—and slow the migration of the Earth's heat out into space. The troposphere would warm, while the upper part of the atmosphere—the stratosphere—would cool.[67]

Ramanathan had *not* argued that the stratospheric cooling was part of a natural cycle. He had argued the reverse: that humans were altering the climate system, causing warming of the troposphere and cooling of the stratosphere. Increasing amounts of CFCs, methane, and CO_2 were likely to cause further stratospheric cooling, so continued human emissions of these gases would produce more stratospheric ozone depletion. It was precisely the opposite of Singer's position. Singer had turned Ramanathan's argument upside down.

He did the same thing to James E. Hansen, director of the Goddard

Institute for Space Studies. In August 1988, Hansen had given dramatic testimony to Congress asserting that "the scientific evidence for the greenhouse effect is overwhelming." "The greenhouse effect is real, it is coming soon, and it will have major effects on all peoples."[68] Singer used a graph from Hansen's presentation, present in the hearing transcripts and published in the *Journal of Geophysical Research*. However, Hansen had not created the graph to assert that the warming trend was part of a natural cycle, but to help show that it wasn't.

Singer neglected to mention Ramanathan and Hansen's arguments, and in doing so misrepresented their larger points: *both* the surface warming and stratospheric cooling trends were direct results of human activities. The ozone hole was anthropogenic from two distinct, but interrelated, standpoints: the excess chlorine came from CFCs, and the cooling effect came from anthropogenic global warming.

Given this, it's hardly surprising that Singer had a hard time getting his letter published. In a 1989 *National Review* article, he complained that it had been rejected by *Science* before being accepted by EOS—the newsletter of the American Geophysical Union. *National Review* was founded by conservative William F. Buckley, and Singer used this platform to launch an attack on the scientific community. In "My Adventures in the Ozone Layer," he cast the scientific community as dominated by self-interest. "It's not difficult to understand some of the motivations behind the drive to regulate CFCs out of existence," he wrote. "For scientists: prestige, more grants for research, press conferences, and newspaper stories. Also the feeling that maybe they are saving the world for future generations."[69] (As if saving the world would be a bad thing!)

Singer alleged that scientists had rushed to judgment. There was a bit of serious illogic here, for if scientists wanted above all to keep their own research programs going, then they would have had no reason to rush to judgment. They would have been better off continuing to insist that more research was needed, rather than saying that there was now sufficient evidence to warrant regulations.

Singer also insisted that Dobson had discovered the ozone hole in 1956, before CFCs had built up significantly, and then concluded by insisting that replacing CFCs was likely to prove difficult and expensive—even dangerous. CFC substitutes "may be toxic, flammable, and corrosive; and they certainly won't work as well. They'll reduce the energy efficiency of appliances such as refrigerators, and they'll deteriorate, requiring frequent replenishment."[70] They *certainly* won't work as well? How could Singer know that, if substitutes hadn't yet been developed? Singer was doing just what he had done for acid rain—insisting that any solution would be difficult and expensive, yet providing scant evidence to support the claim. In fact, he was going further, making bold assertions about the nature of technologies that did not yet exist.

Was Singer's description a fair summary of what the research community actually thought and did during the 1980s? No. It had hardly been "obvious" to everyone in the research community that CFCs were implicated; when the ozone hole was first detected, both solar effects and meteorology were considered and investigated.[71] Singer also ignored the field expeditions and laboratory experiments sponsored by NASA and NOAA. That's an important omission, since the chemical data that clinched the case came from them. The claim that Dobson had actually discovered the hole was completely false. What Dobson *had* discovered was that the seasonal ozone variation in the Antarctic was greater than in the Arctic.[72] That was a significant observation, but the hole was an entirely different matter.

In short, Singer's story had three major themes: the science is incomplete and uncertain; replacing CFCs will be difficult, dangerous, and expensive; and the scientific community is corrupt and motivated by self-interest and political ideology. The first was true, but the adaptive structure of the Montreal Protocol had accounted for it. The second was baseless. As for the third, considering Singer's ties to the Reagan administration and the Heritage Foundation, and considering the venues in which he published, this was surely the pot calling the kettle black. And we now know what happened when CFCs were banned. Non-CFC refrigerants are now available that are more energy efficient—due to excellent engineering and stricter efficiency standards—than the materials they replaced, and they aren't toxic, flammable, or corrosive.[73]

WITH THE AMENDMENTS to the Montreal Protocol adopted in 1992, ratified by the U.S. Senate, and even accepted by the DuPont Corporation, the debate over ozone depletion had come to a practical end. Scientists continued to do research, particularly on the role of ice crystals and other particulates in accelerating the depletion reactions, but primarily because other compounds, including some of the proposed substitutes for CFCs, were suspected of causing problems of their own.

Still, Singer did not give up. In 1990 he had established his own nonprofit organization, the Science and Environmental Policy Project (SEPP), to organize his work. The outfit was initially affiliated with the Washington Institute for Values in Public Policy, which was itself financed by the Reverend Sun Myung Moon's Unification Church.[74] (The Unification Church was known for its passionate anti-Communism, perhaps an attraction to Singer. One of its supporters was Eugene Wigner, the Ph.D. advisor and mentor of Fred Seitz.)[75] The church owned a newspaper, the *Washington Times*, and it also operated a publisher, Paragon House. In the years to come, Singer would use both to expand the reach of his views.

In 1991, Singer reiterated his claim that the science of ozone depletion was too uncertain in the *Washington Times* and *Consumers' Research Magazine*. He also introduced a new argument: that the Ozone Trends Panel was wrong to use the "ground-based rather than the more accurate satellite ozone data."[76] But we've seen that the satellite data had shown *larger* depletions, and that the panel had concluded that the higher satellite-derived depletion rate was an artifact of instrument decay in space (a phenomenon that should have been very familiar to Singer, given his origins in rocket research). If the panel had used the satellite data, Singer no doubt would have attacked them for ignoring the problem of instrument decay.

But whether or not they had any basis in fact, Singer's efforts began to bear fruit. In 1990, Dixy Lee Ray, a zoologist and former chair of the Atomic Energy Commission, as well as former governor of the state of Washington, was the lead author of the book *Trashing the Planet: How Science Can Help Us Deal with Acid Rain, Depletion of Ozone, and Nuclear Waste (Among Other Things)*. Billed as an effort "to separate fact from factoid, to unmask the doom-crying opponents of all progress, and to re-establish a sense of reason and balance with respect to the environment and modern technology," it was a tirade against the environmental movement—and the science that supported it.[77] Ray dismissed energy conservation and renewable energy, attacked toxic chemical "scares" promoted by environmentalists, and constructed a narrative that sedulously omitted the findings of the scientific experts and replaced them with the claims of professional critics and skeptics. Here's what she had to say about ozone.

> Although there is widespread belief that the necessary chloride ion [that damages ozone] comes from chlorofluorocarbon this has not been unequivocally established. On the other hand, the eruption of Mount St. Augustine in 1976 injected 289 billion kilograms of hydrochloric acid directly into the stratosphere. That amount is 570 times the total world production of chlorine and

fluorocarbon compounds in the year 1975. Mount Erebus, which is located just 15 kilometers upwind from McMurdo Sound, has been erupting, constantly, for the last 100 years, ejecting more than 1,000 tons (907,184 kg) of chlorine per day . . . We cannot be sure where the stratospheric chloride comes from, and whether humans have any effect upon it.[78]

Where did she get these claims? Ray cited a 1989 article by Singer in his *Global Climate Change*, which she praised as one of only two "significant, critical contributions" to the subject of ozone depletion and global warming—the other being the National Acid Precipitation Assessment Program, which had nothing to do with either ozone or global warming.[79] If you read Singer's paper, you find that he presented no original data. He had simply cited other papers, without explaining what those papers actually said.

The details about Mt. Erebus and Mt. Augustine can actually be found in two articles, published in 1989, by a man named Rogelio Maduro, in a political magazine called *21st Century Science and Technology*, which is supported by Lyndon LaRouche's organization.[80] In 1992, Maduro would publish a book, *The Hole in the Ozone Scare: The Scientific Evidence that the Sky Isn't Falling*, but the basic argument was already laid out in his 1989 work.[81] Maduro had concluded that the ozone depletion theory was a "fraud" after interviewing Reid Bryson for an article on the "hoax" of global warming. Bryson, an expert on paleoclimate studies using pollen and tree rings— nothing to do with ozone—had told Maduro that Mt. Erebus erupted more chlorine into the atmosphere in a week than CFCs released in a year.

Ray had apparently confused chlorine emission to the atmosphere and chlorine concentration in the stratosphere. Mt. Erebus did produce substantial chlorine emissions, but it did not erupt explosively, so whatever chlorine it released did not get injected into the stratosphere; it would have to have been transported upward by tropospheric winds. Yet the Antarctic data collected by the two NASA/NOAA field expeditions showed very little chlorine in the troposphere and a great deal in the stratosphere. Moreover, balloon measurements showed that the bitterly cold stratospheric air was sinking, not rising, so there was simply no way that air masses carrying materials upward from Mt. Erebus could be the source of the chlorine.

Maduro's claims were published in an obscure source, and they might easily have vanished into obscurity—but for Dixy Lee Ray. When she repeated them in her book, they suddenly gained currency and credibility.

After all, she was a scientist, and had been chairman of the Atomic Energy Commission. Surely she was credible? The press thought so, as the mass media extensively reviewed the *Trashing* book. It sold well enough that Ray expanded it into a 1993 bestseller, *Environmental Overkill*. In addition to repeating the claims of the 1990 book, Ray expanded them, by insisting that CFCs were too heavy to rise into the stratosphere in the first place![82]

Sherry Rowland was disturbed by the rapid spread of this misinformation and dedicated his 1993 AAAS presidential address to combating it.[83] Without naming names, Rowland chided "senior scientists" for helping to spread such erroneous claims. Then he addressed specifics, starting with the idea that CFCs didn't reach the stratosphere. In fact, CFCs had been measured "in literally thousands of stratospheric air samples by dozens of research groups all over the world."[84]

Rowland also addressed the volcano red herring. First, he debunked the 1980 *Science* paper that had argued that a single eruption of Mt. Augustine, Alaska, in 1976 had put as much chlorine into the stratosphere as the entire 1975 CFC production. That claim was based on the chlorine content of ashfall, not on what had actually reached the stratosphere. Rainout would have reduced the amount reaching the stratosphere, but the rain's chemistry hadn't been measured. "No actual evidence was presented in this *Science* paper to show that any hydrogen chloride had really reached the stratosphere in this volcanic plume."[85] He then recounted evidence that the eruption of El Chichón in April 1982 had produced an increase of hydrogen chloride in the stratosphere of less than 10 percent, and that the June 1991 eruption of Pinatubo—a much larger eruption—had increased it even less. Yet hydrogen chloride levels had increased steadily between those two eruptions, despite the lack of any other explosive eruptions during the interceding nine years. This showed conclusively that the chlorine did *not* come from volcanoes.

Rowland traced the next phase of confusion over volcanic effects to Fred Singer's 1989 *National Review* article. The confusion had been amplified by Ray's attributing extremely high chlorine releases to Mt. Augustine.[86] This had been taken as fact by people "who are relying, often unquestioningly, upon such fourth-hand descriptions of the volcano problem, rather than going back to the original literature." Then the error had been broadcast far and wide by a variety of media outlets.[87]

ROWLAND'S ATTEMPT TO CORRECT these errors didn't make a difference. In March 1994, Singer repeated the now-refuted claim that the evidence "suggest[ed] that stratospheric chlorine comes mostly from natural sources."[88] In September 1995, Singer served as a star witness in hearings in the U.S. Congress, sponsored by Republican congressman Dana Rohrabacher—on "scientific integrity." Singer recycled some of his earlier claims and concluded that the committee was being "misled, bamboozled, and otherwise manipulated" by the testimony of Robert Watson, former director of the NASA Upper Atmosphere Research Panel and currently at the Office of Science and Technology Policy. Referring to the issue as "so-called" ozone depletion, he asserted that scientific basis for concern was simply "wrong."[89] In his written statement to the committee, Singer added that there was "no scientific consensus on ozone depletion or its consequences."[90] Just a few weeks later, Sherry Rowland shared the 1995 Nobel Prize in Chemistry with Mario Molina and Paul Crutzen for their work on the understanding of stratospheric ozone chemistry—the highest honor any scientist can achieve—and the clearest possible evidence of broad acceptance and appreciation of one's work.[91]

So Singer attacked the Nobel committee, too. "In awarding the 1995 Nobel Prize in Chemistry to the originators of the stratospheric ozone depletion hypothesis, the Swedish Academy of Sciences has chosen to make a political statement," he began, writing again in the *Washington Times*. Swedish public opinion had supported the "hasty phaseout" of CFCs and even a "putative carbon tax to turn back a global climate warming that has not even been detected yet . . . In short, the country is in the throes of collective environmental hysteria."[92]

Did all of Singer's efforts to discredit mainstream science matter? When asked in 1995 where he got his assessments of ozone depletion, House Majority Leader Tom DeLay, probably the most powerful man in Congress at the time, said, "my assessment is from reading people like Fred Singer."[93]

What Was This Really About?

Everyone is entitled to an opinion. But when a scientist consistently rejects the weight of evidence, and repeats arguments that have been thoroughly rebutted by his colleagues, we are entitled to ask, What is really going on?

From 1988 to 1995, Singer insisted that the ozone research community was misleading the public about even the existence of ozone depletion, let alone its origins. He argued in his 1989 *National Review* article that researchers were doing this to line their own pockets, and those of their graduate students, by scaring public officials who could fund their research.[94] Of course, similar charges might be levelled at Singer. While we don't have access to SEPP's tax returns for the 1990s, in 2007 it netted $226,443, and had accumulated assets of $1.69 million.[95] His skepticism also gained him a huge amount of attention—far more than most scientists ever get for their research, quietly published in academic journals. So if scientists should be discredited for getting money for their research, or for enjoying the limelight, the same argument would logically apply to Singer.

What *was* Singer really up to? We suggest that the best answer comes from his own pen. "And then there are probably those with hidden agendas of their own—not just to 'save the environment' but to change our economic system," he wrote in 1989. "Some of these 'coercive utopians' are socialists, some are technology-hating Luddites; most have a great desire to regulate—on as large a scale as possible."[96] In a 1991 piece on global warming, he reiterated the theme that environmental threats—in this case global warming—were being manufactured by environmentalists based on a "hidden political agenda" against "business, the free market, and the capitalistic system."[97] The true goal of those involved in global warming research was not to stop global warming, but to foster "international action, preferably with lots of treaties and protocols."[98] The "real" agenda of environmentalists—and the scientists who provided the data on which they relied—was to destroy capitalism and replace it with some sort of worldwide utopian Socialism—or perhaps Communism. That echoed a common right-wing refrain in the early 1990s: that environmental regulation was the slippery slope to Socialism. In 1992, columnist George Will encapsulated this view, saying that environmentalism was a "green tree with red roots."[99]

To fight environmental regulation, Singer and Ray told a story in which science was corrupt and scientists could not be trusted. Once planted, this counternarrative did not easily go away. Fred Seitz included its claims in a 1994 Marshall Institute "report" on ozone depletion and climate change, repeating, for example, the Dobson-found-it claim about the Antarctic ozone hole and even implying that CFCs couldn't reach the stratosphere—a claim even a freshmen physics major would know was wrong—much less a former president of the National Academy of Science.[100] Patrick Michaels, an agricultural climatologist at the University of Virginia who had joined Singer in attacking the mainstream view of ozone depletion in

1991 on the pages of the *Washington Times*, reiterated the volcanic argument as late as 2000.[101]

It's not surprising that the Marshall Institute took up Singer's ozone claims, because they shared his passionate anti-Communism. Nor is it surprising that he found willing publication venues. The *Washington Times* and *National Review* were stridently anti-Communist in their editorial views; the *Wall Street Journal*, *Forbes*, and *Fortune* were obviously probusiness and market oriented. The *Wall Street Journal* kept up the drumbeat for several years with articles and editorials having titles such as "Bad Climate in Ozone Debate," and "Ozone, CFCs, and Science Fiction," "The Dreaded Ozone Hole," and, after the Nobel award to Rowland and his colleagues, "Nobel Politicized Award in Chemistry."[102] One of these pieces, an article in February 1993 entitled "Too Many Holes," was written by a man named Kent Jeffreys.[103] In the next chapter we will see how Jeffreys joined forces with Singer to attack the EPA over the science of secondhand smoke.

CHAPTER 5

What's Bad Science? Who Decides?
The Fight over Secondhand Smoke

B Y THE MID-1980S, NEARLY EVERY AMERICAN knew that smoking caused cancer, but still tobacco industry executives successfully promoted and sustained doubt. Scientists continued to play a crucial role in that effort, as men like Dr. Martin Cline provided powerful "expert" testimony when cases went to court.[1] In 1986, a new panic ripped through the industry, much like the one that tobacco salesmen must have felt in 1953 when those first painted mice developed cancer from cigarette tar, and again in 1963 when the industry read the first Surgeon General's report. The cause was a new Surgeon General's report that concluded that secondhand smoke could cause cancer even in otherwise healthy non-smokers. When the EPA took steps to limit indoor smoking, Fred Singer joined forces with the Tobacco Institute to challenge the scientific basis of secondhand smoke's health risks. But they didn't just claim that the data were insufficient; they claimed that the EPA was doing "bad science." To make this claim seem credible, they didn't just fight EPA on secondhand smoke; they began a smear campaign to discredit the EPA in general and tarnish any scientific results that any industry didn't like as "junk."

A Brief History of Secondhand Smoke

Today, we know that secondhand smoke can kill. The U.S. Department of Health and Human Services tells us that "there is no risk-free level of exposure to second-hand smoke: even small amounts . . . can be harmful to people's health."[2] But just as the tobacco industry knew that smoking could cause cancer long before the rest of us did, they knew that secondhand smoke could cause cancer, too. In fact, they knew it well before most independent scientists did.[3] In the 1970s, industry researchers had found that sidestream smoke contained *more* toxic chemicals than mainstream smoke—in part because smoldering cigarettes burn at lower temperatures at which more toxic compounds are created. So they got to work trying to produce less harmful sidestream smoke by improving filters, changing cigarette papers, or adding components to make the cigarettes burn at higher temperatures. They also tried to make cigarettes whose sidestream smoke was not less dangerous, but simply less *visible*.[4]

Those charged with protecting public health were less sanguine. The states were moving actively against tobacco. By 1979, every state except Kentucky and Nevada had some kind of antismoking legislation passed or pending. Many bills were aimed at active smoking—to discourage it by increasing taxes or restricting advertising—but some targeted what was becoming known as the "indoor air quality" issue: the impact of secondhand smoke on bystanders.[5]

New Jersey, for example, had been debating restrictions on public smoking since 1974.[6] This is curious, because in 1974 there was very little published scientific evidence to show that secondhand smoke was dangerous. Perhaps it just seemed like common sense: if smoke was harmful to the person who inhaled it on purpose, wouldn't it also be harmful to the person who inhaled it by accident?

Evidence to support that common sense began to emerge in 1980, when researchers published a paper in the weighty *New England Journal of Medicine* showing that nonsmokers working in smoky offices had decreased lung function—as much as if they were actually light smokers.[7] It was a large study—twenty-one hundred subjects—and it was statistically significant, but the science was heavily criticized. It later turned out that nearly all the critics had links to the tobacco industry, but they still had a point: it was hard to demonstrate just how much passive smoke a person was exposed to. You could make general claims about "smoky" environments, but to make a scientifically robust causal claim, you should, ideally, measure exposure levels and show that the more exposure, the more risk. This is known as a "dose-response" curve. A second study provided it.[8]

Takeshi Hirayama was chief epidemiologist at the National Cancer Center Research Institute in Tokyo, Japan. In 1981, he showed that Japanese women whose husbands smoked had much higher death rates from lung cancer than those whose husbands did not. The study was long-term and big—540 women in twenty-nine different health care districts studied over fourteen years—and showed a clear dose-response curve: the more the husbands smoked, the more the wives died from lung cancer. Spousal drinking had no effect, and the husbands' smoking had no impact on diseases like cervical cancer that you wouldn't expect to be affected by cigarette smoke. The study did exactly what good epidemiology should do: it demonstrated an effect and ruled out other causes. The Japan study also explained a long-standing conundrum: why many women got lung cancer even when they didn't smoke.[9] Hirayama's study was a first-rate piece of science; today it is considered a landmark.

The tobacco industry lambasted its findings. They hired consultants to mount a counterstudy and undermine Hirayama's reputation.[10] One of these consultants was Nathan Mantel, a well-known biostatistician, who claimed that Hirayama had committed a serious statistical error. The Tobacco Institute promoted Mantel's work, convincing the media to present "both sides" of the story. Leading newspapers played into their hands, running articles with headlines such as SCIENTIST DISPUTES FINDINGS OF CANCER RISK TO NONSMOKERS and NEW STUDY CONTRADICTS NON-SMOKERS' RISK. Then the industry ran full-page ads in major newspapers highlighting these headlines.[11]

The "new study" was of course funded by the industry. In private, a different story was unfolding, as industry advisors acknowledged that the Hirayama study was correct. "Hirayama [and his defenders] are correct and Mantel and TI [Tobacco Institute] are wrong," one internal memo acknowledged. Industry scientific advisors "believe Hirayama is a good scientist and that his nonsmoking wives publication was correct," concluded another. Another memo put it even more strongly, saying, "Hirayama was correct, that the TI knew it, and that TI [attacked] Hirayama knowing that the work was correct."[12]

The scientific community knew it, too, and the Hirayama study had a galvanizing effect. Doctors, public health officials, and antitobacco activists began to push for controls on public smoking. By 1984, thirty-seven states and the District of Columbia had passed restrictions on smoking in public places; two years later, the number was up to forty.[13] Congress held hearings on controlling cigarette advertising and restricting sales to minors, and the Civil Aviation Board considered a smoking ban on airplane flights.[14] This would make sense, of course, only if secondhand smoke affected bystanders. In 1986, the Surgeon General declared that it did.

"The question of whether or not tobacco smoke is carcinogenic . . . was conclusively resolved more than 20 years ago," the secretary of Health and Human Services wrote to President George H. W. Bush in a cover letter to the 1986 report. For secondhand smoke the question had now been answered, too: "Involuntary smoking is a cause of disease, including lung cancer, in healthy nonsmokers."[15] Ambient tobacco smoke also caused respiratory illness and decreased lung function in infants and young children and increased the risk of asthma. "As a physician," the secretary concluded, "I believe that parents should refrain from smoking."

The report's Executive Summary was written by Robert Windom, a physician nominated by President Ronald Reagan, who took the results to their logical policy endpoint: "Actions to protect nonsmokers from ETS [environmental tobacco smoke] exposure not only are warranted, but are essential to protect public health."[16] An independent report by the National Research Council that year came to the same conclusion.[17] Smoking was not just a matter of personal preference; it was a serious risk to bystanders, like driving drunk or shouting fire in a crowded theater.

The tobacco industry was worried, very worried. It was one thing to say that smokers accepted uncertain risks in exchange for certain pleasures, but quite another to say that they were killing their friends, neighbors, and even their own children. Philip Morris vice president Ellen Merlo put it this way: "All of us whose livelihoods depend upon tobacco sales—directly or indirectly—must band together into a unified force . . . it's not a question of 'are we going to do well or badly . . . this year?' It's a question of: 'Are we going to be able to survive and continue to make a living in this industry in the years to come?' " The bottom line, she explained, was this: "If smokers can't smoke on the way to work, at work, in stores, banks, restaurants, malls and other public places, they are going to smoke less," and the industry was going to shrink.[18]

Industry disinformation campaigns now took new and creative forms. Sylvester Stallone was paid $500,000 to use Brown and Williamson products in no fewer than five feature films to link smoking with power and strength, rather than sickness and death.[19] The Center for Tobacco Research set up a "special projects" office to deal with secondhand smoke, including the development of countervailing scientific evidence, expert witnesses, and industry-sponsored conferences to challenge the emerging scientific consensus.

Several of these special projects were run though a law firm to shield these efforts from scrutiny using attorney-client privilege.[20] (We already saw how UCLA scientist Martin Cline hid behind attorney-client privilege when testifying as an expert witness, claiming not to work for the tobacco industry, but for a law firm.) Other projects developed legal strategies to suggest that restricting smoking in the workplace would be a form of employment discrimination.[21] The industry promoted the idea of "sick building syndrome" to suggest that headaches and other problems suffered by workers in smoky atmospheres were caused by the buildings, not smoke.[22] They attempted to join forces with antitax groups to resist cigarette excise taxes.[23] And they redoubled their efforts to recruit scientists. Project Whitecoat—as its name suggests—enlisted European scientists to "reverse scientific and popular misconception that ETS [environmental tobacco smoke] is harmful."[24] Once again, the industry was fighting science with science—or at least, scientists.

In 1991, Philip Morris executives outlined four objectives specifically related to secondhand smoke. One was to fight bans on smoking in workplaces and restaurants. A second was to maintain smoking areas in transportation facilities like airports. A third was to promote the idea of "accommodation"—that smokers (like the disabled?) had the right to be accommodated. Atlanta, Georgia, would be targeted to become a "model accommodation city," because of its tradition of Southern hospitality, but there was a (literally fatal) flaw in this argument.[25] Everyone appreciates hospitality, but few would argue that it includes the right to kill your guests. So "Objective #1"—on which all else hinged—was "to maintain the controversy . . . about tobacco smoke in public and scientific forums."[26] The budget for maintaining the controversy was $16 million.

The year that followed was crucial for maintaining the controversy, because the battle had now been joined by the U.S. Environmental Protection Agency.[27] The tobacco industry had promoted the use of the phrase "environmental tobacco smoke" in preference to passive smoking or secondhand smoke—perhaps because it seemed less threatening—but this proved a tactical mistake, because it virtually invited EPA scrutiny. If secondhand smoke was "environmental," then there was no question that it fell under the purview of the *Environmental* Protection Agency. And this meant the prospect of federal regulation—what the industry most dreaded.

In December 1992, the EPA released *Respiratory Health Effects of Passive Smoking*. The report attributed 3,000 lung cancer deaths per year to secondhand smoke, as well as 150,000 to 300,000 cases of bronchitis and pneumonia in infants and young children. Another 200,000 to 1,000,000 children had their asthma aggravated, and ETS also increased the risk of asthma in children who did not already have it. These data were statistically significant and could not be explained away by other causes, risk factors, or chance. Environmental tobacco smoke was a class A—a known human—carcinogen.[28]

Despite this strong central conclusion, the report was in many ways cautious. One potential bombshell was left out of the Executive Summary and press releases. This was the statistically significant correlation between ETS and sudden infant death syndrome (SIDS). The evidence clearly showed that ETS increased the risk of SIDS, but the panel couldn't decide whether that risk was caused by prenatal smoking, postnatal ETS, or both. Several other possible connections and correlations—increased cardiovascular disease in adults, respiratory infections in older children, and more—were also left unresolved, pending further research. But it seemed likely that at least some would be resolved on the side of harm, so the scientists concluded: "The total public health impact from ETS will be greater than that discussed here."[29]

The authors also confronted one important methodological difficulty. To assess risk, you have to compare exposed people with unexposed ones, but since ETS was everywhere, it was difficult, perhaps impossible, to find a truly "unexposed" population. So they decided to focus on studies with high spousal exposure, where effects were most likely to show up clearly. Seventeen (of thirty) studies fit this bill, and every single one showed increased risk, nine at the 95 percent confidence level, and the rest at the 90 percent level.[30] Moreover, among women who smoked, the lung cancer rate was even higher if their husbands also smoked. This showed that ETS added extra risk on top of that carried by smoking itself.

It was a judgment call to focus on high spousal exposure and to accept results at the 90 percent confidence level, but it was a reasoned one, and supported by the "weight of evidence approach" advocated by EPA risk assessment guidelines. In 1983, Congress had commissioned the National Academy of Sciences to review risk assessment in the federal government. The Redbook, as the resulting report came to be known for the color of its cover, asked each federal agency to establish clear and consistent guidelines for risk assessment.[31] The EPA had done this, and concluded that there was no magic bullet of risk assessment—different kinds of studies were useful in different ways—so the best approach was to scrutinize all the available evidence and determine where the weight of the evidence lay.[32]

There was no scientific trump card, either. Animal studies face the obvious difficulty that animals aren't people. Human studies face the difficulty that it is generally unethical to deliberately expose people to known or sus-pected risks. Statistically based epidemiology grapples with the well-known problem that correlation is not causation: some associations occur by chance. Nowadays most human toxic exposures are fairly low, because most of the time most reasonable people (and reasonable employers) try to minimize exposures to substances that we know (or seriously suspect) are harmful. And when the dose is low, the response is typically small, and therefore hard to detect.

However, all of these limitations could be addressed through the weight-of-evidence approach: no one study is perfect, but each can contribute use-ful information. For example, to test if a correlation in humans is causal or coincidental, you can deliberately expose animals in a controlled environ-ment. If the animals show the same effect, and if that effect follows a dose-response curve, then the effect is probably not a coincidence. This is what the EPA now argued for secondhand smoke. Environmental tobacco smoke contains the same chemicals found in direct smoke and these chemicals were known to cause cancer in lab rats. So when the epidemiology revealed increased rates of cancer in the wives of smokers, with a clear dose-response curve, it was reasonable to infer a causal connection.

Consistency and quantity of information were also important consider-ations. On secondhand smoke, the good news (sort of) was that there was plenty of evidence on human exposure and the results were consistent. Lots of smoke produced lots of cancer.[33] Less smoke produced less cancer. The effects were seen in the United States, Germany, and Japan, despite other differences in lifestyle, diet, and the like. The weight of evidence was heavy, indeed.[34] The EPA called it "conclusive."[35]

Who could deny all that? The answer: Both Fred Seitz and Fred Singer.

As we saw in our first chapter, Fred Seitz began working for the tobacco industry in 1979. In 1989, he took up the defense of secondhand smoke. He coordinated a report, "Links between passive smoking and disease," which frankly acknowledged the abundant scientific evidence linking ETS to lung cancer in adults, and to respiratory illness, asthma, and ear infec-tion in children, and even to perinatal death.[36]

Seitz did not suggest, however, that the industry give up the fight. Rather, he suggested that the best way to fight such a heavy weight of evidence was to challenge the weight-of-evidence approach. The idea was to reject "ex-haustive inclusion"—examining all the evidence—and to focus on the "best evidence" instead.[37]

Seitz had a point. Not all scientific studies are created equal, and lumping the good with the bad can cause confusion and error. An epidemiological study with ten thousand people is clearly better than one with ten. But it doesn't take much imagination to see how easily a "best evidence" approach could be biased, excluding studies you don't like and including the ones you do. Seitz's report stressed that inclusion criteria should always be stated up front—such as a preference for studies with "ideal research designs." But medical studies are never conducted under ideal conditions: you cannot put people in cages and control what they eat, drink, and breathe, 24/7. Animals are by definition models for what a researcher is really interested in— people. At best, animal studies are reliable representations or good first approximations, but they can never be considered ideal; Seitz's argument was transparently self-serving. The industry was not charmed, and they took up a different banner instead. It was the banner of "sound science." For this they turned to Fred Singer.

In 1990, Singer had created his Science and Environment Policy Project to "promote 'sound science' in environmental policy."[38] What did it mean to promote "sound science"? The answer is, at least in part, to defend the tobacco industry. By 1993, he was helping the industry to promote the concept of sound science to support science they liked and to discredit as "junk" any science they didn't. He did this in collaboration with APCO Associates, the public relations firm that Philip Morris had hired to help with the secondhand smoke campaign.

Tom Hockaday was an APCO employee, and March 1993 found him working closely with Philip Morris vice president Ellen Merlo to develop scientific articles to defend secondhand smoke and promote the idea that the EPA work was "junk science." "We have been working with Dr. Fred Singer and Dr. Dwight Lee [an economist, holding the Ramsey Chair of Private Enterprise at the University of Georgia], who have authored articles on junk science and indoor air quality," Hockaday explained in a memo to Merlo. "Attached you will find copies of the . . . articles which have been approved by Drs. Singer and Lee." Merlo approved the overall approach but wished that Singer's junk science article had a "more personal introduction." Tom Hockaday reported back that Singer was "adamant that this would not be his style."[39]

What was his style? A full-frontal assault, claiming that the science done at the EPA was "junk." The headline of the article he prepared for APCO read: JUNK SCIENCE AT THE EPA. The EPA was taking "extreme positions not supported by science," he asserted. Claiming that they "could not rule out other factors . . . such as diet, outdoor air pollution, genetics, prior lung disease, etc.," he charged that the EPA had "rig[ged] the numbers" by accepting the 90 percent confidence level instead of a 95 percent one.[40]

Why would the EPA "rig" the numbers? Singer's answer: Controlling smoke would lead toward greater regulation in general. "The litany of questionable crises emanating from the Environmental Protection Agency is by no means confined to these issues. It could just as easily include lead, radon, asbestos, acid rain, global warming, and a host of others." By the early 1990s, every one of these items—lead, radon, asbestos, and global warming—had come under serious scrutiny because of substantial scientific evidence, and in every case that concern has been legitimated by further scientific work. The EPA had a legal obligation to be concerned about these things. But the agency had not called ETS a "crisis." That was Singer's word. EPA had called it a carcinogen and therefore a risk.

Was there any substance to Singer's complaints? The short answer is no. The EPA scientists *had* considered and ruled out other factors. That is what it means to do epidemiology. No one had denied that genetics and lifestyle played a role in health and disease, but the statistical evidence was overwhelming that ETS was an *added* risk. It is not plausible to suppose that Singer did not understand this—he was a highly educated and intelligent man—but the reality wasn't convenient to his motivation. He was not practicing science; he was attacking it. His broader purpose, the historical evidence suggests, was to undermine the EPA in order to stop or delay regulation regarding secondhand smoke.

Consider a handbook the tobacco industry distributed that same year, which drew on Singer's work.[41] *Bad Science: A Resource Book* was a how-to handbook for fact fighters. It contained over two hundred pages of snappy quotes and reprinted editorials, articles, and op-ed pieces that challenged the authority and integrity of science, building to a crescendo in the attack on the EPA's work on secondhand smoke. It also included a list of experts with scientific credentials available to comment on *any* issue about which a think tank or corporation needed a negative sound bite.[42]

Bad Science was a virtual self-help book for regulated industries, and it began with a set of emphatic sound-bite-sized "MESSAGES":

1. *Too often science is manipulated to fulfill a political agenda.*

2. *Government agencies . . . betray the public trust by violating principles of good science in a desire to achieve a political goal.*

3. *No agency is more guilty of adjusting science to support precon-*

ceived public policy prescriptions than the Environmental Protection Agency.

4. *Public policy decisions that are based on bad science impose enormous economic costs on all aspects of society.*

5. *Like many studies before it, EPA's recent report concerning environmental tobacco smoke allows political objectives to guide scientific research.*

6. *Proposals that seek to improve indoor air quality by singling out tobacco smoke only enable bad science to become a poor excuse for enacting new laws and jeopardizing individual liberties.*

Bad, bad science. You can practically see the fingers wagging. Scientists had been bad boys; it was time for them to behave themselves. The tobacco industry would be the daddy who made sure they did. It wasn't just money at stake; it was individual *liberty*. Today, smoking, tomorrow . . . who knew? By protecting smoking, we protected freedom.

As we saw in chapter 3, science really *was* manipulated for political purposes in the case of acid rain, but not by the scientists who had done the research. It was Bill Nierenberg who changed the Executive Summary of the Acid Rain Peer Review Panel, not the EPA, which played no role in the Acid Rain Peer Review. Still, if the best defense is a good offense, the tobacco industry now took the offensive. To anyone who understood the science, their actions were pretty darn offensive, indeed.

Bad Science was divided into six chapters, each one beginning with a list of sound bites entitled "What others are saying." On page 1 an economics professor was quoted: "Crises can be exploited by organized groups to justify government action . . . If a real crisis is not available, an artificial crisis . . . will serve just as well." The professor was Dwight Lee, the paid consultant working alongside Fred Singer, via APCO, for Philip Morris. Another quote claimed that undue regulation cost a family of four $1,800 per year. What was that claim based on? No one knew, because *Bad Science* contained no primary sources or annotations. Nearly all the quotes were assertions presented as facts. "Costly solutions are . . . enacted into law . . . before they are scientifically justified," said one. "Publicly funded scientists may be playing fast and loose with the facts for political reasons," said another. "Many environmental zealots in and out of government . . . have proved themselves quite willing to bend science to the service of their political . . . goals." And so on.

If the quotable quotes were assertions without evidence, so too were many of the articles, often taken from the *Wall Street Journal* and *Investor's Business Daily*, and written by individuals with long histories of defending risky industrial products. Michael Fumento, for example, a syndicated columnist for Scripps Howard papers and a longtime defender of pesticides, asked, "Are Pesticides Really So Bad?" in *Investor's Business Daily*. (Fumento was later fired from Scripps Howard for failing to disclose receiving $60,000 from Monsanto, a chemical corporation whose work he covered in his columns.)[43] "Frontline Perpetuates Pesticide Myth," "Earth Summit Will Shackle the Planet, Not Save It," and other articles from the *Wall Street Journal* variously attacked efforts to control pesticides, stop global warming, and limit the risks of asbestos. A *St. Louis Post-Dispatch* headline declared, SCIENTISTS RIPPED AS ALARMISTS IN ECOLOGY WARNING above an article quoting Candace Crandall—Singer's wife.

If *Bad Science* often quoted "experts" who were paid consultants to regulated industries, sometimes it followed a more sophisticated strategy: reminding readers of the fallibility of science. Reprints from respected media outlets provided well-documented examples of scientific error and malfeasance. "The Science Mob," from the *New Republic*, recounted the David Baltimore case, where a postdoctoral fellow in Baltimore's lab falsified experimental results, and the scientific establishment closed ranks to defend Baltimore—a giant in his field—rather than support the whistleblower who exposed it. Other pieces discussed bias and distortion in medical research caused by industrial financing (the irony of this was unremarked). Several pieces from the *New York Times* focused on the limits of animal studies, while a special issue of *Time*, "Science under Siege," described growing public distrust of science in the face of mistakes like the premature announcement of cold fusion and mismanagement of the Hubble telescope.[44] Collectively, the articles created an impression of science rife with exaggeration, mismanagement, bias, and fraud.

The strategy was nothing if not clever, for these articles were based on real events and real concerns within the scientific community. David Baltimore did dismiss evidence of malfeasance in his lab, animal studies do have serious limits, and science has been corrupted by industry funding. Yet not a single piece reported an actual study demonstrating that these problems were widespread—or any more widespread than in any other place where politics and business intersect. More to the point, not one of these studies showed that assertions of an environmental hazard had later been proven wrong. In fact, no scientific results were *corrected* by any of these articles, because the point wasn't to correct particular scientific mistakes. It was to provide the reader with materials to challenge science in general, as a means to challenge science on any topic. And the topic at issue was secondhand smoke.

Message #3 from *Bad Science* declared, "No agency is more guilty of adjusting science to support preconceived public policy prescriptions than the Environmental Protection Agency." The resource book outlined in chapter and verse the tobacco industry's complaint about EPA on ETS: that their conclusions were politically motivated, that they were based on inadequate science, that the EPA had no right to accept the 90 percent confidence level, and so on. "The EPA report has been widely criticized within the scientific community," the book proclaimed, but in truth very few scientists had criticized the EPA report, except ones linked to the tobacco industry. This was the *Bad Science* strategy in a nutshell: plant complaints in op-ed pieces, in letters to the editor, and in articles in mainstream journals to whom you'd supplied the "facts," and then quote them as if they really were facts. Quote, in fact, yourself. A perfect rhetorical circle. A mass media echo chamber of your own construction.

The phrases "excessive regulation," "over-regulation," and "unnecessary regulation" were liberally sprinkled throughout the book. Many of the quotable quotes came from the Competitive Enterprise Institute (CEI), a think tank promoting "free enterprise and limited government" and dedicated to the conviction that the "best solutions come from people making their own choices in a free marketplace, rather than government intervention."[45] The Institute's "Science Policy Clips and Highlights" compiled articles by Institute staff that had been published in mass media venues such as the *Washington Times*, the *St. Louis Post-Dispatch, Reason, Advertising Age*, and *Insight*. The CEI compilation for January 1993 through April 1994 included "EPA's Bad Science Mars ETS Report," "EPA and the Pesticide Problem," "When Chemophobia Ruled the Land," and "Safety Is a Relative Thing for Cars: Why Not Cigarettes?"[46]

In short, *Bad Science* was a compendium of attacks on science, published in places like the *Washington Times*, and written by staff of the Competitive Enterprise Institute. The articles weren't written by scientists and they didn't appear in peer-reviewed scientific journals. Rather, they appeared in media venues whose readers would be sympathetic to the Competitive Enterprise Institute's laissez-faire ideology.

And that was precisely the point. The goal wasn't to correct scientific mistakes and place regulation on a better footing. It was to undermine regulation by challenging the scientific foundation on which it would be built. It was to pretend that you wanted sound science when really you wanted no science at all—or at least no science that got in your way.

Bad Science lambasted the EPA for not "seek[ing] out the nation's leading scientists [to] conduct a peer-reviewed study" on ETS, but the EPA *had* sought leading scientists and their work *had* been peer-reviewed. Had the EPA commissioned a brand new study, the industry would no doubt have attacked them for wasting taxpayer money on superfluous work. But that was precisely the point: to attack the EPA, because it was just about impossible to defend secondhand smoke any other way. At least, this was what Philip Morris had concluded.

Blaming the Messenger: The Industry Attack on the EPA

Craig Fuller was a former chief of staff to Vice President George H. W. Bush; in 1993 he also worked with Ellen Merlo to defend ETS by attacking the EPA. Desperate times called for desperate measures, and the industry now appeared desperate indeed. In July, Fuller paid $200,000 to a group called Federal Focus, Inc., run by James Tozzi.[47] Tozzi had been an administrator at the Office of Management and Budget in the Reagan administration, and was well-known among public health officials for his resistance to the scientific evidence that aspirin causes Reye's syndrome in children. (Critics charged him with perfecting the strategy of "paralysis by analysis": insisting on more, and more, and more, data in order to avoid doing anything.)[48] After reading Seitz's report on ETS, Tozzi suggested that Federal Focus could channel money to the Marshall Institute for further work on ETS. The Marshall Institute would be good for this, Tozzi suggested, because of its lack of obvious links to Philip Morris: "Possibly PM [Philip Morris] could provide funding, through Federal Focus, to the George C. Marshall Institute . . . they could address the ETS conclusion . . . I think the Marshall Institute will have considerable credibility since it does not take funding from private companies nor the government. It is funded solely through foundations such as Federal Focus . . ."[49]

When *Investor's Business Daily* ran a front-page article favorable to the tobacco industry, Fuller sent a memo to his team saying, "[It] ought to be mailed to every one of our allies and any opinion leaders we can get it to as quickly as possible. It offers a comprehensive review that is among the best I've seen." But this was no coincidence, as Fuller acknowledged, no doubt grinning as he wrote: "(And, I know that it's no coincidence . . . it is very fine work!)" On the bottom of the memo he added in pencil: "This is Tom Borelli's work."[50] (Borelli was Philip Morris's manager of Corporate Scientific Affairs.)[51]

As we saw in chapter 1, the tobacco industry had long tried to make its case for "balance" to writers, editors, and radio and television producers. Now they targeted particular journalists of a "revisionist ilk," whom they considered susceptible to the suggestion that environmentalism had run amok. These included Nicholas Wade, science editor of the *New York Times*, P. J. O'Rourke at *Rolling Stone*, and Gregg Easterbrook, a frequent writer for the *New Republic*. (Wade was the coauthor of a 1983 book, *Betrayers of the Truth*, which asserted that fraud and deceit were endemic to science; the industry saw him as a potential ally, carefully tracking his work and places where he was quoted. Easterbrook we shall meet again in chapter 6.) Other targets for influence were the First Amendment Center, "a media peer group that is respected and has the capability of changing reporters' attitudes on issues on a wholesale basis"; the national meeting of mayors conference and the mayors' regional meetings, who would be approached through the issue of unfunded mandates; and proindustry groups such as the Institute for Regulatory Policy and Citizens for a Sensible Environment.[52] Later the industry would also enlist Rush Limbaugh.[53]

Most of the science upon which the EPA relied was independent—it came from academic researchers and other federal agencies, such as the National Institutes of Health, the Food and Drug Administration, and the Department of the Interior—so attacks on the EPA as a corrupt bureaucracy wouldn't work alone; they'd have to be coupled with attacks on the science itself. "Without a major, concentrated effort to expose the scientific weaknesses of the EPA case, without an effort to build considerable reasonable doubt . . . then virtually all other efforts . . . will be significantly diminished in effectiveness," ran a memo from Philip Morris communications director, Victor Han, to Ellen Merlo.

The EPA was "an agency that is at least misguided and aggressive, at worst corrupt and controlled by environmental terrorists," Han asserted.[54] Since few people were sympathetic to secondhand smoke, attacking the EPA offered "one of the few avenues for inroads." The industry would abandon its defensive posture—defending smokers' right to smoke—and argue instead that "over-regulation" was leading to "out-of-control expenditures of taxpayer money."[55] Much of this would be done through a newsletter called EPA Watch—an "asset" created by Philip Morris through the public relations firm APCO.[56]

Han concluded, "The clock is ticking."[57] This is where EPA Watch came in, as Han, Merlo, Fuller, and their associates developed "a plan for EPA Watch," and to use a man named Bonner Cohen as an "expert on EPA matters." Cohen was associated with the Committee for a Constructive Tomorrow—a Cornucopian group committed to harnessing "the power of the market combined with the applications of safe technologies . . . [to address] the world's pressing concerns." Cohen had written extensively for the *Wall Street Journal, Forbes, Investor's Business Daily, National Review,* and the *Washington Times.*[58] Merlo and Fuller's group resolved to do "whatever can be done to increase his visibility and credibility on matters dealing with the EPA."[59]

No one in 1993 would have argued that the EPA was a perfect agency, or that there weren't some regulations that needed to be revamped; even its supporters had said as much. But the tobacco industry didn't want to make the EPA work better and more sensibly; they wanted to bring it down. "The credibility of EPA is defeatable," Victor Han concluded, "but not on the basis of ETS alone. It must be part of a larger mosaic that concentrates all of the EPA's enemies against it at one time."[60] That mosaic would soon be created.

"Junk science" quickly became the tag line of Steven J. Milloy and a group called TASSC—The Advancement of Sound Science Coalition—whose strategy was not to advance science, but to discredit it. Milloy—who later became a commentator for Fox News—was affiliated with the Cato Institute and had previously been a lobbyist at Multinational Business Services (MBS)—a firm hired by Philip Morris in the early 1990s to assist in the defense of secondhand smoke.[61] (Milloy's supervisor at MBS had been James Tozzi.)

TASCC was launched by APCO Associates, in November 1993, with measures taken to hide the Philip Morris connection.[62] APCO was enlisted because Philip Morris's main PR agency, Burson-Marsteller, was too obviously associated with the tobacco giant.[63] John Boltz, a manager of media affairs at Philip Morris, supplied APCO with a list of sympathetic reporters, but APCO, not Boltz, placed the calls to "remove any possible link to PM [Philip Morris]."[64] The launch would be focused on "receptive" secondary markets, rather than in conventionally attractive cities for PR like New York and Washington, "to avoid cynical reporters from major media" who might be inclined to dig.[65]

Philip Morris executive John C. Lenzi summarized for Ellen Merlo how TASSC had "launched itself" with the help of selected media and sympathetic scientists. "As you know, The Advancement of Sound Science Coalition (TASSC) publicly launched itself . . . with a national, five-city media tour . . . Rather than do a press conference at each site . . . TASSC elected to conduct one-on-ones with interested media . . . This appears to have worked well, particularly when combined with TASSC's effort to highlight a regional 'bad science' problem of interest, and appear with a known and respected member of the local scientific community who is also a member of TASSC . . . In total TASSC created coverage that potentially reached approximately 3 million people."[66]

The launch was particularly successful in Albuquerque. Former New Mexico governor Garrey Carruthers, now TASSC's honorary chairman, was the keynote speaker at the American Farm Bureau Federation state convention in New Mexico; Carruthers used the occasion to introduce "TASSC, its goals, and its objectives." In Denver, the launch was featured in the *Post* and *Denver Business Journal*, and on three radio programs; in San Diego, in the *Union Tribune* and *Daily Transcript*; in Dallas, in various newspapers, radio stations, and at least one television network. Lenzi boasted to Merlo that the people reached included "more than 350,000 by television, 850,000 by radio and more than 1.7 million by print." The launch was deemed sufficiently successful that they budgeted over $500,000 for TASSC efforts in 1994.[67]

Scientific advisors to TASSC included Fred Singer, Fred Seitz, and Michael Fumento—names familiar from both *Bad Science* and from earlier arguments over tobacco, acid rain, and ozone. Richard Lindzen, a distinguished meteorologist at MIT who was a major global warming skeptic and industry expert witness, was also invited to join.[68] The goal, as Craig Fuller put it, was to mobilize as many "third party allies" as possible.[69]

Meanwhile, Milloy wrote articles for the *Wall Street Journal*, the *Washington Times*, and *Investor's Business Daily*, and created a Web site, JunkScience .com, that freely attacked science related to health and environmental issues. It didn't matter who had done the work—the EPA, the World Health Organization, the U.S. National Academy of Sciences, or distinguished scientists at private universities. If the results challenged the safety of a commercial product, Milloy attacked them.

TASSC also ran ads in commercial and campus newspapers across the country, and developed potential congressional testimony on "public health priorities."[70] They also created a "Sound Science in Journalism Award," first granted to *New York Times* reporter Gina Kolata, "who responsibly detailed . . . how science has been distorted and manipulated to fuel litigation" on silicone breast implants.[71] (Kolata has subsequently been heavily criticized by scientists, environmentalists, and her journalism colleagues for a persistent proindustry, protechnology bias, and an overt skepticism about environmental causes of cancer.)[72] Still, despite their managing to place their views in so many media outlets—and even finding a voice through Kolata at the *New York Times*—TASSC faced an uphill battle, as the American people were increasingly turning against smoking, and industry attacks over arcane scientific issues like confidence limits got scant traction. So the industry now launched a flank attack through yet another think tank, this one called the Alexis de Tocqueville Institution.

In the mid-1990s, the Tobacco Institute identified Alexis de Tocqueville as one of many organizations that it would support in its effort to fight higher tobacco taxes; and members of Tocqueville's advisory board—among them Dwight Lee and Fred Singer—had links to the tobacco industry.[73] One industry document described the connection this way: "TI's chief economist works closely with leading figures at the Alexis de Tocqueville Institution (AdTI). Some member companies [also] support the organization. Opinions expressed and promoted by AdTI frequently support industry arguments on economic and other matters."[74]

Officially the mission of the Alexis de Tocqueville Institution is to promote democracy; in 1993 the Institution decided to promote democracy by defending secondhand smoke. "EPA and the Science of Environmental Tobacco Smoke" was written by Fred Singer and Kent Jeffreys.[75] The Tocqueville Institution had anointed Jeffreys with the title of "adjunct scholar," but he was in fact a lawyer affiliated with the Cato Institute, the Competitive Enterprise Institute, and the Republican Party. He was well-known for his attacks on Superfund—the federal fund designed to pay for the cleanup of toxic waste sites—and for his advocacy of "free-market environmentalism." One of his slogans was "behind every tree should stand a private . . . owner." To prevent overfishing, Jeffreys wanted to privatize the oceans.[76]

The defense of secondhand smoke was part of a larger report criticizing the EPA over radon, pesticides, and the Superfund, but the center of it— and the focus of the accompanying press releases—was what Singer and Jeffreys called "Case Study No. 1: Environmental Tobacco Smoke." It be-

gan by accusing the federal government of seeking a ban on smoking—
although there was no pending legislation to do so—and asserting that the
vehicle of the alleged ban would be the EPA. But the EPA had not asked
for a ban, so how did Singer and Jeffreys build their case? By asserting that
"scientific standards were seriously violated in order to produce a report to
ban smoking in public settings."[77] What was the alleged violation? The
EPA panel had assumed a linear dose-response curve. They had assumed
the risk was directly proportional to the exposure.

Singer and Jeffreys argued that the EPA should have assumed a "threshold effect"—that doses below a certain level would have no effect. Citing the
old adage "the dose makes the poison," they insisted that there might be a
threshold value below which no harm occurred. Since the EPA had failed to
provide proof that this wasn't so, the linear-dose response assumption was
"flawed."[78]

A memo from the Tobacco Institute to the members of its Executive
Committee in August 1994 described the report's release at a press conference held by two members of Congress—Peter Geren, Democrat from
Texas, and John Mica, Republican from Florida—joined by the executive
director of the Alexis de Tocqueville Institution and "co-authors Dr. S. Fred
Singer and Kent Jeffreys." Singer stressed how money was being wasted
on "phantom" environmental problems; Jeffreys focused on the Clinton
administration's "lying to or withholding information" from Congress,
implying the EPA was doing the same. He concluded by invoking the hob-
goblin of absolute proof: "I can't prove that ETS is not a risk of lung cancer,
but EPA can't prove that it is."[79]

Were any of these charges true? Should the EPA have insisted on 95
percent confidence limits? Should they have used a threshold? Were scientific standards violated? Was this bad science? And how's an ordinary person
to judge?

Scientists are confident they know bad science when they see it. It's science that is obviously fraudulent—when data have been invented, fudged,
or manipulated. Bad science is where data have been cherry-picked—
when some data have been deliberately left out—or it's impossible for the
reader to understand the steps that were taken to produce or analyze the
data. It is a set of claims that can't be tested, claims that are based on
samples that are too small, and claims that don't follow from the evidence
provided. And science is bad—or at least weak—when proponents of a position jump to conclusions on insufficient or inconsistent data. (As we saw
in chapter 4, Sherwood Rowland had used his AAAS presidential lecture

to show how Dixy Lee Ray, Fred Seitz, and Fred Singer had relied on bad science to challenge ozone depletion; they had made demonstrably false assertions and ignored widely available, published evidence.) But while these scientific criteria may be clear in principle, knowing when they apply in practice is a judgment call. For this scientists rely on peer review. Peer review is a topic that is impossible to make sexy, but it's crucial to understand, because it is what makes science science—and not just a form of *opinion*.

The idea is simple: no scientific claim can be considered legitimate until it has undergone critical scrutiny by other experts. At minimum, peer reviewers look for obvious mistakes in data gathering, analysis, and interpretation. Usually they go further, addressing the quality and quantity of data, the reasoning linking the evidence to its interpretation, the mathematical formulae or computer simulations used to analyze and interpret the data, and even the prior reputation of the claimant. (If the person is thought to do sloppy work, or has previously been involved in spurious claims, he or she can expect to attract tougher scrutiny.)

Scientific journals submit all papers to peer review. Typically three experts are asked to comment. If the reviewers are very divided, the editor may seek additional voices, and he may weigh in his judgment as well. Many papers go through two or more rounds, as authors try to correct mistakes and address concerns raised by the reviewers. If they fail, the paper will be rejected, and the authors go back to the drawing board—or try another, less prestigious, journal. Conferences are usually less strict, which is why conference papers are generally not considered serious— and generally do not count in academic circles for promotion and tenure— until published in peer-reviewed journals. (This is also why the industry could exploit an apparent loophole by sponsoring their own conferences and publishing their proceedings.) The reviewers must also be real experts—they must know enough to be able to judge the methods used and the claims made—and they must not have a close relationship, either personal or professional, with the person whose work is being judged. Editors spend considerable time finding people who meet these criteria. And this is all done for free. Scientists review papers as part of a communal system in which everyone is expected to review other people's papers, with the understanding that others will in turn review theirs.

The EPA report on passive smoking was reviewed not just by three experts, but by an entire panel commissioned by the EPA's Science Advisory Board: nine experts and nine consultants, aided by staff members from the Advisory Board.[80] Unlike Singer (a physicist), Jeffreys (a lawyer), and Milloy (a lobbyist), these were true experts: a professor of medicine at Yale University; a senior staff scientist at the Lawrence Berkeley Laboratory; the chief of Air and Industrial Hygiene for the California Department of Health, and six others, all medical doctors or Ph.D. scientists. And they reviewed it not once, but twice. What did these experts have to say about the EPA's report? "The Committee concurs with the judgment of EPA that environmental tobacco smoke should be classified as a Class A carcinogen."[81]

Typically, reviewers are skeptics. They challenge scientists on the claims they make, often demanding more evidence, more clarification, more persuasive arguments. The reviewers of the draft EPA report did request more discussion of certain matters: the uncertainties and confounding effects, the limits of using spousal exposure as a surrogate for total ETS exposure, and the recent work on ETS and respiratory disorders in children. But they did so not because they thought the report had overstated the case. On the contrary, their major concern was that the report had *understated* the risks. Its conclusions were not too strong, but too weak.

The major issue involved the epidemiological data. Ill effects of chemicals in the environment are detected through epidemiology: statistical studies of affected populations. If a chemical is very toxic, or exposures are very high, then ill effects are easy to detect: lots of people get sick, far more than you'd otherwise expect for a population group of that type. But if the chemical is only mildly harmful, or exposures are low, then the task is much harder. Only a few people get sick, and it's hard to say for sure that the observed effect isn't just random variation.

How do you judge epidemiological evidence when there's only a modest effect? You judge it in light of what else you know about the issue. If strong epidemiology is a red flag, then weak epidemiology is a pink one. Imagine placing both against a wall: a white one if you know nothing else (a blank slate, if you will), a black one if you already have good reason to think there's a problem. If the wall is white, the pink flag barely shows up, but if it's black, then you've got no problem seeing that flag. ETS was a pink flag against a black wall.

Here's why. Secondhand smoke is quickly diluted in the air, so most people's exposures are low, and epidemiology is a weak tool with which to detect effects: a pink flag. But scientists already knew that active smoking causes cancer, and that passive smoking introduces the same toxins into the lungs. That was the black wall.[82] The reviewers put it this way: "The causality of the connection between direct inhalation of tobacco smoke and excess risk of lung cancer cannot be in doubt . . . and ETS resembles mainstream tobacco smoke in terms of particle size distribution and composition of carcinogens, co-carcinogens, and tumor producers."[83] So even if the statistical effects were modest, there was good reason to believe that they were real. The reviewers wanted the EPA panel to make this explicit, "with each step in the argument . . . carefully addressed."[84]

The reviewers especially found the report too weak on its discussion of the impact of ETS on children. They "found the evidence for respiratory health effects in children to be stronger and more persuasive" than stated, and suggested that the panel consider the possibility that "the impact of ETS on respiratory effects in children may have much greater public health significance than the impact of ETS on lung cancer in nonsmokers."[85] In other words, while 3,000 additional adult lung cancer deaths per year was a serious public health concern, 150,000 to 300,000 cases of bronchitis and pneumonia in infants and young children was even worse.

The panel revised their report in light of the peer review, and five months later it was reviewed a second time. The panel found the overall assessment of risk to children to be still "on the conservative side."[86] On the central question of labeling ETS as a class A carcinogen, "the Committee was unanimous in endorsing this classification."[87]

Here's what the reviewers did *not* criticize: They did not reject the use of spousal smoking as surrogate for exposure or the studies from other countries, which they considered appropriately included as part of the "totality of evidence." They did not criticize the 90 percent confidence limit or the linear dose-response model. And they did not suggest that EPA should have presumed a threshold effect. On the contrary, they noted the "clear dose-related association of lung cancer risk with exposure to [mainstream] smoke," accepting that a similar relationship would likely apply to sidestream smoke.[88]

Why didn't the peer reviewers address the issue of confidence limits? This was a major point of Singer and Jeffrey's contention, so we might expect the peer reviewers to have at least mentioned it. The answer is simple. There's nothing magic about 95 percent. It could be 80 percent. It could be 51 percent. In Vegas if you play a game with 51 percent odds in your favor, you'll still come out ahead if you play long enough.

The 95 percent confidence level is a social convention, a value judgment. And the value it reflects is one that says that the worst mistake a scientist can make is to fool herself: to think an effect is real when it is not. Statisticians call this a type 1 error. You can think of it as being gullible, naïve, or having undue faith in your own ideas.[89] To avoid it, scientists place the burden of proof on the person claiming a cause and effect. But there's another kind of error—type 2—where you miss effects that are really there. You can think of that as being excessively skeptical or overly cautious. Conventional statistics is set up to be skeptical and avoid type 1 errors. The 95 percent confidence standard means that there is only 1 chance in 20 that you believe something that isn't true. That is a very high bar. It reflects a scientific worldview in which skepticism is a virtue, credulity is not.[90] As one Web site puts it, "A type I error is often considered to be more serious, and therefore more important to avoid, than a type II error."[91] In fact, some statisticians claim that type 2 errors aren't really errors at all, just missed opportunities.[92]

Is a type 1 error more serious than a type 2? Maybe yes, maybe no. It depends on your point of view. The fear of type 1 errors asks us to play dumb. That makes sense when we really *don't* know what's going on in the world—as in the early stages of a scientific investigation. This preference also makes sense in a court of law, where we presume innocence to protect citizens from oppressive governments and overzealous prosecutors. However, when applied to evaluating environmental hazards, the fear of gullibility can make us excessively skeptical and insufficiently cautious. It places the burden of proof on the *victim*—rather than, for example, the manufacturer of a harmful product—and we may fail to protect some people who are really getting hurt.[93]

And what if we aren't dumb? What if we already have strong, independent evidence to support a cause-and-effect relationship? Let's say you know *how* a particular chemical is harmful, for example, that it has been shown to interfere with cell function in laboratory mice. Then you might argue that it is reasonable to accept a lower statistical threshold when examining effects in people, because you already have good reason to believe that the observed effect is not just chance. This is exactly what the ETS reviewers did argue. Even if 90 percent is less stringent than 95 percent, it still means that there is a 9 in 10 chance that the observed results did not occur by chance. Think of it this way. If you were nine-tenths sure about a crossword puzzle answer, wouldn't you write it in?[94]

"The extent of the consistency defies attribution to chance," the EPA stressed when the final report was released.[95] Consistency—not any arbitrary significance level—is the real gold standard of scientific evidence, and this was the key point that Singer and Jeffreys had obfuscated. It was true that some of the included studies were small, and alone could not prove a causal connection, but when you looked at *all* the studies, you found that twenty-four of thirty showed increased risk associated with increased exposure—and the odds of that happening by chance were less than 1 in 1,000.

What about the threshold effect? Why didn't this come up in the peer review, either? The answer here is simple, too: the reviewers did not need to comment because the panel had followed EPA guidelines.[96] One chemist who has worked closely with the EPA for decades put it this way. "Linear dose-response is the 'official' EPA default [position]. If there is sufficient evidence for a non-linear mode of action then that is used. Otherwise, it is linear."[97]

This is true, but it isn't just EPA guidelines; it's normal scientific practice, too. The logic is twofold. One reason derives from centuries of scientific practice and the principle known as Ockham's razor. Use the simplest theory that accounts for the evidence. Just as a well-designed machine has no unnecessary parts, a well-designed theory does not introduce additional assumptions that are not supported by evidence. If you have evidence for complications like threshold effects at low doses (or amplifying effects at high doses) then of course you pay attention, but absent evidence you don't make complications up.

The second reason is just common sense. If something is harmful, then more exposure means more risk. At least, that is what one would expect. However, not all poisons work this way. Some do show threshold effects: up to a point, your body can deal with it. Certain substances, including some vitamins and minerals, are poisonous at high doses but actually helpful or even essential at low ones. This effect has a scientific name: hormesis.[98] But as a rule of thumb, if a little of something is known to be bad, a lot is probably worse, and if a lot of something is known to be bad, then a little is probably not great either. And while Ronald Reagan infamously claimed that ketchup was a vegetable, no one, not even Fred Singer, would claim that cigarette smoke was a vitamin.

How did the EPA defend itself against these attacks? In normal scientific practice, the mere fact of withstanding peer review is the first line of defense, but Singer and Jeffreys had misrepresented the peer review process, claiming that the EPA report had been widely criticized in the scientific community, ignoring that the report had not only been unanimously endorsed by the independent experts, but that those experts had encouraged EPA to make it *stronger*. So the EPA established a Web site: Setting the Record Straight: Secondhand Smoke is a Preventable Health Risk. The site said everything that needed to be said, so it's worth quoting in full:

> A recent high profile advertising and public relations campaign by the tobacco industry may confuse the American public about the risks of secondhand smoke. EPA believes it's time to set the record straight about an indisputable fact: secondhand smoke is a real and preventable health risk.
>
> EPA absolutely stands by its scientific and well documented report. The report was the subject of an extensive open review . . . by EPA's Science Advisory Board (SAB), a panel of independent scientific experts. Virtually every one of the arguments about lung cancer advanced by the tobacco industry and its consultants was addressed by the SAB. The panel concurred in the methodology and unanimously endorsed the conclusions of the final report. The report has also been endorsed by the U.S. Department of Health and Human Services, the National Cancer Institute, the Surgeon General, and many major health organizations.

The criticism had come not from the scientific community, but from the tobacco industry and groups and individuals funded by it. The peer review panel endorsed the EPA conclusions, and so had every other relevant major agency and organization. As for the 90 percent confidence limit, this was "a standard and appropriate statistical procedure" given the prior evidence, and had been used in many other EPA cancer risk assessments when there was similarly strong prior evidence; there was nothing special or irregular about how the EPA had treated secondhand smoke. Moreover, in the portions of the report dealing with other respiratory effects, where there wasn't as much prior evidence, 95 percent confidence intervals *were* used.

Singer and Jeffreys had focused their attention on cancer risk, but the bombshell of the report was the danger to children. "The tobacco industry neither acknowledges nor disputes EPA's conclusions on respiratory effects on children. It focuses instead on EPA's findings on lung cancer," the EPA noted. This silence was telling, as both the peer review panel and the secretary of Health and Human Services highlighted the impact on children as the most important finding—as no doubt most of the public would, too. It was one thing for adults to choose to take risks, another thing to impose those risks on children (or anyone else). The EPA wisely refocused attention on this crucial distinction: "Having a choice to take a risk for themselves should not permit smokers to impose a risk on others."[99] This was the crux of the issue. But we've found no evidence that the mass media paid any attention to the EPA Web site. And what impact *could* a lonely Web site have against a multimillion-dollar disinformation campaign?

The EPA made clear that the fuss about confidence limits was a red herring, but what about the threshold issue? Was there any substance to Singer's insistence that there might be a threshold effect for secondhand smoke? The EPA's answer was simple: "There is no evidence that this threshold exists."[100] So where did Singer get the idea from? Did he just make it up? Should the EPA have considered the threshold effect in analyzing smoking?

In the report with Jeffreys, Singer was promulgating an old adage: that the dose makes the poison. Where did that come from, anyway? The answer is Paracelsus, a Renaissance medic who died in 1541. Singer and Jeffreys were challenging the EPA with a five-hundred-year-old aphorism.[101] While it's possible that Singer was reading Latin medical texts, it seems more likely that he got the argument from a contemporary debate about radioactivity.

Many Japanese citizens exposed to the devastation of the atomic bomb developed cancer in later years—many, but not all. What protected the resistant survivors? Some scientists argued for a threshold effect: that up to a certain point, radiation does not cause cancer. People who were far enough away from the blast, or protected by thick walls or metal plates, might have received exposures below the level required to cause cancer.

This was a reasonable argument, because radiation is a natural phenomenon to which we are all exposed every day. Many ordinary elements, including carbon, potassium, and uranium, have naturally occurring radioactive varieties, found in rocks, minerals, soils, and even in the air. Cosmic rays from outer space add a bit more to this "natural background" radiation. While the natural background varies from place to place, it's always present, so it stands to reason that living creatures might be used to it. Having evolved on a planet that has always had background radiation, we may have evolved natural defenses against it. So the concept of a permissible, safe, or "threshold" dose gained currency, and this was used to set standards in industries where workers were exposed to radiation, such as uranium mining and nuclear power generation.

Some people went even further, arguing for radiation hormesis—that small doses of radiation were actually good for you. One of these people was Chauncey Starr, the physicist with the Electric Power Research Institute who we met in chapter 3 writing to George Keyworth and Bill Nierenberg to persuade them that "public anxiety" was being unnecessarily inflamed about acid rain.[102] (We will meet Starr again in chapter 6.)

By the 1970s, the threshold concept was being used by all sorts of people to defend all sorts of hazardous materials. This was illogical, because the threshold argument was about *natural* hazards—like background radiation and trace metals that occur in soils—but that didn't stop some people from using it to defend unnatural ones, too.

In 1973, Emil Mrak, a former chancellor of the University of California, Davis, was invited to the Philip Morris laboratories to speak about food safety. Mrak was dubious about the alleged dangers of DDT and other artificial pesticides, and he used the threshold argument to defend them. "Is there a level below which compounds have no effect?" he asked rhetorically, referring to chemicals in the environments suspected of causing cancer. Most cancer experts said no—meaning no, there wasn't evidence to support that claim, because if a chemical is hazardous at some dose, then the only dose that is guaranteed to produce no harm is no dose—but Mrak rejected this, arguing hyperbolically, "If this is the case, we can start right out by outlawing almost everything."[103] He ended his speech with the reductio ad absurdum that, if you didn't embrace the threshold concept, then you ended up concluding that "everything is harmful."[104]

Mrak was pulling a rhetorical switcheroo because it wasn't *environmentalists* who argued everything was harmful; it was the *tobacco industry*. The industry insisted that everything from crossing the street to riding a bicycle was harmful, so tobacco should be viewed as just one of the routine risks that people accept by living life. The menace of daily life, some industry apologists called it.[105] Life is dangerous. So is tobacco. Get used to it.

So the tobacco industry argued. But there's a world of difference between risks we choose to accept in exchange for rewards we want—like driving a car, drinking alcohol, or having unprotected sex—and having those risks imposed upon us against our will. There's also a world of difference between the idea that evolution has equipped humans with some immunity to natural hazards and the idea that we somehow have immunity to something we'd never been exposed to in two million years of evolution. The secondhand smoke debate was crucial precisely because the risk wasn't a choice and it wasn't natural. It was a man-made risk that was being imposed without consent.

The very fact that Singer was recycling arguments from earlier debates about nuclear power and pesticides—alongside Singer's previous activities to defend acid rain and CFCs—suggests that none of this was really about the science of secondhand smoke. Singer simply was not an expert on every one of these issues. Modern science is too complex and specialized for that.

For the tobacco industry, of course, the goal was to protect profits. Indeed, in 1995, Philip Morris reported record profits; *USA Today* reported that "the Marlboro Man continues to ride high."[106] Philip Morris was the highest-yielding stock on the Dow Jones Industrial Average that year, and *Money* magazine noted that while "uncertainties created . . . by smoking liability lawsuits" continued to keep a lid on stock values, "cigarette makers have never had a judgment go against them."[107] Philip Morris was determined to maintain that winning streak.[108]

But what about the scientists who helped their effort? What was this about for Fred Singer, Fred Seitz, and the other scientists who made common cause with the tobacco industry?

One answer has already emerged in our discussion of acid rain and ozone depletion: these scientists, and the think tanks that helped to promote their views, were implacably hostile to regulation. Regulation was the road to Socialism—the very thing the Cold War was fought to defeat. This hostility to regulation was part of a larger political ideology, stated explicitly in a document developed by a British organization called FOREST— Freedom Organisation for the Right to Enjoy Smoking Tobacco. And that was the ideology of the free market. It was free market fundamentalism.

Using Tobacco to Defend Free Enterprise

FOREST was a British group that purported to be a grassroots organization, defending the rights of smokers. In fact, it was the creation of the British Tobacco Advisory Council, an industry group that served much the same function in the United Kingdom as the Tobacco Institute did in the United States.[109] Its chair was Sir Christopher Foxley-Norris, a retired Royal Air Force commander (and confirmed smoker) who had fought in the Battle of Britain. One industry memo recounted in the late 1970s that Foxley-Norris had approached British tobacco executives about "adopting a more robust public stance" in the face of "increasing interference by Government and other do-gooding bodies in many aspects of people's private lives."[110]

More than three thousand documents in the Tobacco Legacy Documents Library detail FOREST's activities.[111] FOREST organized campaigns to defend smoking, particularly in the workplace, and to challenge the scientific evidence that secondhand smoke was dangerous. They launched an attack on the London Science Museum for an exhibit on passive smoking that they labeled "junk science," and issued a "Good Smoker's Airline Guide" steering readers to smoke-friendly airlines and encouraging them to boycott British Airways for its smoking ban. In 1997, FOREST made plans for a pair of research conferences designed to convince business executives that "anti-smoking policies can have serious consequences for staff morale, commercial viability, and public relations."[112] They conducted campaigns to fight smoking bans in hotels and pubs, to challenge antismoking education in British schools, and to defend the rights of smokers to adopt children. FOREST also sought to fund research to highlight the social and economic costs of smoking restrictions and high tobacco taxes.[113]

A 1994 FOREST report entitled "Through the Smokescreen of Science: The Dangers of Politically Corrupted Science for Democratic Public Policy" claimed much the same thing as Fred Singer had: that science was being rigged to advance a political agenda. Whether or not that was true, this report made clear that the inverse was certainly true: science was being attacked to advance their agenda, the defense of free market capitalism.

The introduction to the report was written by Lord Harris of High Cross, the economist who ran the British Institute of Economic Affairs and who is widely considered the architect of Thatcherism. An avowed free market ideologue, Harris idolized Adam Smith; his nemesis was John Maynard Keynes. In one of her first appointments, Margaret Thatcher had made Harris a peer of the realm, but he allegedly declined a coat of arms on the grounds that the invisible hand could not be blazoned.[114]

The Lord Harris laid out the stakes on page 1. Public health officials were "puritan paternalists . . . who see other men's lives as the proper end product of their own activity." Antismoking scientists were perverting science "on the age-old pretext that the end, namely banning smoking, justifies any means, including . . . systematic selection or suppression of the evidence." But if their tactics were Communistic—with the ends justifying the means—they were also somehow Nazis, perpetrating "scientific deception worthy of the late Herr Goebbels."[115]

The crucial issue was freedom. "Non-smokers have as much to lose as smokers if they acquiesce in the prostitution of science . . . to justify . . . depriv[ing] free men and women of inexpedient freedoms," Lord Harris warned. "Smoking is only the first target. Beware!" The real aim was to control men's and women's lives. "There is little likelihood that we [will] end up being more healthy—only less free."[116] The same argument was reiterated in the body of the report, which repeatedly stressed that the defense of smoking was a defense of individual liberty. Smoking critics were health paternalists, moving toward the view that "the State is justified in attempting to manipulate and coerce."[117]

This was the ideological core. Indeed, Fred Singer had said virtually the same thing in his attack on the EPA: "If we do not carefully delineate the government's role in regulating [danger] . . . there is essentially no limit to how much government can ultimately control our lives."[118]

Perhaps a man like Foxley-Norris, a hero of the Battle of Britain, could be forgiven for worrying about the specter of totalitarianism. After all, the Nazis *had* been the first government to actively discourage smoking. But forty-nine years had passed since the end of World War II—and for Harris, Singer, and their friends in the Reagan and Thatcher administrations, men who fought not the Second World War but the first Cold War—the enemy was not Nazism but Communism. Anti-Communism had launched the weapons and rocketry programs that launched the careers of Singer, Seitz, and Nierenberg, and anti-Communism had underlain their politics since the days of *Sputnik*. Their defense of freedom was a defense against Soviet Communism. But somehow, somewhere, defending America against the Soviet threat had transmogrified into defending the tobacco industry against the U.S. Environmental Protection Agency.

We saw in chapter 2 how Russell Seitz, a cousin of Frederick Seitz, had been enlisted by the Marshall Institute to attack not only Carl Sagan but the entire scientific community over the issue of nuclear winter, and to insist that the United States could triumph in a nuclear exchange with the Soviet Union—and win the Cold War. In the mid-1990s, the younger Seitz took up the defense of secondhand smoke, and he did it in the fashion of a Cold Warrior, too.

Seitz was affiliated with the John M. Olin Institute for Strategic Studies at Harvard University, so why would a researcher at an Institute for Strategic Studies defend secondhand smoke? An answer is suggested by looking a bit more closely at the Olin affiliation. The Olin Institute was funded by the John M. Olin Foundation, which, like the Cato and the Competitive Enterprise institutes, promoted free market ideas. (Its president was William Simon, secretary of treasury in the Nixon administration.)[119] The foundation had funded numerous conservative and Libertarian think tanks, including the American Enterprise Institute, the Heritage Foundation, the Hoover Institution, and, through the Olin Center, they funded Russell Seitz.[120]

In an article in *Forbes* magazine, Seitz argued that rather than trying to control smoking, the government should fund research into making a safe cigarette. After all, the government funded all kinds of other safety devices, many of dubious value, so why not put at least some money into creating a safe cigarette? "Vast sums have been spent to good effect on reducing auto emissions and on curing—as well as preventing—AIDS."[121] Why not do the same for cigarettes?

The real culprit in smoking, Seitz argued, was smoke, and this was "no more wanted by smokers than coffee grounds by cappuccino addicts or a hangover by drinkers of red wine." So Seitz suggested that the U.S. government should figure out how to remove the smoke from cigarettes. "Only one-tenth of one percent of a cigarette is nicotine, and it should not take a rocket scientist to devise a means to volatilizing that small drop of active ingredient without generating a thousand times its weight in burning leaves."[122] Seitz was proposing that the government should spend taxpayer money figuring out how to safely deliver nicotine—an addictive and toxic substance—to the American people.

This sort of exigent approach might make sense for methadone since it helps people get off heroin, whose dangers to individuals and society are both grave *and* immediate. But what public good would be served by the government deliberately enabling people to continue to smoke?

The answer was to preserve smokers' right to smoke, and Seitz suggested that smokers should use their liberty to insist on it. "The nation's 50 million smokers remain at liberty to vote *en bloc* for a fussbudget-free Congress. Are the polls ready to accommodate smokers?"

Liberty was of course a keyword of the Cold War. We were free; the Soviet people were not. We cherished liberty; they did not. We believed in liberty and justice for all, and we fought to defend it. When Lt. General Daniel O. Graham (who had served on Team B in the SDI debate) wrote to Bill Nierenberg in 1984 asking him to help Bob Jastrow defend SDI, Liberty—with a capital L—was his catchword, too. This was *their* chance, Graham argued, to recapitulate the work of the Founding Fathers and "secure the blessings of Liberty to ourselves and our posterity."[123]

Russell Seitz and the defenders of tobacco invoked liberty, too. But as the philosopher Isaiah Berlin sagely pointed out, liberty for wolves means death to lambs.[124] Our society has always understood that freedoms are never absolute. This is what we mean by the rule of law. No one gets to do just whatever he feels like doing, whenever he feels like doing it. I don't have the right to yell fire in a crowded theater; your right to throw a punch ends at my nose. All freedoms have their limits, and none more obviously than the freedom to kill other people, either directly with guns and knives, or indirectly with dangerous goods. Secondhand smoke was an indirect danger that killed people.

The EPA was saying no more and no less than this. It was saying that protecting lambs required the government to control the wolves—and government control was what the Cold Warriors most feared. It was what they had spent their lives fighting.

Marxists were often criticized for believing that the ends justified the means, yet these old Cold Warriors were now the ones using ends to justify means—attacking science in the name of freedom. Suppressing evidence. Misrepresenting what their colleagues had done and said. Taking quotes out of context. Making allegations that were unsupported by evidence. One claim in particular was repeated several times in the FOREST report on secondhand smoke: that of a prominent epidemiologist who had allegedly said of the EPA work, "Yes it's rotten science, but it's in a worthy cause."[125] Did any one actually ever say that? Maybe yes, maybe no—there's no way to tell, because it was given without attribution. It's not the sort of thing that scientists typically say, but even if it were true, so what? It would just be the opinion of one man—and hardly evidence of a conspiracy to undermine the free market.

Like TASSC, FOREST's strategy was to insist that science was being used as a cover for an ideological program. The whole antismoking argument—with its "totalitarian flavour"—would be seen as transparently coercive, they insisted, were it not for the veneer of respectability provided by science. "In a world in which science is increasingly the source of both truth and value the scientific character of health paternalism is decisive," Lord Harris declaimed.[126] It was so decisive, in fact, that it had to be attacked. As the FOREST report put it, "Everything therefore depends on science. And with so much at stake, the pressure to adjust, shave, create, ignore, reevaluate, even manipulate, is enormous."[127] Indeed.

If a reader had any remaining doubt that the objective of FOREST was to undermine science as a "source of truth and value," he would only have needed to turn to the report's appendix, written by Professor Christie

Davies—a sociologist who wrote extensively for the *Daily Telegraph* and *Wall Street Journal* and compared cigarette smoking to drinking tea and eating chocolates.[128] British American Tobacco described him as "one of the most senior and well respected right-wing sociologists in the UK . . . A radical free marketer, he carries sound ideological baggage when it comes to issues of risk and personal freedom."[129]

Davies's appendix was a veritable tirade against "the state control of science." Old-timers would have recognized this as a reprisal of 1930s arguments against Marxist scientists, who were fairly widespread in Britain in those days, but how many British scientists were still Marxists in the 1990s? Not many, but still Davies went on, offering a manifesto for resisting "a system with a potential for repression now that is greater than it has ever been in the past. And this is far more dangerous than any form of tobacco smoking."[130]

FOREST made fact fighting into a cause célèbre: capitalism vs. Socialism, and the way science was (allegedly) being used to push the latter. "In a capitalist society individual economic pressure groups such as the tobacco industry do not have the same kind of power [as state bureaucrats and their scientific lackeys]. Rather power lies with a 'new class' of civil servants." It was class warfare, only the underclass was the tobacco industry.

The final flourish of the FOREST report was a bibliography of attacks on science—four pages on secondhand smoke and three more on "fraud, corruption and politicization" reminiscent of *Bad Science: A Resource Book*. Just about every potential threat to human or environmental health was included: acid rain, ozone depletion, and global warming (all fraudulent scares); pesticides, asbestos, chlorine, nuclear power, genetic engineering, biotechnology, and electromagnetic radiation from power lines (all harmless). Some surprising topics were included too: AIDS and the "myth" of heterosexual transmission, "allegedly disappearing species," forestry (an attack on environmental management), and food, drink, and lifestyle (defending the safety of alcohol and fatty foods—and straining credibility to its breaking point with a defense of British food). Sections were included on "science in general" and "environmentalism in general." Evidently, there wasn't a scientific or environmental claim that couldn't be attacked. Articles ranged from the plausible (Malcolm Gladwell on "Risk, Regulation and Biotechnology" in the *American Spectator*) to the ridiculous ("Is British food bad for you?"). It was all of a piece. If you believed in capitalism, you had to attack science, because science had revealed the hazards that capitalism had brought in its wake.

The biggest hazard of them all—one that could truly affect the entire planet—was just at that moment coming to public attention: global warming. Global warming would become the mother of all environmental issues, because it struck at the very root of economic activity: the use of energy. So perhaps not surprisingly, the same people who had questioned acid rain, doubted the ozone hole, and defended tobacco now attacked the scientific evidence of global warming.

CHAPTER 6

The Denial of Global Warming

ANY AMERICANS HAVE THE impression that global warming is
something that scientists have only recently realized was important. In 2004, *Discover* magazine ran an article on the top
science stories of the year, one of which was the emergence of a scientific
consensus over the reality of global warming. *National Geographic* similarly declared 2004 the year that global warming "got respect."[1]

Many scientists felt that respect was overdue: as early as 1995, the leading international organization on climate, the Intergovernmental Panel on
Climate Change (IPCC), had concluded that human activities were affecting global climate. By 2001, IPCC's Third Assessment Report stated that
the evidence was strong and getting stronger, and in 2007, the Fourth Assessment called global warming "unequivocal."[2] Major scientific organizations and prominent scientists around the globe have repeatedly ratified
the IPCC conclusion.[3] Today, all but a tiny handful of climate scientists are
convinced that Earth's climate is heating up, and that human activities are
the dominant cause.

Yet many Americans remained skeptical. A public opinion poll reported
in *Time* magazine in 2006 found that just over half (56 percent) of Americans thought that average global temperatures had risen—despite the fact
that virtually all climate scientists thought so.[4] An ABC News poll that
year reported that 85 percent of Americans believed that global warming
was occurring, but more than half did not think that the science was settled; 64 percent of Americans perceived "a lot of disagreement among
scientists." The Pew Center for the People and the Press gave the number

believing that there is "solid evidence the Earth is warming" as 71 percent in 2008, but in 2009, the answer to that same question was only 57 percent.[5]

The doubts and confusion of the American people are particularly peculiar when put into historical perspective, for scientific research on carbon dioxide and climate has been going on for 150 years. In the mid-nineteenth century, Irish experimentalist John Tyndall first established that CO_2 is a greenhouse gas—meaning that it traps heat and keeps it from escaping to outer space. He understood this as a fact about our planet, with no particular social or political implications. This changed in the early twentieth century, when Swedish geochemist Svante Arrhenius realized that CO_2 released to the atmosphere by burning fossil fuels could alter the Earth's climate, and British engineer Guy Callendar compiled the first empirical evidence that the "greenhouse effect" might already be detectable. In the 1960s, American scientists started to warn our political leaders that this could be a real problem, and at least some of them—including Lyndon Johnson—heard the message. Yet they failed to act on it.[6]

There are many reasons why the United States has failed to act on global warming, but at least one is the confusion raised by Bill Nierenberg, Fred Seitz, and Fred Singer.

1979: A Seminal Year for Climate

In 1965, the President's Science Advisory Committee asked Roger Revelle, then director of the Scripps Institution of Oceanography, to write a summary of the potential impacts of carbon dioxide–induced warming. Revelle had been interested in global climate for some time, and in the late 1950s had obtained funding for his colleague, chemist Charles David Keeling, to measure CO_2 systematically. (This work would produce the Keeling curve—showing CO_2's steady increase over time—for which Keeling would win the National Medal of Science and be made famous by Al Gore in *An Inconvenient Truth*.) Revelle knew that there was a lot about the problem that wasn't well understood, so he focused his essay on the impact he considered most certain: sea level rise.[7] He also made a forecast: "By the year 2000 there will be about 25% more CO_2 in our atmosphere than at present [and] this will modify the heat balance of the atmosphere to such an extent that marked changes in climate . . . could occur."[8]

The report made it to the Office of the President, and Lyndon Johnson mentioned it in a Special Message to Congress later that year: "This generation has altered the composition of the atmosphere on a global scale through . . . a steady increase in carbon dioxide from the burning of fossil fuels."[9] But with the war in Vietnam going badly, civil rights workers being murdered in Mississippi, and the surgeon general declaring that smoking was hazardous to your health, Johnson had more pressing things to worry about. Nor was it easy to get Richard Nixon's focus a few years later. Nixon undertook a number of important environmentally oriented reforms, including creating the Environmental Protection Agency, but during his administration climate concerns were focused on the SST project and the potential climate impact of its water vapor emissions, not CO_2.

Yet, while CO_2 didn't get much attention in the 1970s, climate did, as drought-related famines in Africa and Asia drew attention to the vulnerability of world food supplies. The Soviet Union had a series of crop failures that forced the humiliated nation to buy grain on the world market, and six African nations south of the Sahel (the semi-arid region south of the Sahara) suffered a devastating drought that continued through much of the 1970s.[10] These famines didn't just hurt poor Africans and Asians; they also caused skyrocketing food prices worldwide.

The famines were also noticed by the Jasons, a committee of elite scientists, mostly physicists, first gathered in the early 1960s to advise the U.S. government on national security issues.[11] The Jasons have long been an independent, self-confident group, and especially in its early days the committee members often told the government what they thought it needed to know. But the Jasons also respond to requests, and in 1977, the Department of Energy asked them to review the DOE research programs related to CO_2. The Jasons decided to look at carbon dioxide and climate.

Their report began with a recognition of the acute sensitivity of agriculture, and thus society in general, to even small changes in climate: "The Sahelian drought and the Soviet grain failure . . . illustrate the fragility of the world's crop producing capacity, particularly in those marginal areas where small alterations in temperature and precipitation can bring about major changes in total productivity."[12]

Over two summers they developed a climate model, which showed that doubling the carbon dioxide concentration of the atmosphere from its preindustrial level (about 270 ppm) would result in "an increase of average surface temperature of 2.4 C." Perhaps more worrying than the average temperature increase was the prospect of "polar amplification"—that warming would be greater, maybe a lot greater, at the poles. In their model, the poles warmed by 10°C to 12°C—a colossal amount.[13]

None of this was new. Professional climate modelers had already published papers that said pretty much the same thing, and in 1977, Robert M. White, the head of the National Oceanic and Atmospheric Administration (and later head of the National Academy of Engineering) had headed a committee for the National Research Council that warned of the serious impacts of unimpeded climate change: "We now understand that industrial wastes, such as carbon dioxide released during the burning of fossil fuels, can have consequences for climate that pose a considerable threat to future society . . . The scientific problems are formidable, the technological problems, unprecedented, and the potential economic and social impacts, ominous."[14]

But what matters in science is not the same as what matters in politics, and while the Jason study found nothing new, the fact that it *was* a Jason study "stimulated some excitement in White House circles."[15] Still, the Jasons were mostly physicists, not climate scientists. They included a couple of geophysicists, one of whom had a long-standing interest in climate, but none claimed climate as their central area of active research. So Frank Press, President Carter's science advisor, asked the National Academy of Sciences president Philip Handler to empanel a review of the Jason study. Handler turned to MIT professor Jule Charney.

One of the founders of modern numerical atmospheric modeling, and perhaps the most revered meteorologist in America, Charney assembled a panel of eight other scientists at the Academy's summer study facility in Woods Hole, Massachusetts. Charney also decided to go a bit beyond reviewing what the Jasons had done, inviting two leading climate modelers—Syukuro Manabe from the Geophysical Fluid Dynamics Laboratory and James E. Hansen at the Goddard Institute for Space Studies—to present the results of their new three-dimensional climate models. These were the state of the art—with a lot more detail and complexity than the Jason model—yet their results were basically the same. The key question in climate modeling is "sensitivity"—how sensitive the climate is to changing levels of CO_2. If you double, triple, or even quadruple CO_2, what average global temperature change would you expect? The state-of-the-art answer, for the convenient case of doubling CO_2, was "near 3 C with a probable error of 1.5 C."[16] That meant that total warming might be as little as 1.5°C or as much as 4°C, but either way, there was warming, and the most likely value was about 3°C. If you more than doubled CO_2, you'd probably get more than 3°C of warming.

There were, however, natural processes that might act as a brake on warming. The panel spent some time thinking about such "negative feedbacks," but concluded they wouldn't prevent a substantial warming. "We have examined with care all known negative feedback mechanisms, such as increase in low or middle cloud amount, and have concluded that the oversimplifications and inaccuracies in the models are not likely to have vitiated the principal conclusions that there will be appreciable warming."[17] The devil was not in the details. It was in the main story. CO_2 was a greenhouse gas. It trapped heat. So if you increased CO_2, the Earth would warm up. It wasn't quite that simple—clouds, winds, and ocean circulation did complicate matters—but those complications were "second-order effects"—things that make a difference in the second decimal place, but not the first. The report concluded, "If carbon dioxide continues to increase, the study group finds no reason to doubt that climate changes will result and no reason to believe that these changes will be negligible."[18]

How soon would these changes occur? Charney's group couldn't say, in part because that depended on how the oceans absorbed heat. The climate models had "swamp oceans," meaning they provided moisture to the atmosphere but did not hold or transport heat, so they weren't realistic. What would happen in real life? Everyone understood that the oceans have a huge "thermal inertia"—meaning that they take a very long time to heat up. Exactly how long depended in part on how well mixed they are, because the more well mixed the oceans are, the more heat would be distributed into the deep waters, and the slower the warming of the atmosphere would be. Scientists use the word "sink" to describe processes that remove components from natural systems; the oceans are almost literally a heat sink, as heat in effect sinks to the bottom of the sea.

The available evidence suggested that ocean mixing was sufficient to delay the Earth's atmospheric warming for several decades.[19] Greenhouse gases would start to alter the atmosphere immediately—they already had—but it would take decades before the effects would be pronounced enough for people to really see and feel. This had very serious consequences: it meant that you might not be able to prove that warming was under way, even though it really was, and by the time you could prove it, it would be too late to stop it.

One Jason recalls being asked by colleagues, "When you go to Washington and tell them that the CO_2 will double in 50 years and will have major impacts on the planet, what do they say?" His reply? "They . . . ask me to come back in forty-nine years."[20] But in forty-nine years it would be too late. We would be, as scientists would later say, "committed" to the warming—although "sentenced" might have been a better word.

Verner E. Suomi, chairman of the National Academy's Climate Research Board, tried to explain this crucial point in his foreword to the Charney report: "The ocean, the great and ponderous flywheel of the global climate system, may be expected to slow the course of observable climatic change. A wait-and-see policy may mean waiting until it is too late."[21] Suomi realized that this conclusion might be "disturbing to policymakers."[22] He was right.

Organizing Delay:
The Second and Third Academy Assessments

Before Charney's study had even been published, the White House Office of Science and Technology started asking the National Academy of Sciences for more information.[23] There was a host of questions about anthropogenic climate warming that Charney hadn't asked, let alone tried to answer. Prominent among them was quantification of the time frame. When would measurable change occur? "Decades" was a pretty loose estimate. What specific effects would follow? Policy makers wanted answers.

The next Academy study to address the anthropogenic warming problem produced only a letter, not a full scientific assessment, but it was nonetheless influential. Chaired by economist Thomas Schelling, famous for his work in game theory (and for which he would later win the Nobel Prize in Economics), the committee included Roger Revelle, Bill Nierenberg, and McGeorge Bundy, the national security advisor to presidents Kennedy and Johnson. Their letter report was submitted in April 1980.

Schelling focused on what warming would mean socially and politically, an aspect of the problem that was scarcely studied, much less understood. So his letter to the Academy focused on uncertainties, although he stressed not just the social scientific uncertainties, but the physical scientific ones as well. Because there were enormous uncertainties about both climate change and its potential costs, policy makers should do nothing yet, he argued, but fund more research. Moreover, Schelling wasn't certain that all the effects of warming would be bad. "The credible range of effects is extremely broad," he wrote. "By the middle of the next century, we may have a climate almost as different from today's as today's is from the peak of the last major glaciation. At the other extreme, we may only experience noticeable but not necessarily unfavorable effects around mid-century or later."[24] No one really knew.

Climate change wouldn't produce *new* kinds of climate, Schelling argued, but would simply change the distribution of climatic zones on Earth. This suggested an idea that climate skeptics would echo for the next three decades: that we could continue to burn fossil fuels without restriction and deal with the consequences through migration and adaptation. Schelling noted that past human migrations "to and throughout the new world subjected large numbers of people—together with their livestock, food crops, and culture—to drastically changed climate."[25]

Schelling acknowledged that these historic migrations occurred in eras with few or no national boundaries, very *unlike* the present, but he nevertheless suggested that adaptation would be the best response. We had time—Charney's group had said so—and during that time the cost of fossil fuel would probably go up, and so usage would go down. The slowing rate of fossil fuel use "will make adaptation to climate change easier and may permit more absorption of carbon into non-atmospheric sinks. It will also permit conversion to alternative energy sources at a lower cumulative carbon dioxide concentration, and it is likely that the sooner we begin the transition from fossil fuels the easier the transition will be." All this, he suggested, would happen naturally as market forces kicked in, so there was no need for regulation now.

Considering all the other uncertainties that Schelling emphasized, his faith in the free market could have been viewed as surprising, and his predictions have turned out to be entirely wrong: fossil fuel use has risen dramatically over the past three decades even as global warming has accelerated. But if his prediction were true, then there would be no need for government action. So this panel of worthies did not recommend a program of emissions reduction that might be phased in over time, despite their own acknowledgment that the sooner we began the transition the easier it would be. Instead, they counseled research:

> In view of the uncertainties, controversies, and complex linkages surrounding the carbon dioxide issue, and the possibility that some of the greatest uncertainties will be reduced within the decade, it seems to most of us that *the near-term emphasis should be*

on research, with as low a political profile as possible . . . We do not
know enough to address most of these questions right now. We
believe that we can learn faster than the problem can develop.[26]

At least one scientist close to this work wasn't sure this prescription was
right. John Perry, the chief staff officer for the Academy's Climate Research Board—and a meteorologist in his own right—was following the
arguments closely, and penned an article for the journal *Climatic Change*.
Its title gave away its argument, "Energy and Climate: Today's Problem, Not
Tomorrow's."[27]

Everyone was focusing on doubling CO_2 in their models and analyses,
but Perry sagely pointed out that this was just a convenient point of comparison. "Physically, a doubling of carbon dioxide is no magic threshold,"
he noted. "If we have good reason to believe that a 100 percent increase in
carbon dioxide will produce significant impacts on climate, then we must
have equally good reason to suspect that even the small increase we have
already produced may have subtly altered our climate," he concluded. "Climate change is not a matter for the next century; we are most probably doing it right now."[28] Schelling's group had expressed the hope that we could
"learn faster than the problem can develop." Perry countered, "The problem is already upon us; we must learn very quickly indeed."[29] Perry would
be proven right, but Schelling's view would prevail politically. Indeed, it
provided the kernel of the emerging skeptics' argument, and the eventual
basis for the Reagan administration to push the problem off the political
agenda entirely.

Congress was also looking into climate change. The 1978 National
Climate Act had established a national climate research program, and Connecticut senator Abraham Ribicoff was planning to introduce an amendment to fund a closer look at CO_2. It's a cliché that scientists always say that
more research is needed, but Ribicoff concluded that more research *was*
needed.[30] President Jimmy Carter was proposing a major effort to increase
U.S. energy independence by developing "synfuels"—liquid fuels made
from coal, oil shales, and tar sands—and scientific experts had warned that
this could accelerate CO_2 accumulation. Ribicoff's amendment authorized
the National Academy of Sciences to undertake a comprehensive study of
CO_2 and climate.[31] While the formal charge to the new committee was not
formulated until June of the following year, a committee was already in place
by October 1980, with Bill Nierenberg as its chair.

Nierenberg seems to have done a certain amount of groundwork, if not actual lobbying, for the job. In August 1979, as the Charney group was compiling its conclusions, John Perry had already been pondering the follow-up. Following normal Academy patterns, Perry suggested to members of the Climate Research Board that the new committee should not undertake new research, but simply review the adequacy and conclusions of existing work.[32] Nierenberg disagreed and argued for a much broader view. He thought the Academy should undertake a comprehensive, integrated assessment of all aspects of the problem, and that the members of the committee should be chosen with more than the usual care.[33] They were. They included Tom Schelling, and another who would support his views, Yale economist William Nordhaus.

Most National Academy reports are written collectively, reviewed by all the committee members, and then reviewed again by outside reviewers. Changes are made by the authors of the various sections and by the chairperson, and the report is accepted and signed by all the authors. An Executive Summary, or synthesis, sometimes written by the chairperson, sometimes by Academy staff, is also reviewed to ensure that it accurately reflects the contents of the study. That didn't happen here. The Carbon Dioxide Assessment Committee—chaired by Bill Nierenberg—could not agree on an integrated assessment, so they settled for chapters that were individually authored and signed. The result, *Changing Climate: Report of the Carbon Dioxide Assessment Committee*, was really two reports—five chapters detailing the likelihood of anthropogenic climate change written by natural scientists, and two chapters on emissions and climate impacts by economists—which presented very different impressions of the problem. The synthesis sided with the economists, not the natural scientists.

The chapters written by the natural scientists were broadly consistent with what other natural scientists had already said. No one challenged the basic claim that warming would occur, with serious physical and biological ramifications. Revelle's chapter on sea level rise warned of the possible disintegration of the West Antarctic Ice Sheet, which "would release about 2 million km^3 of ice before the remaining half of the ice sheet began to float. The resulting worldwide rise in sea level would be between 5 and 6 m[eters]."[34] The likely result: "The oceans would flood all existing port facilities and other low-lying coastal structures, extensive sections of the heavily farmed and densely populated river deltas of the world, major portions of the state[s] of Florida and Louisiana, and large areas of many of the world's major cities."[35]

How quickly could such a disaster occur? Total disintegration of that ice sheet would take a long time, perhaps two hundred to five hundred years, but smaller effects might begin much sooner. If temperature increases of 2°C to 3°C were achieved by midcentury, thermal expansion alone would produce seventy centimeters of sea level rise, to which one could add another two meters by 2050 or so if the ice sheet began to fail. Whether fast or slow, "disintegration of the West Antarctic Ice Sheet would have . . . far-reaching consequences."[36]

Other chapters addressed the impacts on climate, water availability, marine ecosystems, and more. The physical scientists allowed that many details were unclear—more research *was* needed—but they broadly agreed that the issue was very serious. When the chapters were boiled down to their essence, the overall conclusion was the same as before: CO_2 had increased due to human activities, CO_2 will continue to increase unless changes are made, and these increases will affect weather, agriculture, and ecosystems. None of the physical scientists suggested that accumulating CO_2 was not a problem, or that we should simply wait and see.

But that's precisely what the economists' chapters, as well as the synthesis, argued. The report's first chapter, written by Nordhaus, National Research Council staff member Jesse Ausubel, and a consultant named Gary Yohe (an economics professor at Wesleyan University), focused on future energy use and carbon dioxide emissions. The long and detailed chapter began by acknowledging the "widespread agreement that anthropogenic carbon dioxide emissions have been rising steadily, primarily driven by the combustion of fossil fuels." Their focus, however, was not so much on what was known, but on what was not known: the "enormous uncertainty" beyond 2000, and the "even greater uncertainty" about the "social and economic impacts of possible future trajectories of carbon dioxide."[37]

Using a probabilistic scenario analysis, they projected atmospheric CO_2 levels to 2100, using various assumptions regarding energy use, costs, and increased economic efficiencies. The range of possible outcomes was large, but they considered the most likely scenario to be CO_2 doubling by 2065.[38] The economists acknowledged the "substantial probability that doubling will occur much more quickly," including a 27 percent chance that it would occur by 2050, and admitted that it was "unwise to dismiss the possibility that a doubling may occur in the first half of the twenty-first century." Yet they did just that.

What could be done to stop climate change? According to Nordhaus, not much. The most effective action would be to impose a large permanent carbon tax, but that would be hard to implement and enforce.

> A significant reduction in the concentration of CO_2 will require very stringent policies, such as hefty taxes on fossil fuels . . . The strategies suggested later [in the report] by Schelling—climate modification or simply adaptation to a high CO_2 and high temperature world—are likely to be more economical ways of adjusting . . . Whether the imponderable side effects on society— on coastlines and agriculture, on life in high latitudes, on human health, and simply the unforeseen—will in the end prove more costly than a stringent abatement of greenhouse gases, we do not now know.[39]

Rather than confront their own caveat that changes might happen much sooner than their model predicted—and thus be much more costly than prevention—the economists assumed that serious changes were so far off as to be essentially discountable.

Schelling picked up the thread of this argument in the final chapter of the report, where the economists' reframing of the climate question became explicit. Natural scientists were not worried about climate change per se—because scientists knew climate was naturally variable—but about rapid, unidirectional change forced by carbon dioxide. Such change would seriously challenge ecosystems that couldn't adapt in only a few decades, as well as human infrastructure. But Schelling rejected this view, insisting that the real issue *was* climate change and that the impact of carbon dioxide needed to be assessed together with "other climate-changing activities," such as dust, land use changes, and natural variability. It was wrong to single out CO_2 for special consideration.

Common sense might suggest that if carbon dioxide is the cause of climate change, then controlling it is the obvious solution, but Schelling rejected this view, too. He insisted that it was a mistake to assume a "preference for . . . dealing with causes rather than symptoms . . . It would be wrong to commit ourselves to the principle that if fossil fuels and carbon dioxide are where the problem arises, that must also be where the solution lies."[40] It might be best just to treat the symptoms through deliberate weather modification or to adapt.

Schelling's attempt to ignore the cause of global warming was pretty peculiar. It was equivalent to arguing that medical researchers shouldn't try to cure cancer, because that would be too expensive, and in any case people in the future might decide that dying from cancer is not so bad. But it was based on an ordinary economic principle—the same principle invoked by Fred Singer when discussing acid rain—namely, discounting. A dollar today is worth more to us than a dollar tomorrow and a lot more than a dollar a century from now, so we can "discount" faraway costs. This is what Schelling was doing, presuming that the changes under consideration were "beyond the lifetimes of contemporary decision-makers."[41] Not only did we not know how much energy future populations would use, and therefore how much CO_2 they would produce, we didn't know how they would live, how mobile they would be, what technologies they would have at their disposal, or even what climates they might prefer.

Schelling had a point: if changes were a century away, then it would be impossible to predict how troubling they would be. Perhaps by 2100 everyone would be living indoors, with agriculture pursued in controlled hydroponic environments. The rub was that most of the physical scientists on the panel did *not* think that trouble was more than a century away. Most of them thought that significant changes were much closer, and that carbon dioxide *was* the problem.

So Nierenberg's committee had produced a report with two quite different views: the physical scientists viewed accumulating CO_2 as a serious problem; the economists argued that it wasn't. And the latter view framed the report—providing its first and last chapters. A fair synthesis might have laid out the conflicting views and tried to reconcile them or at least account for the differences. But this synthesis didn't. It followed the position advocated by Nordhaus and Schelling. It did not disagree with the scientific facts as laid out by Charney, the Jasons, and all the other physical scientists who had looked at the question, but it rejected the interpretation of those facts as a problem. "Viewed in terms of energy, global pollution, and worldwide environmental damage, the 'CO_2 problem' appears intractable," the synthesis explained, but "viewed as a problem of changes in local environmental factors—rainfall, river flow, sea level—the myriad of individual incremental problems take their place among the other stresses to which nations and individuals adapt."[42]

Some climatic effects—like serious sea level rise—might make some areas of the world uninhabitable, but this could be addressed through migration. Nierenberg stressed that people had often migrated in the past, and when they did, they often had to adapt to new climates. "Not only have people moved" Nierenberg noted, "but they have taken with them their horses, dogs, children, technologies, crops, livestock, and hobbies. It is extraordinary how adaptable people can be."[43] Thus Nierenberg's argument was the same as Schelling's had been in 1980: research, not policy action, was necessary, and that research should take the lowest possible political profile. Vern Suomi had admonished that a "wait and see" attitude was likely to be untenable, but that's exactly what Nierenberg's committee recommended.

The fact is, historical mass migrations had been accompanied by massive suffering, and typically people moved under duress and threat of violence. So Nierenberg's cavalier tone, and suggestion that these migrations were essentially benign, flew in the face of historical evidence. At least one reviewer recognized this. Alvin Weinberg, a physicist who had led the Oak Ridge National Laboratory for nearly twenty years, wrote a scathing eight-page critique. Weinberg was one of the first physicists to recognize the potential severity of global warming, arguing in 1974 that climate impacts might limit our use of fossil fuels before they were even close to running out.[44] This perspective meshed with his advocacy of nuclear power, which he believed was the only energy source that could enable better living conditions for all humanity, an opinion he and Nierenberg shared. But Weinberg was outraged by what he read in Nierenberg's report.

The report was "so seriously flawed in its underlying analysis and in its conclusions," Weinberg wrote, that he hardly knew where to begin. The report flew in the face of virtually every other scientific analysis of the issue, yet presented almost no evidence to support its radical recommendation to do nothing. Improvements in irrigated agriculture would no doubt occur, but could they be put in place fast enough and on a sufficient scale, particularly in poor countries? The report provided no evidence. As for migration, "does the Committee really believe that the United States or Western Europe or Canada would accept the huge influx of refugees from poor countries that have suffered a drastic shift in rainfall pattern?," Weinberg demanded. "I can't for the life of me see how historic migrations, which generally have taken place when political boundaries were far more permeable than they are now, can tell us anything about migrations 75 to 100 years from now when large areas lose their capacity to support people. Surely there will be times of trouble then."[45]

Weinberg wasn't alone in realizing that the claims made in the synthesis were not supported by the analysis presented in the body of the report. Two other reviewers made the same point, although with less passion.[46]

Yet these reviewers were also ignored. How was it possible for the reviewers' comments to be ignored, and for a report to be issued in which the synthesis was at odds with the report it claimed to synthesize and in which major claims were unsupported by evidence? One senior scientist many years later answered this way: "Academy review was much more lax in those days." But why didn't anyone object after the report was released? This same scientist: "We knew it was garbage so we just ignored it."[47]

But the Nierenberg report didn't go out with the morning trash. It was used by the White House to counter scientific work being done by the Environmental Protection Agency. The EPA prepared two reports of its own, both of which concluded that global warming would be serious, and that the nation should take immediate action to reduce coal use.[48] When the EPA reports came out, White House Science Advisor George Keyworth used Nierenberg's report to refute them. In his monthly report for October prepared for Ed Meese, Keyworth wrote, "The Science Advisor has discredited the EPA reports . . . and cited the NAS report as the best current assessment of the CO_2 issue. The press seems to have discounted the EPA alarmism and has taken the conservative NAS position as the wisest."[49]

Keyworth was right. The press would indeed take the "conservative" position. A *New York Times* reporter put it this way: "The Academy found that since there is no politically or economically realistic way of heading off the greenhouse effect, strategies must be prepared to adapt to a 'high temperature world.' "[50] But the Academy hadn't *found* that; the committee had *asserted* it. And it wasn't the Academy; it was Bill Nierenberg and a handful of economists.

Was it just coincidence—a meeting of minds—that Nierenberg gave the White House just what it wanted? The historical record suggests not. In meetings with the Climate Research Board, Energy Department officials had told Academy members that they "did not approve of . . . speculative, alarmist, 'wolf-crying' scenarios."[51] They simply wanted "guidance on the on-going research program."[52] Tom Pestorius, the senior policy analyst at the White House Office of Science and Technology who was a White House liaison to the Acid Rain Peer Review, was involved here, too. There was no need for alarm, he told John Perry, who reported this back to Nierenberg's committee, because "technology will ultimately be the answer to the problems of providing energy and protecting the environment."[53]

Nierenberg's CO_2 and climate report pioneered all the major themes behind later efforts to block greenhouse gas regulation, save one. Nierenberg didn't deny the legitimacy of climate science. He simply ignored it in favor of the claims made by economists: that treating symptoms rather than causes would be less expensive, that new technology would solve the problems that might appear so long as government didn't interfere, and that if technology couldn't solve all the problems, we could just migrate. In the two decades to come, these claims would be heard again and again.

But just as Alvin Weinberg hadn't bought these arguments, not all economists did, either. A handful of economists in the late 1960s had realized that free market economics, focused as it was on consumption growth, was inherently destructive to the natural environment and to the ecosystems on which we all depend. The Earth doesn't have infinite resources, and, as we saw in chapter 3 with acid rain, it doesn't have an infinite ability to withstand pollution. Nierenberg hadn't put any of *these* economists on his panel. So just as Nierenberg had built his Executive Summary around a one-sided view of climate change, he'd built it around a one-sided view of economics.

Nierenberg gave the administration everything it wanted: a report that presented a united front rather than the real differences of opinion between the social and physical scientists, insisted that no action was needed now, and concluded that technology would solve any problems that did, in the future, emerge. The government did not need to do anything—except fund research.

Meeting the "Greenhouse Effect" with the "White House Effect"

Two crucial developments during the presidential campaign year of 1988 changed climate science forever. The first was the creation of the Intergovernmental Panel on Climate Change. The second was the announcement by climate modeler James E. Hansen, director of the Goddard Institute for Space Studies, that anthropogenic global warming had begun. An organized campaign of denial began the following year, and soon ensnared the entire climate science community.

In November 1987 Colorado senator Tim Wirth had sponsored a hearing on climate in which Hansen had testified, but it had been widely ignored by the nation's media establishment.[54] A drought was setting in across the United States, however, and by the following summer, the nation was in crisis. The year 1988 proved to be one of the hottest and driest in U.S. history. As 40 percent of the nation's counties were affected, and as crops failed, livestock died, and food prices rose, people were beginning to wonder if perhaps global warming was not so far off after all. Popular and media interest in climate soared. In June, Wirth tried again. Senator J. Bennett Johnston of Louisiana delivered the opening statement of the hearing:

> Today, as we experience 101°[F] temperatures in Washington, DC, and the soil moisture across the midwest is ruining the soybean crops, the corn crops, the cotton crops, when we're having emergency meetings of the Members of the Congress in order to figure out how to deal with this emergency, then the words of Dr. Manabe and other witnesses who told us about the greenhouse effect are becoming not just concern, but alarm.[55]

Hansen was the star of the show. He testified about some new research at the Goddard Institute for Space Studies, showing that there had been a warming since 1980 of just about half a degree Celsius—or one degree Fahrenheit—relative to the 1950–1980 average. The probability that this could be explained by natural events was only 1 percent. "The global warming is now large enough that we can ascribe with a high degree of confidence a cause and effect relationship to the greenhouse effect," Hansen told the committee.[56]

His team had also modeled the increase of carbon dioxide and other trace gases according to three "emissions scenarios." The scenarios were not intended to be predictions of the actual course of human carbon emissions; they were what-if scenarios bracketing likely rates of future emissions and their consequences. One scenario imagined rapid reduction of fossil fuel use after 2000, which reduced future warming. The other two—more realistic scenarios—raised the Earth's global mean temperature rapidly. Within twenty years, it would be higher than at any time since the warmest previous interglacial period then known, which ended about 120,000 years ago.[57]

This time, major newspapers across the country covered the hearings. The *New York Times* put Hansen's testimony on the front page; suddenly he was the leading advocate for doing something about the global warming.[58] Some colleagues, uncomfortable with all the media attention—and maybe a bit jealous, too—attacked Hansen for going too far, thinking he had discounted the significant uncertainties that still remained. On the other hand, Hansen had captured attention as no one else had. Moreover, most of the scientific community *did* believe that one could not endlessly raise atmospheric concentrations of greenhouse gases without a climatic response. It was basic physics. Still, Hansen's claim of detection was unexpected, and seemed perhaps premature.[59]

DURING THE FIVE-YEAR INTERREGNUM between the release of the Nierenberg report and Hansen's powerful testimony, atmospheric scientists had been busy with other things. They had discovered the Antarctic ozone hole, investigated it, and explained its cause. They had also demonstrated the existence of global ozone depletion through the work of the Ozone Trends Panel. Certain scientists, including NASA's Bob Watson, began to think that something like the Ozone Trends Panel was needed for global warming, too. This became the Intergovernmental Panel on Climate Change.

Bert Bolin, the man who had first warned about acid rain in Europe, thought that Hansen's temperature data hadn't been "scrutinized well enough," and accepted the task.[60] He divided the panel into three working groups. The first would produce a report reflecting the state of climate science. The second would assess the potential environmental and socio-economic impacts. The third would formulate a set of possible responses. The scientists set themselves a deadline of 1990 for their first assessment: a very short time given their intent to involve more than three hundred scientists from twenty-five nations.[61]

The political pressure generated by the June hearings also caused presidential candidate, and sitting vice president, George H. W. Bush to promise to counter the "greenhouse effect with the White House effect" by bringing the power of the presidency to bear on the problem.[62] After his inauguration as forty-first president of the United States in January 1989, he sent his secretary of state, James Baker, to the first IPCC meeting, and had the Federal Coordinating Council for Science, Engineering, and Technology's Committee on Earth Sciences outline a proposed U.S. Global Climate Change Research initiative for the fiscal year 1990 budget.[63] It was welcomed in the U.S. Senate, where the Committee on Commerce, Science, and Transportation had prepared a bill proposing the same thing: the National Global Change Research Act of 1989.[64] The United States, it seemed, was preparing to deal with anthropogenic climate change. As Gus Speth later recalled, "We thought we were on track to make real changes."[65] He underestimated the challenge.

Blaming the Sun

In 1984 Bill Nierenberg retired as director of the Scripps Institution of Oceanography, and joined the Board of Directors of the George C. Marshall Institute. As we saw earlier, Robert Jastrow had established the Institute to defend President Reagan's Strategic Defense Initiative against attack by other scientists. But by 1989, the enemy that justified SDI was rapidly disappearing. The Warsaw Pact had fallen apart, the Soviet Union itself was disintegrating, and the end of the Cold War was in sight. The Institute might have disbanded—its raison d'être disappeared—but instead, the old Cold Warriors decided to fight on. The new enemy? Environmental "alarmists." In 1989—the very year the Berlin Wall fell—the Marshall Institute issued its first report attacking climate science. Within a few years, they would be attacking climate scientists as well.

Their initial strategy wasn't to deny the fact of global warming, but to blame it on the Sun. They circulated an unpublished "white paper," generated by Jastrow, Seitz, and Nierenberg and published as a small book the following year, entitled "Global Warming: What Does the Science Tell Us?"[66] Echoing the tobacco industry strategy, they claimed that the report would set the record straight on global warming. The Institute's Washington office staff contacted the White House to request the opportunity to present it. Nierenberg gave the briefing himself, to members of the Office of Cabinet Affairs, the Office of Policy Development, the Council of Economic Advisers, and the Office of Management and Budget.[67]

The briefing had a big impact, stopping the positive momentum that had been building in the Bush administration. "I was impressed with the report," said one member of the cabinet affairs office. "Everyone has read it. Everyone takes it seriously." Another ruminated, "It is well worth listening to. They are eminent scientists. I was impressed."[68] White House chief of staff John Sununu—a nuclear engineer by training—was particularly taken. Stanford University's Stephen Schneider lamented, "Sununu is holding the report up like a cross to a vampire, fending off greenhouse warming."[69] Meanwhile, no one had invited Bert Bolin to the White House. Perhaps he hadn't known to ask to be invited.

The central claim of the Marshall Institute report was that the warming that Hansen and others had found didn't track the historical increase in CO_2. The majority of the warming had been prior to 1940—prior to the majority of the carbon dioxide emissions. Then there was a cooling trend through 1975, and a return to warming. Since the warming didn't parallel the increase in CO_2, it must have been caused, they claimed, by the Sun.[70]

Drawing on sunspot and carbon-14 data from tree rings, they argued that the Sun had entered a period of higher energy output during the nineteenth century, and that this solar output increase (of about 0.3 percent) was responsible for the climate warming to date. They also contended that the data showed a two-hundred-year cycle, so the warming trend was almost over, and things would soon begin to cool off. "If the correlation between solar activity and global temperatures also continues, a trend toward a cooler planet can also be expected in the 21st century as a result of natural forces of climate change."[71]

Had there been cooling between 1940 and 1975? Yes, but the Marshall report misrepresented it. The Institute's source for their diagram was an article by Hansen's team, so it looked eminently credible.[72] It looked like they were relying on peer-reviewed science. But Jastrow, Nierenberg, and Seitz had cherry-picked the data—using only one diagram out of six that were relevant. They had shown their readers only the top piece of figure 5 (see next page). What Hansen and his group had done was to explore the role of various "forcings"—the different causes of climate change. One was greenhouse gases, a second was volcanoes, and the third was the Sun. Hansen's team had done what scientists are supposed to do—objectively considered all the known possible causes.

Then they asked, What cause or combination of causes best explains the observations? The answer was all of the above. "CO_2 + volcanoes + Sun" fit the observational record best. The Sun did make a difference, but greenhouse gases did, too. The observed climate of the twentieth century was a product of all three forcings, but since Jastrow, Seitz, and Nierenberg had shown their readers only the top portion of Hansen's figure, they'd made it appear as if only the Sun mattered. The warming prior to 1940 probably was the effect of a nineteenth-century increase in solar output, but not the increase that had started in the mid-1970s. There hadn't been any solar output increase in the mid-twentieth century, so only CO_2 explained the recent warming.

There was an even larger problem with the Marshall analysis that climate modeler Steven Schneider pointed out. If Jastrow and company were right that the climate was extremely sensitive to small changes in solar

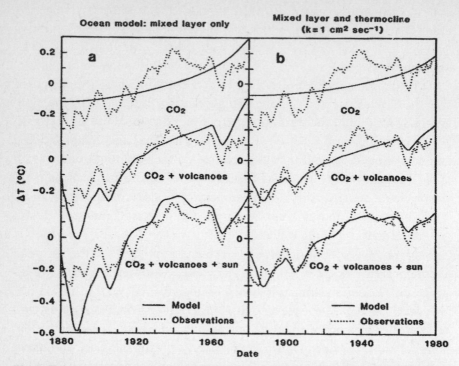

Fig. 5, Global temperature trend obtained from climate model with sensitivity 2.8°C for doubled CO_2. The results in (a) are based on a 100-m mixed-layer ocean for heat capacity; those in (b) include diffusion of heat into the thermocline to 1000 m. The forcings by CO_2, volcanoes, and the sun are based on Broecker (25), Lamb (27), and Hoyt (48). Mean ΔT is zero for observations and model.

This set of charts was part of an article by James E. Hansen at the Goddard Institute for Space Studies, showing (left side) model results for an "Earth" with only very shallow oceans exchanging heat with the atmosphere, and (right side) oceans with much deeper mixing of heat. Hansen's team argued that the bottom right image best reflected the behavior of the real Earth—with ocean mixing to 1,000 meter depth, solar irradiance, volcanic dust and aerosols, and CO_2 all playing roles. The Marshall Institute's version included only the top left portion of the diagram, leaving the impression that CO_2 didn't matter. *From J. Hansen et al., "Climate Impact of Increasing Atmospheric Carbon Dioxide,"* Science *(28 August 1981): 963. Reprinted with permission from AAAS.*

output, then it meant that the climate would also be extremely sensitive to small changes in greenhouse gases. Schneider argued,

> If only a few tenths of a percent change in solar energy were re-
> sponsible for the [observed] .5 C long trend in climate over the
> past century, then this would suggest a planet that is relatively
> sensitive to small energy inputs. The Marshall Institute simply
> can't have it both ways: they can't argue on the one hand that small
> changes in solar energy output can cause large temperature
> changes, but that comparable changes in the energy input from
> greenhouse gases will not also produce comparable large signals.
> Either the system is sensitive to large scale radiative forcing or it
> is not.[73]

Sensitivity cuts both ways. And as physicists, Jastrow, Seitz, and Nierenberg would of course have known this.

The Intergovernmental Panel on Climate Change published its first assessment of the state of climate science in May 1990. It reiterated the result that was by now familiar to anyone who had been following the issue:
unrestricted fossil fuel use would produce a "rate of increase of global
mean temperature during the next century of about .3°C per decade; this is
greater than that seen over the past 10,000 years."[74] Global warming from
greenhouse gases would produce changes unlike what humans had ever
seen before.

The IPCC explicitly addressed—and rejected—the Marshall Institute
argument for blaming the Sun. The upper limits on solar variability, they
explained, are "small compared with greenhouse forcing and even if such
a change occurred over the next few decades, it would be swamped by the
enhanced greenhouse effect."[75]

But the IPCC's refutation didn't matter to the Marshall Institute. In 1991,
they reiterated their argument in a longer version, and in October 1992 Bill
Nierenberg took it on the road to the World Petroleum Congress in Buenos
Aires, where he launched a full frontal attack on the IPCC. Nierenberg insisted that global temperatures would increase at most by 1°C by the end of
the twenty-first century, based on a straight linear projection of twentieth-
century warming. Bert Bolin confronted him directly, pointing out that
greenhouse gas emissions were increasing *exponentially*, not linearly. Add to
this the time lag induced by the oceans—which Jule Charney had warned
about a decade earlier—and warming would accelerate over time.

In his memoir, Bolin called Nierenberg's conclusion "simply wrong."[76] A less polite man would have said something far worse. If Nierenberg had been a journalist, one might suppose he was just confused. But Nierenberg was no journalist; one longtime associate at Scripps once said she never knew a man who was more careful in choosing what he worked on and how he worked on it.[77] Meanwhile, the Cato Institute distributed an uncorrected version of the graph printed in the original Marshall Institute white paper—the one that showed only the top part of Hansen's graph.[78] Given all the efforts the climate scientists had made to set the record straight, it's not plausible that this was simply a mistake.

Moreover, they were proud of the results. In a February 1991 letter to the vice president of the American Petroleum Institute, Robert Jastrow crowed, "It is generally considered in the scientific community that the Marshall report was responsible for the Administration's opposition to carbon taxes and restrictions on fossil fuel consumption." Quoting *New Scientist* magazine, he reported that the Marshall Institute "is still the controlling influence in the White House."[79]

Fred Singer would push their efforts one step further.

The Attack on Roger Revelle

While Jastrow, Seitz, and Nierenberg were broadcasting their "blame the Sun" claim, Fred Singer was preparing to attack climate science in a different way: by claiming that Roger Revelle had changed his mind about global warming. In addition to his role in helping to launch the Keeling Curve, Revelle had played another crucial role in the history of climate science, as mentor to Al Gore. Gore had studied with Revelle in the 1960s at Harvard, and it was well-known that Gore's concern about climate change stemmed from his tutelage under Revelle. If Revelle no longer considered global warming worrisome, this would be news indeed. It would also embarrass Gore, who was running his 1992 presidential campaign on environmental themes.

On February 19, 1990, the eighty-one-year-old Revelle had presented a paper entitled "What Can We Do About Climate Change?" at the American Association for the Advancement of Science meeting in New Orleans. Research and observations over the next ten to twenty years "should give us a much better idea of the likely magnitude of atmospheric and oceanic warming during the twenty-first century," he noted.[80] In the meantime, there were six approaches that could be taken to reduce future warming: emphasizing natural gas over coal and oil, conservation, substitution of non-fossil energy sources, carbon sequestration by stimulating phytoplankton production, increasing atmospheric reflection through artificial intervention, and expanding forests. Revelle had lately developed an interest in the possibility that high-latitude (or "boreal") forests might expand as the Earth warmed, removing carbon dioxide from the atmosphere and preventing some of the warming. He thought this expansion might remove 2.7 billion tons of carbon per year, roughly half the total contributed by fossil fuel combustion each year.[81] This wouldn't be negligible—it might even be the negative feedback that the Charney panel had looked for but never found—and he thought more research *was* needed.

Revelle's discussion of mitigation strategies—conservation, nuclear power, boreal forests, etc.—would have made no sense if he didn't think there was something to mitigate against. Read in full, his talk clearly demonstrates that he believed the prudent step was to begin to switch to nuclear power and natural gas and improve energy conservation, while continuing research. Like all good scientists, Revelle was careful not to overstate his claims. He knew as well as anyone that there were still important uncertainties, and perhaps because he was intrigued by the prospect that boreal forests might delay warming significantly, he'd started his talk with this potentially ambiguous statement: "There is a good but by no means certain chance that the world's average climate will become significantly warmer during the next century."[82]

That gave Fred Singer the opening he needed. Singer approached Revelle after the talk about collaborating on an article for the *Washington Post*. The historical record doesn't tell us exactly what the article was supposed to be about, and had Revelle stayed healthy, he might have left a fuller record. But on his way back to La Jolla, Revelle suffered a massive heart attack. He went straight from the airport to the hospital, where he underwent a triple-bypass operation.

Revelle didn't recover quickly. After finally returning home in March, he was forced back to the hospital for an emergency hernia operation. Then he contracted a severe infection and spent another six weeks in the hospital. When he finally returned home in May, he was so weak that his personal secretary, Christa Beran, and Justin Lancaster, a graduate student with whom Revelle was teaching, arranged to limit his appointments to under a half hour.[83] Famous for his energy, Revelle was now falling asleep while dictating letters. He was not well.

The title of the paper that Singer would later publish, with Revelle as coauthor, was "What To Do about Greenhouse Warming: Look Before You Leap," but, given the state of his health, it's not clear how closely Revelle was able to look at the various drafts that Singer sent him, or how closely he checked that Singer had made the changes he suggested. Revelle had never been good at saying no to people; one of Revelle's closest colleagues, oceanographer Walter Munk, admits that "Roger often leapt before he looked."[84]

What we do know from Revelle's papers at Scripps Institution of Oceanography is that Singer sent three drafts of their proposed article during March, while Revelle was still in the hospital. We also know that something about the paper clearly bothered Revelle. Christa Beran later recalled that whenever Singer sent him a draft, Revelle buried it under piles of paper on his desk. When Singer called, Beran would dig up the draft and put it on top, and Revelle would bury it again. Beran wondered why, and Revelle, she recalled later in a legal affidavit, told her, "Some people don't think Fred Singer is a very good scientist."[85]

Singer had made himself an unpopular figure in the scientific mainstream by attacking fellow scientists over acid rain and ozone, so perhaps after having said yes to Singer at the AAAS meeting in New Orleans, Revelle was regretting it, hoping that if he ignored the paper, it would go away. But Singer was not one to go away.

While Singer was trying to get Revelle to review the drafts, he published an article on his own in the journal *Environmental Science and Technology*, with essentially the same title, "What To Do about Greenhouse Warming." Singer echoed the Marshall Institute's arguments, implying that scientists just didn't know what had caused the warming of the twentieth century. "There is major uncertainty and disagreement about whether this increase [in CO_2] has caused a change in the climate during the past 100 years; observations simply don't fit the theory," he insisted. Of course there was disagreement—the Marshall Institute had generated it—but not among climate scientists. The IPCC had clearly stated that the unrestricted fossil fuel use would produce a "rate of increase of global mean temperature during the next century of about .3 C per decade; this is greater than that seen over the past 10,000 years."[86] Singer rejected this, asserting instead that "the scientific base for [greenhouse warming] includes some facts, lots of uncertainty, and just plain ignorance." He concluded emphatically, "*The scientific base for a greenhouse warming is too uncertain to justify drastic action at this time.*"[87] This, of course, was precisely what he had said about acid rain. And ozone depletion. It was easy to see why many working scientists didn't like Fred Singer. He routinely rejected their conclusions, suggesting that he knew better than they did.

In February 1991, Singer visited Scripps. In one multihour meeting, Singer and Revelle went over the paper, which was already set in galleys. There was at least one point of contention between the two, and it was a big one: what *was* the climate sensitivity to carbon dioxide? The galleys that Singer gave to Revelle to review asserted, "Assume what we regard as the most likely outcome: A modest average warming in the next century of less than one degree Celsius, well below the normal year to year variation."[88]

This was completely inconsistent with what the Jasons had said, what Charney's panel had said, and what the IPCC had said. No one in the climate community was asserting that the climate change from increased greenhouses gases would be no different from normal year-to-year variation. In fact, the IPCC had said just the opposite. Revelle apparently crossed out "less than one degree" and wrote in the margin next to it: "one to three degrees."[89]

This might not seem like a big difference, but it was. One to three degrees fell within the mainstream view, and clearly *outside* the range of the natural climate variability of the past few hundred years. This was the key point: would warming lead us into a new man-made climate regime, unlike anything we had seen before? Revelle (and thousands of climate scientists) said yes; Singer said no.

Singer finessed the disagreement by dropping numbers altogether. The sentence as published read, "Assume what we regard as the most likely outcome: A modest average warming in the next century well below the normal year to year variation."[90] The paper contradicted what Revelle had written in the margin, and asserted that there was no likelihood of significant warming. What little change would occur would be not noticeably different from natural variation. Singer had prevailed, and it looked as if Revelle had agreed.

The paper was published later that year in *Cosmos*, the journal of the elite Washington Cosmos Club, founded in 1878 (and that only opened its doors to women in 1988 when forced to by the threat of an antidiscrimination suit). Revelle was listed as second author.[91] There was also a third author: Chauncey Starr, the physicist we met in chapter 3 casting doubt on the reality of acid rain, and in chapter 5 arguing for radiation hormesis— that radiation is good for you.[92]

Did Roger Revelle agree to this final version? We will never know for sure, because in July, Revelle suffered a fatal heart attack, but it's hard to believe that he would have—at least, not if he were in good health and clear of mind—and no one close to him did believe that he had.

Scientists already knew from paleoclimate data that the lowest possible climate sensitivity to doubled CO_2 was 1.5°C. We knew from the geological record that CO_2 levels had varied in the past, and temperatures had varied in a manner consistent with an overall sensitivity of not less than 1.5°C for CO_2 doubling. Revelle—a geologist by training—knew this very well. He had cotaught a course at Scripps with Justin Lancaster that included discussion of this natural climate variation.

Lancaster later recalled that Revelle was embarrassed when the *Cosmos* paper was published.[93] But *Cosmos* wasn't a scientific journal—it wasn't peer reviewed—and it didn't have a very high circulation. Few scientists would have seen the article, much less paid much attention to it, so even had he been in good health, Revelle might well have just let it drop. Perhaps he would have thought it was "garbage" and just ignored it.

But as the 1992 election campaign got under way, the *Cosmos* article was not ignored. It was used to attack Senator Al Gore. The first salvo seems to have fired by Gregg Easterbrook in the July issue of the *New Republic,* and reiterated in August in the *Independent.* Criticizing Gore's new book, *Earth in the Balance*, Easterbrook sniffed indignantly that Gore had failed to mention that "before his death last year, Revelle published a paper that concludes, *'the scientific base for a greenhouse warming is too uncertain to justify drastic action at this time.'* "[94]

Those were *Singer's* words, not Revelle's. Singer had used them in his stand-alone 1990 paper, and again in 1991, in a book chapter questioning the existence of global warming and attacking the Intergovernmental Panel on Climate Change.[95] Revelle had said nothing like that in his AAAS talk. Moreover, it's customary in both academic and journalistic circles to credit the lead author of a paper. That, of course, was Fred Singer. Easterbrook might just as well have said he was quoting Chauncey Starr. Either Easterbrook was being sloppy or he was exploiting the Revelle connection for political purposes. After all, it was Revelle, not Singer or Starr, who was Gore's mentor.

Easterbrook's attack was picked up by conservative columnist George Will, who repeated it almost verbatim in a September 1992 column. "Gore knows that his former mentor at Harvard, Roger Revelle, who died last year, concluded: 'The scientific base for greenhouse warming is too uncertain to justify drastic action at this time. There is little risk in delaying policy responses.'"[96] From there, it became part of the only vice-presidential debate of the campaign. Retired admiral James B. Stockdale, the running mate of Ross Perot, attacked Gore with the claim, again using the statement that had originated in Singer's 1990 article.[97]

The use of Revelle's name to attack Al Gore infuriated the Revelle family, as well as his colleagues at Scripps. Revelle's daughter, Carolyn Hufbauer, protested Will's attack in an op-ed published just before the vice-presidential debate, September 13.[98] Two of Revelle's closest colleagues at Scripps, oceanographer Walter Munk and physicist Edward Frieman, agreed with Hufbauer that Revelle's views were being misrepresented. They wrote a letter to *Cosmos*, but the journal declined to publish it, so they published it in the journal *Oceanography*, along with the text of Revelle's AAAS paper.[99] (Yet again, unscientific claims were being circulated broadly, but the scientists' refutation of them was published where only fellow scientists would see it.)

Munk and Frieman explained that the *Cosmos* paper hadn't been written by Revelle at all. "S. Fred Singer wrote the paper," they explained, suggesting that "as a courtesy, [Singer] added Roger as a co-author based upon his willingness to review the manuscript and advise on aspects relating to sea-level rise."[100]

More than a decade later, Munk was still angry about what he referred to as "Singer's betrayal of Roger."[101] But the person who fought longest and hardest to defend Revelle's legacy—and paid the highest price—was Justin Lancaster. In that last year of Revelle's life, Lancaster had seen him on nearly a daily basis. The two had taught a class together, and they shared a commitment to addressing policy questions. (This was something that most of the scientists at Scripps weren't actually interested in; they just wanted to do pure science.) Lancaster felt he knew Revelle's views as well as anyone.

Lancaster and his thesis advisor, Dave Keeling, wrote a letter to the *New Republic* challenging the Easterbrook article, but it was never published. For a second time, scientists close to Revelle were attempting to refute the misrepresentation, but their attempts to set the record straight were rejected by the journals that had published the misrepresentation in the first place. So Lancaster did what Munk and Frieman had done. He turned to the scientific community, who he figured did care about the truth. At the time, Lancaster was serving on the editorial board for a volume titled *A Global Warming Forum*, and Singer intended to republish the *Cosmos* piece there. Lancaster tried to get Singer to remove Revelle's name from it, but Singer refused. A struggle among Singer, Lancaster, and the volume's editorial staff ensued as Lancaster tried to remove Revelle's name from the article; when the volume was finally published in 1993, it contained a footnote on the first page point-ing readers to Revelle's AAAS paper, now published in *Oceanography*.[102]

In October, Harvard held a memorial symposium for Revelle, the same month that the vice-presidential debate placed the *Cosmos* dispute in the national light. Originally the organizers had planned to have Singer pres-ent the now-infamous paper, but they'd also invited Walter Munk and the Revelle family. Given their objections to the piece, the organizers removed Singer from the program, hoping to prevent a confrontation. But it didn't work. Singer went anyway.

Walter Munk and Justin Lancaster complained about the *Cosmos* article, Munk apparently in his introductory remarks, and Lancaster in a statement that read in part: "Revelle did not write the *Cosmos* article and was reluctant to join it. Pressured rather unfairly at a very weak moment while recover-ing from heart surgery, Revelle finally gave in to the lead author." The chairman of the symposium allowed Singer to respond. Singer denied having pressured Revelle, insisting that the *Cosmos* paper was based on Revelle's AAAS paper, and he attacked Munk and Lancaster for their "poli-tically inspired misrepresentations."[103]

Singer neglected to mention that the key sentence of the *Cosmos* paper—the one that had been loudly quoted in the press—came from his own 1990 paper. But Singer wasn't content with having made a scene at a sym-posium that was supposed to be celebrating Revelle's life and work. As Lan-caster continued to publicly dispute Revelle's coauthorship of the paper, Singer filed a libel lawsuit against him. Lancaster had little money and fewer resources, but he tried to fight Singer, insisting that the facts were on his side. The only other person who could corroborate Lancaster's account, Revelle's secretary, Christa Beran, did. It wasn't enough. Singer's pockets were deeper than Lancaster's, and in 1994, Lancaster accepted a settlement that forced him to retract his claim that Revelle hadn't really been a coau-thor, put him under a ten-year gag order, and sealed all the court docu-ments.[104] (In 2007, he spoke to us. He now also has a Web site.)[105]

What did Roger Revelle really believe about global warming in 1991? We have looked closely at the records in Revelle's papers at Scripps, and can find only one other statement of his thoughts at the time. It's a short, apparently unpublished, introduction to a November 1990 meeting on cli-mate variability. Revelle wrote:

> There is good reason to expect that because of the increase of
> greenhouse gases in the atmosphere there will be a climate

warming. How big that warming will be is . . . very difficult to say. Probably somewhere between 2 and 5 degrees centigrade at the latitudes of the United States, probably a greater change in average temperature at higher latitudes and a lesser change at lower latitudes . . . Whatever climate change there is will have a profound effect on some aspects of water resources.[106]

The documentary record clearly shows that Roger Revelle did not change his mind. He believed that global warming was coming and it would have serious impacts on water resources. This, of course, is precisely what his colleagues said then and continue to say today. He also believed that the best way to address it was to shift our energy sources. Nowhere did he ever suggest that he considered that a "drastic" action. It seems to us, in fact, that he considered it pretty darn obvious.

The rest of the world did too, as leaders of governments and NGOs made plans to convene in Rio de Janeiro for the U.N. Earth Summit. In June 1992, 108 heads of state, 2,400 representatives of nongovernmental organizations, and more than 10,000 on-site journalists converged in Rio, along with 17,000 other individuals who would convene in a parallel NGO forum, to address the problem of anthropogenic climate change. Yet it was unclear whether President Bush would even attend. At the last minute, President George H. W. Bush flew to Rio de Janeiro to sign the U.N. Framework Convention on Climate Change, which committed its signatories to preventing "dangerous anthropogenic interference in the climate system."[107] President Bush then pledged to translate the written document into "concrete action to protect the planet."[108] By March 1994, 192 countries had signed on to the Framework Convention, and it came into force.

Like the Vienna Convention for the Protection of the Ozone Layer, the U.N. Framework Convention on Climate Change had no real teeth: it set no binding limits on emissions. It was an agreement in principle. Real limits would be determined later, in a protocol that would be eventually signed in Kyoto, Japan. And with the threat that real limitations would soon be enforced, the merchants of doubt redoubled their efforts.

Doubling Down on Denial

Despite the best efforts of Jastrow, Seitz, Nierenberg, and Singer to create doubt, the scientific debate over the detection of global warming was reaching closure. By 1992, Hansen's 1988 claim that warming was detectable no longer seemed bold. It seemed prescient. The only remaining issue really was whether we could prove that the warming was caused by human activities. As scientists had acknowledged many times, there are many causes of climate change, so the key question was how to sort out these various causes. Now that warming had been *detected*, could it be definitively *attributed* to humans?

"Detection and attribution studies" work by considering how warming caused by greenhouse gases might be different from warming caused by the Sun—or other natural forces. They use statistical tests to compare climate model output with real-life data. These studies were the most threatening to the so-called skeptics because they spoke directly to the issue of *causality*: to the social question of whether or not humans were to blame, and to the regulatory question of whether or not greenhouse gases need to be controlled. As these studies began to appear in the peer-reviewed literature, it's not surprising that Singer and his colleagues tried to undermine them. Having taken on the patriarch of climate change research, they went after one of its rising young stars: Benjamin Santer of the Program for Climate Model Diagnosis and Intercomparison at the Lawrence Livermore National Laboratory.

Santer had done his Ph.D. work in the 1980s at the University of East Anglia, England, where he had compared climate model results to observational data, using so-called Monte Carlo methods to make a rigorous statistical analysis.[109] Until this point, model comparisons had been mostly done qualitatively. Scientists looked at maps of model output and compared them to maps of real-life observations to identify similarities and differences. Santer and his Ph.D. supervisor, Tom Wigley (the director of the Climatic Research Unit at East Anglia, U.K.), thought statistical analysis offered more to climate science than such qualitative comparisons. Besides, other parameters—detailed patterns of surface pressure, precipitation, and humidity—might actually provide better tests of the models than global mean temperature. Depending on the driving force—greenhouse gases, volcanic dust, or the Sun—you'd expect different changes in some of these parameters.[110]

After finishing his thesis with Wigley, Santer was invited to the Max Planck Institute for Meteorology in Hamburg. One of the Institute directors was Klaus Hasselmann, a physicist who spent much of his spare time working on unification theory: the effort to merge the four known fundamental forces in the universe into a single field at extremely high energies like those that theoretically existed at the universe's first few moments of existence. This was pretty far from climate science, but Hasselmann had also made a number of major scientific contributions to climate questions. One of those was a paper in 1979 that proposed a new detection and attribution technique called "optimal fingerprinting."[111] The idea was derived from signal processing theory, and the paper was so technical, so elegant, and so laden with dense tensor field mathematics, that Santer at first didn't get it. Santer recalls it as "a thing of beauty. It was many years ahead of its time. I was just too dumb to understand it."[112]

Hasselmann's key insight was that climate scientists faced the same basic problem as communications engineers: how to detect a weak signal—the thing you're interested in—amid lots of noise that you don't care about. In climate science, the noise is caused by phenomena that are internal to the climate system, such as El Niño. The "signal" is something caused by things that are external to the Earth's natural climate system: the Sun, volcanic dust, or man-made greenhouse gases. Engineers had worked for a century to develop mathematical techniques to sort out signals from noise, but they were largely unknown to climate scientists. They also aren't simple to master.

Santer got started, but progress was slow. The results of his Ph.D. thesis hadn't been all that encouraging, either. He and Wigley had shown that some of the models used for the IPCC's first assessment had large errors in surface pressures; it was what scientists call a "negative result." Still, it was important to point out such errors, and based on this and his preliminary work with Hasselmann, he was offered a position at the Climate Model Intercomparison project at the Lawrence Livermore National Laboratory, in California. The program's founder, Lawrence Gates, believed that if models were to be used for policy purposes—and they obviously would be if climate policy were to be based partly on model forecasts—it was important to evaluate them to see whether they were reliable or not. Gates pioneered the idea of "benchmark experiments"—getting climate model centers around the world to perform exactly the same calculation with their models—to permit scientists to rule out differences in model design as an explanation for the differences in model performance. (Model benchmarking was a radical idea at the time; now it is standard procedure.) Gates also argued for making the results of these experiments widely available, so that model diagnosis became an activity of the entire climate science community, not just the responsibility of the modelers themselves—who might not be entirely objective. The lab, in other words, was trying to make modeling more rigorous, more objective, and more transparent.

Santer had the good fortune to arrive at the lab not only in the middle of one of the first major model intercomparison projects, but also at a time when Livermore colleagues Karl Taylor and Joyce Penner were performing an innovative set of climate model experiments that considered not only greenhouses gases, which cause warming, but also sulfate aerosol particles, which generally cause cooling. The Taylor and Penner experiments clearly showed that human influences on climate were complex: changes in CO_2 and sulfate aerosols had distinctly different climate fingerprints.

Fingerprinting proved to be a powerful tool for studying cause-and-effect relationships. Up to that point, much of the scientific argument about the causes of climate change had gone like this: if greenhouse gases increased, then you would expect temperatures to increase, too. They had. So the prediction had come true—textbook scientific method. The problem with the textbook method, however, is that it's logically fallacious. Just because a prediction comes true doesn't mean the hypothesis that generated it is correct. Other causes could produce the same effect. To prove that greenhouse gases had caused climate change, you'd have to find some aspect of it that was *different* than if the cause were the Sun or volcanoes. You needed a pattern that was unique.

We saw in chapter 4 that V. Ramanathan, a prominent atmospheric scientist, had suggested one: the vertical structure of temperature.[113] If warming were caused by the Sun, then you'd expect the whole atmosphere to warm up. If warming were caused by greenhouse gases, however, the effect on the atmosphere would be different, and distinctive. Greenhouse gases trap heat in the lower atmosphere (so it warms up), while the reduced heat flow into the upper atmosphere causes it to cool. Collaborating with colleagues at the Max Planck Institute, and six other research institutions around the world, Santer started to look at the vertical variation of temperature.[114] Before they'd finished the work, Santer was asked to become the convening lead author for "Detection of Climate Change and Attribution of Causes," chapter 8 of the second IPCC assessment.

Nowadays, there's a lot of prestige associated with the IPCC, since they shared the 2007 Nobel Peace Prize, but back in 1994 most scientists considered it a distraction from their "real" work—doing basic research—and Bert Bolin was having a difficult time finding someone to take the lead on the detection and attribution chapter. In the spring of 1994, after some of the other chapters were already started, Tim Barnett of the Scripps Institution of Oceanography called Santer to ask if he'd be willing to do it. Barnett had been one of two lead authors of the equivalent chapter in the First Assessment, and he convinced Santer that it would be a feather in his cap. Santer signed on.

The job of the convening lead author (this position is now called the coordinating lead author) is to produce an assessment of some aspect of climate science based on "the best scientific and technical information available."[115] This involves working together with other "lead authors" and "contributing authors" to agree on the structure and scope of the future chapter. Individual scientists are then assigned the task of drafting different sections of the chapter. Once all sections are drafted, the convening lead author and the lead authors attempt to hammer out a complete draft that's acceptable to the entire group. Santer's chapter ultimately had four lead authors, including his old mentor at East Anglia, Tom Wigley, Tim Barnett, and thirty-two additional contributing authors—in other words, thirty-six of the world's top climate scientists.[116]

The chapter 8 author group met in Livermore, California, in August 1994 to identify the key scientific areas that needed to be addressed. There were a total of twenty participants (from the United States, Canada, the United Kingdom, Germany, and Kenya). After this initial meeting, most of the author group's discussion took place by e-mail. Then, in October through November, Santer attended the first of three so-called drafting sessions, involving the lead authors and the convening lead authors of all chapters of the IPCC Working Group I Report.

The first drafting session convened in Sigtuna, Sweden, and Santer encountered his first challenge: a disagreement over whether the chapter should include a discussion of model and observational uncertainties. Since the topic was covered in other chapters, some authors thought it would be redundant to do it here, but Santer didn't think readers would search other chapters to find it, and in any case his panel would have no control over what was said in those chapters. Santer prevailed, and the published version contained about six pages of discussion of model and observational uncertainties.

Shortly after the Sigtuna meeting, chapter 8 went through an initial round of peer review. The "zeroth" draft was sent out to roughly twenty scientific experts in detection and attribution work, to all scientific contributors to the chapter, and to the lead authors of all other chapters of the report. After updating their chapters in response to the peer review comments, the

IPCC lead authors met for a second drafting session in March 1995 in the British seaside resort of Brighton. In May, a complete draft of the entire IPCC Working Group I Report, as well the Summary for Policymakers, was submitted for full "country review" by the governments participating in the IPCC. The governments chose reviewers—a mixture of scientists and laypeople—who were supposed to provide comments to the lead authors prior to the third drafting session, in Asheville, North Carolina, in July, but because Santer had been chosen so late as the convening lead author, this schedule didn't quite work out for his group. Santer arrived in Asheville having yet to receive the government reviewers' comments.

At the Asheville meeting, Santer presented the results of his fingerprint study of changes in the vertical structure of atmospheric temperatures, which by this point had been submitted to *Nature*.[117] One scientist present at the meeting reported that Santer's presentation electrified the audience; it was "mind-boggling to a lot of the scientists there."[118] It looked like Santer and his colleagues might just have proved the human impact on climate.

After Asheville, all chapters were revised in response to the country review—all, that is, except chapter 8, because Santer was still awaiting the comments. The final stage in the process was the IPCC plenary meeting, scheduled to start in Madrid on November 27. In October, drafts of the Working Group I Report and the Summary for Policymakers had been sent to all the government delegates to the Madrid meeting. When Santer arrived at the Madrid meeting, he was handed a sheaf of comments— including comments from the U.S. government—that he had never seen before.

Meanwhile, sometime in September, a draft of the entire Working Group I Report was leaked. The central message of chapter 8, that the anthropogenic fingerprint had been found, drew widespread attention.[119] "In an important shift of scientific judgment, experts advising the world's governments on climate change are saying for the first time that human activity is a likely cause of the warming of the global atmosphere," the *New York Times* declared on its front page. This, of course, wasn't quite right. Scientists had been saying for a long time that human activity was a *likely* cause of warming. They were now saying that it was *demonstrated*. The *New York Times* didn't get it. But the skeptics did, and they went on the attack.

Two weeks before the plenary session in Madrid, the Republican majority in the U.S. Congress launched a preemptive strike. In a set of hearings held in November, they repeatedly questioned the scientific basis for concern. The star witness was another well-known contrarian, Patrick J.

Michaels, who had completed his Ph.D. at the University of Wisconsin–Madison in 1979, building models relating climate change to crop yields. In 1980, he was appointed state climatologist of Virginia by Republican governor John Dalton (although many years later Michaels was forced to forego that title when it was shown that Dalton had acted without legal authority).[120] In the 1980s, Michaels had published scientific work on the climate sensitivity of various crops and ecosystems, but by the early 1990s, he was mainly known not for mainstream science, but his contrarian views.[121] He had joined Fred Singer in publicly attacking the mainstream view of ozone depletion in a series of columns in the *Washington Times*.[122] He produced a quarterly newsletter called the *World Climate Review*, funded at least in part by fossil fuel interests, and used it as a platform to attack mainstream climate science. The *Review* was circulated free to members of the Society for Environmental Journalism, ensuring that its claims got wide attention.[123] In the early 1990s, he had worked as a consultant to the Western Fuels Association—a coal mining industry group—to promote the idea that burning fossil fuels was good, because it would lead to higher crop yields as increased atmospheric CO_2 led to increased photosynthesis and therefore increased agricultural productivity.[124]

In the Republican hearings, Michaels was presented as an expert who somehow knew more than all the scientists working within the IPCC umbrella. His own personal analysis of the difference between a model prediction of greenhouse gas–induced warming and atmospheric temperatures derived from NOAA's weather satellites showed, he claimed, that the IPCC climate models had heavily overpredicted global warming and could not be trusted. He complained in the hearing that while he'd made many critical comments on the various chapters of the IPCC report, his comments had been ignored, resulting "in not one discernable change in the text of the IPCC drafts."[125]

Congressman George E. Brown Jr. of California asked Jerry Mahlman, director of NOAA's Geophysical Fluid Dynamics Laboratory (GFDL), to respond to Michaels's claims. The particular model study that Michaels attacked was the work of GFDL scientist Syukuro Manabe—probably the world's most respected climate modeler—the man who, along with Jim Hansen, had presented his work to the Charney committee back in 1979. Mahlman explained that Michaels's analysis contained an elementary flaw. Manabe's study was designed to investigate the impact of CO_2 on climate, and had deliberately omitted other factors—including volcanic dust. However, there had been a set of large volcanic eruptions in the early 1990s, most famously Mt. Pinatubo in 1992. The satellite measurements obviously did incorporate these other real world phenomena, so naturally, they'd be different from the model results.

"The bottom line," Mahlman concluded, "is that there is no logical basis for a direct comparison of this GFDL model experiment with that of [satellite] data sets or any other data set."[126] A legitimate comparison between models and observations could only be carried out when the models and observations examined the same things. It was obvious why the IPCC had ignored Michaels's complaints.

The hearing wasn't very successful at getting press attention, receiving no notice from the *New York Times*, the *Washington Post*, or even the *Washington Times*. Among major newspapers, only the *Boston Globe* seems to have bothered covering it. It wasn't exactly news by late 1995 that the Republican congressional leadership opposed environmental protection; there had been discussion that year of repealing the Clean Water Act, one of the cornerstones of American environmental improvement. But the lack of press attention didn't matter; the hearing had the desired effect of reinforcing the Republican majority's do-nothing attitude. Writing to Fred Seitz after the hearing, Nierenberg said, "I doubt that Congress will do anything foolish. I can also tell you that at least one high-level corporate advisor is advising boards that the issue is politically dead. Happy holiday."[127]

Santer presented the findings in chapter 8 on November 27, 1995, the first day of the plenary session (and the same day Nierenberg proclaimed the issue politically dead in his letter to Seitz). The chapter was immediately opposed by the Saudi Arabian and Kuwaiti delegates. In the words of the *New York Times*'s reporter, these oil-rich states "made common cause with American industry lobbyists to try to weaken the conclusions emerging from Chapter 8."[128] The lone Kenyan delegate, Santer remembers, "thought there should not be a detection and attribution chapter at all."[129] Then the chairman of a fossil fuel industry group, the Global Climate Coalition, and automobile industry representatives monopolized the rest of the afternoon.[130] Finally the IPCC chairman, Britain's Sir John Houghton, closed the discussion and appointed an ad hoc drafting group to work out the disagreements and to address all of the late government comments. The working group included the lead authors, and delegates from the United States, Britain, Australia, Canada, New Zealand, the Netherlands, Saudi Arabia, Kuwait, and the lone Kenyan.

A portion of the ad hoc group hammered out an acceptable language. Steve Schneider convinced the Kenyan that there really was a scientific basis for the chapter's central conclusion that anthropogenic climate change had been detected.[131] But the Saudis never sent a representative to the ad hoc sessions, and when Santer presented the revised draft, the Saudi head delegate protested all over again. A bit of a shouting match ensued, and Houghton had to intervene, effectively tabling the issue while the working group finished negotiating the Summary for Policymakers. There the entire issue boiled down to a single sentence, in fact a single *adjective*, drawn from Santer's chapter: "The balance of evidence suggests that there is a [blank] human influence on global climate."[132]

What should the adjective be? Santer and Wigley wanted "appreciable." This was unacceptable to the Saudi delegate, but it was too strong for Bert Bolin, too. One participant recalls the group trying about twenty-eight different words before Bolin suggested "discernible." That clicked, and the outcome of the Madrid meeting was this sentence: "The balance of evidence suggests that there is a discernible human influence on global climate."[133] This line would be quoted repeatedly in the years to come.

With the Summary for Policymakers settled, the individual chapters had to be revised in the light of all the late review comments, and Houghton instructed the lead authors to make the necessary changes after the meeting.[134] Santer went from Madrid to the Hadley Center in Bracknell, England, where he made the changes in long-distance collaboration with Wigley and Barnett. The most significant of these changes was structural. The draft chapter 8 had summary statements at both beginning and end of the chapter, but none of the other chapters did. They only had summaries at the beginning. Therefore, Santer had been instructed to remove the summary statement at the end of the chapter so that it would have the same structure as the rest of the chapters. That, Santer remembered years later, was a fateful decision, as critics would later attack him for "removing material."[135]

Then Fred Singer launched an attack. In a letter to *Science* on February 2, 1996, four months before formal release of the Working Group I Report, Singer presented a litany of complaints. The Summary for Policymakers, he claimed, ignored satellite data that showed "no warming at all, but actually a slight cooling." On this basis he claimed that the climate models, which all showed warming, were wrong. The IPCC had violated one of its "major rules" by including the fingerprinting work, because "the research had not yet, to my knowledge, appeared in the peer-reviewed literature." The panel had also ignored an "authoritative U.S. government report" that had found the twenty-first-century warming might be as little as 0.5°C, making global warming a nonproblem. (Singer didn't cite the report.)

Finally, he concluded, "The mystery is why some insist in making it into a problem, a crisis, or a catastrophe—'the greatest global challenge facing mankind.' "[136]

Tom Wigley responded to Singer's criticisms in March. Rejecting the "no warming" claim entirely, he simply stated, "This is not supported by the data; the trend from 1946 to 1995 is .3 C. As shown in chapter 8 of the full report (figure 8.4) there is no inconsistency between the observed temperature record and model simulations." There were some differences between measurements made with satellites and measurements made with "radiosondes"—instruments on balloons, with radios attached to transmit the results—but climate scientists didn't expect them to perfectly track each other; the reasons were explained in both chapters 3 and 8. "There are good physical reasons to expect differences between these two climate indicators," Wigley noted, because they were in different places measuring somewhat different things.

The claim that the pattern recognition studies violated the IPCC's rules was wrong on two counts. First, Wigley explained, the IPCC allowed use of material from outside the peer-reviewed journals as long as it was accessible to reviewers. This was to ensure the report was "up to date" when published. Moreover, the specific work Singer referred to "on the increasing correlation between the expected greenhouse-aerosol pattern and observed temperature changes, *is* in the peer-reviewed literature."[137]

Moreover, Singer was again creating a straw man. "Singer refers to the [Summary for Policymakers] as saying that global warming is 'the greatest global challenge facing mankind,' " Wigley and his coauthors wrote. "We do not know the origin of this statement—it does not appear in any of the IPCC documents. Further, it is the sort of extreme statement that most involved with the IPCC would not support."[138]

Wigley was right. The IPCC had not described global warming as the "greatest global challenge facing mankind." The words Singer attributed to the IPCC don't appear in either the Working Group I Report or in its Summary for Policymakers. Singer was putting words into other people's mouths—and then using those words to discredit them.

The IPCC had in fact bent over backward *not* to use alarmist terms. Bert Bolin had deliberately imposed a policy of extreme conservatism of language; witness his rejection of "appreciable" in favor of "discernible." The opposition of the Saudi and Kuwaiti delegations had ensured only least common denominator statements. Everyone involved had seen how the process led to a conservative estimation of the threat. What was Singer's response to this refutation of his allegations? He provided the missing citation for his claim that there would be only a 0.5°C warming in the twenty-first century.[139]

The IPCC had contracted with Cambridge University Press to publish the Working Group 1 Report, scheduled to appear in the United States in June 1996. In May, Santer and Wigley presented their chapter at a briefing in the Rayburn House Office Building on Capitol Hill, organized by the American Meteorological Society and the U.S. Global Change Research Program. The two scientists were now challenged by William O'Keefe of the American Petroleum Institute and by Donald Pearlman, an industry lobbyist and registered foreign agent of several oil-producing nations.[140] O'Keefe and Pearlman accused them of "secretly altering the IPCC report, suppressing dissent by other scientists, and eliminating references to scientific uncertainties."[141]

"Who made these changes to the chapter? Who authorized these changes? Why were they made?" Pearlman demanded. "Pearlman got up and in my face, turned beet red and [started] screaming at me," Santer recalls. AMS officer Anthony Socci "finally separated us, but Pearlman kept following me around."[142] Santer explained that he'd been required by IPCC procedures to make the changes in response to the government comments and discussions at Madrid, and the chapter had never been out of his control, but the truth did not satisfy the opposition.[143]

The Global Climate Coalition meanwhile had circulated a report entitled "The IPCC: Institutionalized Scientific Cleansing" to reporters, members of Congress, and some scientists. By chance, anthropologist Myanna Lahsen interviewed Nierenberg about his "skepticism" about global warming two weeks before the Working Group I Report was published, and found that he had a copy of the coalition report. He had evidently accepted its veracity, even though there was no way to compare its claims against the real chapter 8 (since the latter had not yet been released). He quoted its claims to Lahsen, telling her that the revisions had "just altered the whole meaning of the document. Without permission of the authors." Moreover, he claimed, "Anything that would imply the current status of knowledge is so poor that you can't do anything is struck out."[144] That was hardly true; Santer's panel had included six pages of discussion of uncertainty in the final text. But Bill Nierenberg knew all about altering scientific reports for political reasons, so perhaps he followed the adage that the best defense is offense. Or perhaps he was guilty of "mirror imaging," as Team B had accused the CIA of in 1976: assuming that his opponents thought and operated the way he did.

Then Fred Seitz took the attack to the national media. In a letter published in the *Wall Street Journal* on June 12, 1996, he accused Ben Santer of fraud. "In my more than 60 years as a member of the American scientific community, including my services as president of the National Academy of Sciences and the American Physical Society, I have never witnessed a more disturbing corruption of the peer-review process than the events that led to this IPCC report." Seitz repeated the Global Climate Coalition's charges that unauthorized changes to chapter 8 had been made after its acceptance in Madrid. "Few of these changes were merely cosmetic; nearly all worked to remove hints of the skepticism with which many scientists regard claims that human activities are having a major impact on climate in general and on global warming in particular," Seitz claimed. If the IPCC couldn't follow its own procedures, he concluded, it should be abandoned and governments should look for "more reliable sources of advice to governments on this important question."[145] Presumably, he meant the George C. Marshall Institute, of which he was still chairman of the board.

Santer immediately drafted a letter to the *Journal*, which forty of the other IPCC lead authors signed. Santer explained what had happened, how he had been instructed by Houghton to make the changes, and why the changes were late in coming. At first the *Journal* wouldn't publish it. After three tries, Santer finally got a call from the *Journal*'s letters editor, and the letter was finally published on June 25. Santer's reply had been heavily edited, and the names of the forty other cosigners deleted.

What the *Journal* allowed Santer to say was that he had been required to make the changes "in response to written review comments received in October and November 1995 from governments, individual scientists, and non-government organizations during plenary sessions of the Madrid meeting." This was peer review—the very process that Seitz, as a research scientist, had been a part of all his life. Only it was extended to include comments and queries from governments and NGOs as well as scientific experts. But the changes didn't affect the bottom line conclusion.

Santer also pointed out that Seitz wasn't a climate scientist, hadn't been involved in creating the IPCC report, hadn't attended the Madrid meeting, and hadn't seen the hundreds of review comments to which Santer had to respond. In other words, his claims were just hearsay.[146]

Bert Bolin and Sir John Houghton also responded with a long letter defending Santer and the IPCC process. "Frederick Seitz's article is completely without foundation," they replied unequivocally. "It makes serious allegations about the Intergovernmental Panel on Climate Change and about the scientists who have contributed to its work which have no basis in fact. Mr. Seitz does not state the source of his material, and we note for the record that he did not check his facts either with the IPCC officers or with any of the scientists involved."[147]

Well, that's what they'd wanted it to say, but the *Journal* edited that statement out, too, along with three more paragraphs explaining the drafting process in some detail. The *Journal* allowed them to say only that:

> ... [in] accordance with IPCC Procedures, the changes to the draft of Chapter 8 were under the full scientific control of its convening Lead Author, Benjamin Santer. No one could have been more thorough and honest in undertaking that task. As the responsible officers of the IPCC, we are completely satisfied that the changes incorporated in the revised version were made with the sole purpose of producing the best possible and most clearly explained assessment of the science and were not in any way motivated by any political or other considerations.[148]

We know how the *Journal* edited the letters because Seitz's attack and the *Journal*'s weakening of the response so offended the officials of the American Meteorological Society and of the University Corporation for Atmospheric Research that their boards agreed to publish an "Open Letter to Ben Santer" in the *Bulletin of the American Meteorological Society*, where they republished the letters in their entirety, showing how the *Journal* had edited them. They voiced their support of Santer and the effort it had taken all the authors to put the report together, and categorically rejected Seitz's attack as having "no place in the scientific debate about issues related to global change."[149] They began, finally, to realize what they were up against.

> [There] appear[ed] to be a concerted and systematic effort by some individuals to undermine and discredit the scientific process that has led many scientists working on understanding climate to conclude that there is a very real possibility that humans are modifying Earth's climate on a global scale. Rather than carrying out a legitimate scientific debate through the peer-reviewed

literature, they are waging in the public media a vocal campaign against scientific results with which they disagree.[150]

But the attack was far from over. On July 11, the *Wall Street Journal* published three more letters reprising the charges, one from Fred Seitz, one from Fred Singer, and one from Hugh Ellsaesser. (Ellsaesser was a retired geophysicist at the Lawrence Livermore National Laboratory who previously had questioned the evidence of the ozone hole. He served in the mid-1990s on the Marshall Institute's Scientific Advisory Board, and in 1995 wrote a report for the Heartland Institute on *The Misuse of Science in Environmental Management*.) Singer and Seitz simply repeated the charges they'd already made, but Singer also took the opportunity to turn the IPCC's caution against it. The IPCC had bent over backward to be judicious, arguing at length to choose just the right, reasonable adjective—"discernible." Singer dismissed the IPCC conclusion as "feeble," at the same time insisting illogically that it was being used to frighten politicians into believing that a climate catastrophe is about to happen.[151]

Santer and Bolin responded a second time to the attacks in letters the *Journal* published July 23—prompting another attack by Singer.[152] This time, the *Journal* wouldn't publish it, and Singer circulated it by e-mail instead. Santer responded by e-mail, too. There was, Singer maintained, no "evidence for a current warming trend." According to Singer, chapter 8 had been based primarily on Santer's "unpublished work," and the panel should have included as a lead author "Professor Patrick J. Michaels, who, at the time, had published the only refereed paper on the subject" of climate fingerprinting. And he repeated the charge of "scientific cleansing." Santer rejected all of Singer's charges. Chapter 8 was based on more than 130 references, not just Santer's two papers. The claim that Michaels had published the only "refereed paper on the subject" of pattern-based recognition before mid-1995 was incorrect: Hasselmann's theoretical paper on the subject was published in 1979, and Tim Barnett and Mike Schlesinger had published a "real-world" fingerprint study as early as 1987. Michaels *had* been invited to be a contributing author to chapter 8 but had refused. Finally, Santer noted, chapter 8 contained several paragraphs discussing Michaels's paper, but when Wigley had approached Michaels for comments, "Prof. Michaels did not respond."[153]

Singer's claims were not only false, but had been *shown* to be false. Still, he wasn't finished repeating them. Now he would claim that Fred Seitz was the real victim of the whole affair.

In November, Singer penned an article for the *Washington Times* entitled "Global Warming Disinformation?" By this time, the IPCC report had
been published and available for months, so Singer could have seen for
himself that chapter 8 contained six pages of discussions of model and observational uncertainties, as Santer had insisted it should all along. Still,
Singer repeated the claim that chapter 8 had been edited to remove uncertainties, and then asserted that "Seitz, one of the nation's most respected
scientists, was attacked for factually reporting the revisions made by the
IPCC leadership, which clearly affected the sense of the report!"[154]

Joined by Bill Nierenberg, Patrick Michaels, and a new ally—MIT meteorologist Richard Lindzen—Singer then attacked the AMS/UCAR Open
Letter. After repeating the refuted charges of "substantial and substantive"
deletions of uncertainty, Singer cast the deletions as a conspiracy that Santer
was now trying to cover up. "Santer . . . has not been forthcoming in reveal-
ing who instructed him to make such revisions and who approved them after they were made. He has, however, told others privately that he was asked
[prevailed upon?] to do so by IPCC co-chairman John Houghton." To Singer
and his co-authors on the letter, this was evidence of political meddling in
the chapter. He continued, "You may not have seen the 15 November [1995]
letter from the State Department instructing Dr. Houghton to 'prevail
upon' chapter authors 'to modify their texts in an appropriate manner following discussion in Madrid.' "[155] Singer's presentation of it as some sort of
clandestine conspiracy was absurd: Bolin and Houghton had already identified themselves months before as the source of Santer's instructions.

In her 1999 analysis, Myanna Lahsen pinned Singer's efforts to "en-
velop the IPCC in an aura of secrecy and unaccountability" to a common
American conservative rhetoric of political suppression.[156] As we have
seen in previous chapters, if anyone was meddling in the scientific assessment and peer review process, it was the political right wing, not the left.
It wasn't the Sierra Club that tried to pressure the National Academy of
Sciences over the 1983 Carbon Dioxide Assessment; it was officials from
the Department of Energy under Ronald Reagan. It wasn't Environmental
Defense that worked with Bill Nierenberg to alter the Executive Summary
of the 1983 Acid Rain Peer Review Panel; it was the White House Office of
Science and Technology Policy. And it was the *Wall Street Journal* spread-
ing the attack on Santer and the IPCC, not *Mother Jones*.

The over-the-top attacks on Santer began to have consequences for
Nierenberg. In April, Nierenberg had invited Tom Wigley to a conference
he wanted to hold at Scripps that November on the costs and benefits of

global warming, but Wigley smelled a rat. "I have decided to withdraw from your November meeting," he wrote. "The reason for this is the letter you co-signed which appeared in BAMS [*Bulletin of the American Meteorological Society*]. I have no desire to cooperate with anyone who endorses such an unmitigated collection of distortions and misinformation."[157]

Nierenberg tried flattery to keep Wigley on board. "The personally difficult part for me is that your work, Klaus [Hasselmann's] work, and [Bill] Nordhaus' work have had the most influence (and still do) on my thinking." He lamented the rift that was developing in the climate science community over the ongoing public attacks, but then followed Singer's lead in imputing conspiracy, this time in the scientific journals. "I remind you in this instance of something that touched you personally about which I only had the slightest information from the gossip columns and some hallway talk. I was told that you faced great opposition in getting your *Nature* paper published. That great pressure was put on you."[158]

Wigley evidently had no idea what *Nature* paper Nierenberg was talking about. "It seems that you have not only NOT been influenced by, but actually disagree with (or are unaware of) the vast bulk of my scientific work: in particular the work on detection, which the BAMS letter you co-signed has unfairly, unjustifiably, unscientifically and incorrectly criticized." Wigley also rejected the imputation that *Nature* had pressured him. "To which paper are you referring? I have published 22 papers in that particular journal. No matter which, you shouldn't take any notice of what you hear in 'gossip columns and hallway talk.'" He concluded, "So Bill, what I said in my previous email stands. Your 17 April 'response' gives me no reason to change my mind—just the opposite. The BAMS letter makes it quite clear that you think my IPCC detection work with Ben Santer was distorted for political motives. I am surprised, therefore, that you would even want a person like me to attend a meeting of yours. I still think you are being duplicitous, and I still suspect your motives."[159]

Wigley wasn't the only one to begin to understand what Nierenberg was really up to. Klaus Hasselmann also wrote to Nierenberg: "I have followed the attacks on Ben Santer during the last year and found them to be grossly unfair and clearly politically motivated. In a letter I wrote to the *Wall Street Journal* (which was not published, with many other similar letters) I pointed out that it was ridiculous to imply that the conclusions of Chapter 8 had been willfully or unintentionally altered against the will of the Madrid delegates."[160] Hasselmann was still willing to come to a meeting about the costs and benefits of global warming—a subject that interested him greatly—but he wouldn't come to a meeting with a political agenda. "In view of the pronounced political colouring of the BAMS letter I am not convinced at this point that the concerns of Tom Wigley are not justified."[161]

Perhaps after so many years as Svengali, Bill Nierenberg did not realize that this time he had gone too far. Nierenberg, despite his intellect, really didn't seem to understand that by participating in this assault on Ben Santer, he was attacking the entire community of climate modelers. By signing on to Singer's letter, he marked himself in their eyes as a political actor, not a scientific one. Nierenberg's comment that he feared the polarization of the community was both perceptive and blinkered; the climate science community was most definitely becoming polarized, but it was due to his own actions, and those of a small network of doubt-mongers.

WE MIGHT DISMISS this whole story as just infighting within the scientific community, except that the Marshall Institute claims were taken seriously in the Bush White House and published in the *Wall Street Journal*, where they would have been read by millions of educated people. Members of Congress also took them seriously. Proposing a bill to reduce climate research funding by more than a third in 1995, Congressman Dana Rohrabacher called it "trendy science that is propped up by liberal/left politics rather than good science."[162] In July 2003, Senator James Inhofe called global warming "the greatest hoax ever perpetrated on the American people."[163] As late as 2007, Vice President Richard Cheney commented in a television interview, "Where there does not appear to be a consensus, where it begins to break down, is the extent to which that's part of a normal cycle versus the extent to which it's caused by man, greenhouse gases, et cetera"—exactly the question Santer had answered a decade before.[164] How did such a small group come to have such a powerful voice?

We take it for granted that great individuals—Gandhi, Kennedy, Martin Luther King—can have great positive impacts on the world. But we are loath to believe the same about negative impacts—unless the individuals are obvious monsters like Hitler or Stalin. But small numbers of people can have large, negative impacts, especially if they are organized, determined, and have access to power.

Seitz, Jastrow, Nierenberg, and Singer had access to power—all the way to the White House—by virtue of their positions as physicists who had won the Cold War. They used this power to support their political agenda, even though it meant attacking science and their fellow scientists, evidently believing that their larger end justified their means. Perhaps this, too, was part of their professional legacy. During the Manhattan Project, and throughout the Cold War, for security reasons many scientists had to hide the true nature of their work. All weapons projects were secret, but so were many other projects that dealt with rocketry, missile launching and targeting, navigation, underwater acoustics, marine geology, bathymetry, seismology, weather modification; the list goes on and on.[165] These secret projects frequently had "cover stories" that scientists could share with colleagues, friends, and families, and sometimes the cover stories were true in part. But they weren't the whole truth, and sometimes they weren't true at all. After the Cold War, most scientists were relieved to be freed of the burdens of secrecy and misrepresentation, but Seitz, Singer, and Nierenberg continued to act as if the Cold War had not ended.

Whatever the reasons and justifications of our protagonists, there's another crucial element to our story. It's how the mass media became complicit, as a wide spectrum of the media—not just obviously right-wing newspapers like the *Washington Times*, but mainstream outlets, too—felt obligated to treat these issues as scientific controversies. Journalists were constantly pressured to grant the professional deniers equal status—and equal time and newsprint space—and they did. Eugene Linden, once an environment reporter for *Time* magazine, commented in his book *Winds of Change* that "members of the media found themselves hounded by experts who conflated scientific diffidence with scientific uncertainty, and who wrote outraged letters to the editor when a report didn't include their dissent."[166] Editors evidently succumbed to this pressure, and reporting on climate in the United States became biased *toward* the skeptics and deniers because of it.

We've noted how the notion of balance was enshrined in the Fairness Doctrine, and it may make sense for political news in a two-party system (although not in a multiparty system). But it doesn't reflect the way science works. In an active scientific debate, there can be many sides. But once a scientific issue is closed, there's only one "side." Imagine providing "balance" to the issue of whether the Earth orbits the Sun, whether continents move, or whether DNA carries genetic information. These matters were long ago settled in scientists' minds. Nobody can publish an article in a scientific journal claiming the Sun orbits the Earth, and for the same reason, you can't publish an article in a peer-reviewed journal claiming there's no global warming. Probably well-informed professional science journalists wouldn't publish it either. But ordinary journalists repeatedly did.

In 2004, one of us showed that scientists had a consensus about the reality of global warming and its human causes—and had since the mid-1990s. Yet throughout this time period, the mass media presented global warming and its cause as a major debate. By coincidence, another study also published in 2004 analyzed media stories about global warming from 1988 to 2002. Max and Jules Boykoff found that "balanced" articles—ones that gave equal time to the majority view among climate scientists as well as to deniers of global warming—represented nearly 53 percent of media stories. Another 35 percent of articles presented the correct majority position among climate scientists, while still giving space to the deniers.[167] The authors conclude that this "balanced" coverage is a form of "informational bias," that the ideal of balance leads journalists to give minority views more credence than they deserve.

This divergence between the state of the science and how it was presented in the major media helped make it easy for our government to do nothing about global warming. Gus Speth had thought in 1988 that there was real momentum toward taking action. By the mid-1990s, that policy momentum had not just fizzled; it had evaporated. In July 1997, three months before the Kyoto Protocol was finalized, U.S. senators Robert Byrd and Charles Hagel introduced a resolution blocking its adoption.[168] Byrd-Hagel passed the Senate by a vote of 97–0. Scientifically, global warming was an established fact. Politically, global warming was dead.

CHAPTER 7

Denial Rides Again: The Revisionist Attack on Rachel Carson

RACHEL CARSON IS AN AMERICAN HERO—the courageous woman who in the early 1960s called our attention to the harms of indiscriminate pesticide use. In *Silent Spring*, a beautiful book about a dreadful topic, Carson explained how pesticides were accumulating in the food chain, damaging the natural environment, and threatening even the symbol of American freedom: the bald eagle. Although the pesticide industry tried to paint her as a hysterical female, her work was affirmed by the President's Science Advisory Committee, and in 1972, the EPA concluded that the scientific evidence was sufficient to warrant the banning of the pesticide DDT in America.

Most historians, we included, consider this a success story. A serious problem was brought to public attention by an articulate spokesperson, and, acting on the advice of acknowledged experts, our government took appropriate action. Moreover, the banning of DDT, which took place under a Republican administration, had widespread public and bipartisan political support.[1] The policy allowed for exceptions, including the sale of DDT to the World Health Organization for use in countries with endemic malaria, and for public health emergencies here at home. It was sensible policy, based on solid science.

Fast-forward to 2007. The Internet is flooded with the assertion that Carson was a mass murderer, worse than Hitler. Carson killed more people than the Nazis. She had blood on her hands, posthumously. Why? Because *Silent Spring* led to the banning of DDT, without which millions of Africans died of malaria. The Competitive Enterprise Institute—whom we encountered in previous chapters defending tobacco and doubting the reality of global warming—now tells us that "Rachel was wrong." "Millions of people around the world suffer the painful and often deadly effects of malaria because one person sounded a false alarm," their site asserts. "That person is Rachel Carson."[2]

Other conservative and Libertarian think tanks sound a similar cry. The American Enterprise Institute argues that DDT was "probably the single most valuable chemical ever synthesized to prevent disease," but was unnecessarily banned because of hysteria generated by Carson's influence.[3] The Cato Institute tells us that DDT is making a comeback.[4] And the Heartland Institute posts an article defending DDT by Bonner Cohen, the man who created EPA Watch for Philip Morris back in the mid-1990s.[5] (Heartland also has extensive, continuing programs to challenge climate science.)[6]

The stories we've told so far in this book involve the creation of doubt and the spread of disinformation by individuals and groups attempting to prevent regulation of tobacco, CFCs, pollution from coal-fired power plants, and greenhouse gases. They involve fighting facts that demonstrate the harms that these products and pollutants induce in order to stave off regulation. At first, the Carson case seems slightly different from these earlier ones, because by 2007 DDT had been banned in the United States for more than thirty years. This horse was long out of the barn, so why try to reopen a thirty-year-old debate?

Sometimes reopening an old debate can serve present purposes. In the 1950s, the tobacco industry realized that they could protect their product by casting doubt on the science and insisting the dangers of smoking were unproven. In the 1990s, they realized that if you could convince people that science in general was unreliable, then you didn't have to argue the merits of any particular case, particularly one—like the defense of second-hand smoke—that had no scientific merit. In the demonizing of Rachel Carson, free marketeers realized that if you could convince people that an example of successful government regulation wasn't, in fact, successful— that it was actually a mistake—you could strengthen the argument against regulation in general.

Silent Spring *and the President's Science Advisory Committee*

DDT was invented in 1873, but got little attention until 1940, when Swiss chemist Paul Müller, working for a Swiss chemical firm, resynthesized it.

Field trials demonstrated its efficacy against numerous pests, including mosquitoes and lice, leading to the realization that DDT could be used to stop the spread of deadly insect-borne diseases like malaria and typhus.[7] The timing was fortunate, because supplies of the pesticide conventionally used against lice—pyrethrum, derived from chrysanthemums—were in short supply and wartime demand was great. In the latter part of World War II, DDT was widely used in Italian and African campaigns, as well as in some parts of the Pacific. Military strategists credited it with saving many lives.[8]

DDT seemed to be a miracle chemical. It killed insects immediately and almost entirely, yet seemed to have no adverse effects on the troops. It was easy to use: soldiers could apply it to their skin and clothing, or it could be mixed with oils and sprayed from airplanes. And it was cheap. In 1948 Müller was awarded the Nobel Prize for Physiology or Medicine for the value of DDT in disease control.[9]

After the war, DDT use expanded, particularly in agriculture. DDT was clearly less immediately toxic than the arsenic-based pesticides that had been previously widely used, and spraying from airplanes was much less expensive than the older methods of disease eradication, such as draining swamps, eliminating sources of open water near buildings, and clearing brush.[10] Across America, pest control districts switched to spraying. State and local governments began using it too, and even ordinary homeowners. Farmers began to use DDT as the U.S. government sold surplus warplanes cheaply and farmers turned them into crop dusters.[11]

Everyone believed that DDT was safe. One documentary from the period shows schoolchildren happily eating their lunches at picnic benches as DDT is sprayed around them.[12] But adverse effects were starting to be noticed. Among the first to recognize damage were biologists at the U.S. Fish and Wildlife Service, where Carson, a biologist, had worked. As she began to investigate, she found that there were numerous case reports of damage to birds and fish after DDT application. There was also some circumstantial evidence that DDT and other widely used pesticides might be doing harm to humans, too. But as with the early evidence of acid rain, most of these descriptions had been published in obscure places, in reports of the Fish and Wildlife Service or specialized journals of wildlife biology. Few people knew about any of this until Carson began to write about it.

. . .

CARSON WAS AN eloquent writer who had already achieved success and the respect of the scientific community with her earlier book, *The Sea Around Us*. As *Silent Spring* neared completion, it was serialized in the *New Yorker*, so by the time it was published in 1962, its basic message was already out: DDT, the supposed miracle chemical, was no miracle at all.

Carson documented at great length both the anecdotal and systematic scientific evidence that DDT and other pesticides were doing great harm.[13] She reported on death to fish in regions that had been sprayed for pest control, on birds dying on college campuses and in suburban neighborhoods, and on spraying campaigns in Michigan and Illinois that had destroyed squirrel populations and the pets of people unfortunate enough to have been outside during the spraying or that had gone out soon after. The pesticides destroyed beneficial species, too. Spraying DDT in New Brunswick to save evergreens from a budworm infestation destroyed the bugs upon which local salmon relied, and the fish starved. DDT also killed useful insects, vital to pollinating flowers and food crops.

Silent Spring wasn't just about DDT—it was about the indiscriminate use of pesticides in general—but DDT was a particular focus for Carson, as it was for her biology colleagues, because of the evidence of bioaccumulation. Other pesticides broke down quickly in the natural environment, but DDT was very persistent, accumulating up the food chain. Because it was so long lasting, it continued to be concentrated in the tissues of the animals and insects that it didn't kill—long after spraying campaigns were over—so when those animals were eaten, the effects rippled through the ecosystem. One of its most alarming effects—interference in the reproductive systems of eagles and falcons—occurred not by direct exposure, but by those predators eating small rodents that had eaten things with DDT in or on them.

Precisely because DDT was so effective, it unbalanced ecosystems. During spraying to prevent the spread of Dutch elm disease by beetles, DDT accelerated the beetles' spread by destroying the natural predators that previously helped to keep those beetles in check.[14] Spraying in the Helena National Forest to protect trees from budworms caused an outbreak of the spider mite, which further damaged the trees. (It also hurt birds that depended on the forest's insect population.)[15] Carson remarked that populations recovered in one portion of the region because it was only sprayed once in a single year; other parts of the region had experienced continual spraying, and populations in those areas didn't recover.

What about people? The two other most commonly sprayed insecticides, aldrin and dieldrin, were already known to be toxic to humans and other mammals at high doses; so it was reasonable to suppose that DDT might show similar effects. Laboratory rats fed DDT had smaller litters and higher infant mortality than control subjects. Even if DDT were perfectly safe to people in the short run, it might not be in the long run.

Historians have suggested that *Silent Spring* was to environmentalism what *Uncle Tom's Cabin* was to abolitionism: the spark for a new public consciousness.[16] Yet almost as soon as *Silent Spring* came out, the pesticide industry went on the attack. They called Carson hysterical and emotional. They claimed that the science behind her work was anecdotal, unproven, inadequate, and wrong. They threatened Carson's publisher with lawsuits.[17]

Of course, not all scientists agreed with Carson, particularly chemists, who tended to believe pesticides were safe if used properly, and food scientists who appreciated the value of DDT in improving agricultural productivity. One of these skeptics was Emil Mrak, chancellor of the University of California, Davis, who testified to the U.S. Congress that Carson's conclusion that pesticides were "affecting biological systems in nature and may eventually affect human health [was] contrary to the present body of scientific knowledge."[18] Most biologists did not agree with Mrak, however, and the personal attacks on Carson backfired. The publicity and furor caused sales of *Silent Spring* to skyrocket, while the obvious sexism of calling a highly trained biologist and world-class writer "hysterical"—in the age of rising feminist consciousness—led many to rally to her defense. Even President John Kennedy spoke in reverent tones of "Miss Carson's work."[19]

But what about the science? *Silent Spring* was well written, but did Carson have the science right? To answer that question, President Kennedy turned to the leading group of scientific experts in America at that time—the President's Science Advisory Committee (PSAC, pronounced peasack). Established in the 1950s, and mostly populated by physicists, PSAC had mainly considered issues related to nuclear weapons and warfare, but in 1962 the president asked his advisors to guide him on DDT.

In the early 1960s, few systematic studies of the cumulative environmental effects of DDT had been done, in part because DDT had been used primarily as a military technology under exigent conditions.[20] Some government scientists had warned of DDT's hazards, but their studies were mostly classified or buried in government file cabinets; few people knew of their findings. After the war, safety considerations were largely brushed aside as DDT was lionized and Müller awarded the Nobel Prize.[21] In any case, pesticide regulation in the United States was based on assuring efficacy and controlling residues on food, not on environmental impact. Food production in the postwar United States was a great success story— American farmers were producing more food than ever at lower and lower prices—so if DDT had played a role in that as well, it showed how success- ful the chemical was.

So PSAC had a difficult charge: to contrast the obvious, rapid benefits of pesticide use in disease control and food production with the subtle, long- term, poorly understood risks to humans and nature. They also had to sort out a multitude of acknowledged scientific uncertainties. These gray areas included the gap between data on acute exposure (whose risks were not dis- puted) and chronic effects; a lack of information on synergistic impacts; the worry that existing data underreported adverse effects (because doctors weren't trained to recognize low-level pesticide poisoning and rarely did); and the familiar problem of extrapolating from experiments on lab rats to people.[22] They also had to address the difficulties of predicting long-term effects based on the few existing clinical studies.[23]

Despite these difficulties, PSAC came to a clear conclusion: it was time for immediate action to restrain pesticide use. The evidence of damage to wildlife was clear and compelling, even in cases of "programs carried out exactly as planned," and these harms would sooner or later spread to hu- mans.[24] "Precisely because pesticide chemicals are designed to kill or meta- bolically upset some living target organism, they are potentially dangerous to other living organisms," the panel concluded logically enough. "The haz- ards resulting from their use dictate rapid strengthening of interim meas- ures until such time as we have realized a comprehensive program for controlling environmental pollution."[25]

In the years to come, the U.S. government developed just such a pro- gram, as bipartisan majorities in Congress passed the Clean Air and Clean Water Acts and established a number of agencies, such as the National In- stitute for Environmental Health Sciences, to address environmental issues. This effort culminated in 1970 in the establishment of the U.S. Environ- mental Protection Agency. In 1972—ten years after the publication of *Silent Spring* and at least three more national-level science assessments—the En- vironmental Protection Agency under President Richard Nixon banned the use of DDT in the United States.[26] There was no rush to judgment against DDT: it took three presidencies to enact the ban. Science was not the cause of that policy—political will was—but the scientific facts supported it.

The Kennedy PSAC report, *Use of Pesticides: A Report of the President's Science Advisory Committee*, is notable in hindsight as much for what it did not do as for what it did. The scientists did not claim that the hazards of persistent pesticides were "proven," "demonstrated," "certain," or even well understood; they simply concluded that the weight of evidence was sufficient to warrant policy action to control DDT. Environmental concerns other than pesticides might be more serious, they acknowledged, but that was no reason to deflect or distract attention from the issue with which they were charged. They did not dismiss alternatives to pesticide use, such as biological pest control, and they did not accuse Carson of harboring a hidden agenda. Nor did they let a lack of scientific understanding of the mechanisms of pesticide damage stop them from accepting the empirical evidence of it. Most important, while calling for more study, they didn't stall or hedge; they called for action.

The committee placed the burden of proof—or at least a substantial weight of it—on those who argued that persistent pesticides were safe, and explicitly invoked the standard of reasonable doubt. The legal phrase "reasonable doubt" suggests that they were guided by existing legal frameworks, such as the landmark federal Food, Drug, and Cosmetic Act (1938), which placed the burden of proof on manufacturers to demonstrate the safety of their products, and the Miller Amendment to that act (1954), which extended the act's reach to pesticides.[27] Manufacturers had not demonstrated the safety of DDT, and reasonable people now had reason to doubt it.[28]

Both science and democracy worked as they were supposed to. Independent scientific experts summarized the evidence. Polls showed that the public supported strong legislation to protect the environment.[29] Gordon MacDonald, a member of President Nixon's Council on Environmental Quality, recalled that Nixon supported the creation of the EPA not because he was a visionary environmentalist, but because he knew that the environment would be an important issue in the 1972 presidential election.[30] Our leaders acted in concert with both science and the will of the people.

Does the story end there? No, for as we began to explain above, Carson has now become the victim of a shrill revisionist attack. "Rachel was wrong," claims the Web site of the Competitive Enterprise Institute.[31] "Fifty million dead," claims another."[32] "More deaths likely," insists a third.[33] Why? Because malaria has not been eradicated, and it would have been, these critics insist, had the United States not succumbed to environmental hysteria. There was no good scientific evidence to support the DDT ban, they say, and DDT was the only effective means to kill the mosquitoes that carry the malarial parasite.[34] Banning it was "the worst crime of the century."[35]

In his bestselling book, *The Skeptical Environmentalist*, Danish economist Bjørn Lomborg (listed by *Time* as one of the one hundred most influential people in 2004) echoed the accusation that Carson's argument was more emotional than rational, insisting that more lives were saved by disease control and improved food supply than were ever lost to DDT. Thomas Sowell, a conservative writer associated with the Hoover Institution, insists "there has not been a mass murderer executed in the past half-century who has been responsible for as many deaths of human beings as the sainted Rachel Carson."[36] Others have compared Carson to Stalin and Hitler.[37]

One might ignore these venomous claims except that they have been repeated in mainstream newspapers. In 2007, the *San Francisco Examiner* ran an op-ed piece alleging that "Carson was wrong, and millions of people continue to pay the price."[38] The *Wall Street Journal* argued that Carson's work led to the attitude that "environmental controls were more important than the lives of human beings."[39] The *New York Times* has run several articles and op-ed pieces doubting the wisdom of U.S. action on DDT.[40] "What the World Needs Now Is DDT" ran the title of a Sunday *New York Times Magazine* piece in 2004. "No one concerned about the environmental damage of DDT set out to kill African children," the article began, but their deaths happened all the same. "*Silent Spring* is now killing African children because of its persistence in the public mind."[41]

One of the anti-Carson voices at the *New York Times* is the "science" columnist John Tierney, who in 2007 argued that *Silent Spring* was a "hodgepodge of science and junk science" and that the person who actually got the science right in the 1960s was I. L. Baldwin, a professor of agricultural bacteriology at the University of Wisconsin. No one listened to him, Tierney insisted, because Baldwin didn't scare people. His calm demeanor was no match for Carson's "rhetoric," which "still drowns out real science."[42]

Is Tierney right? Was Carson wrong? What does real science—and real history—tell us? It tells us that Carson—and the President's Science Advisory Committee and the U.S Environmental Protection Agency and President Richard Nixon—were not wrong about DDT.

After DDT's demonstrated successes in World War II, the United States and the World Health Assembly launched a Global Malaria Eradication Campaign (1955–1969). It was not based on large outdoor spraying campaigns—the principal target of Carson's indictment—but primarily on indoor spraying of household walls and surfaces with DDT (and dieldrin). The U.S. Centers for Disease Control summarizes the results: "The campaign did not achieve its stated objective." Endemic malaria was eliminated in developed nations, mainly in Europe and Australia, and sharply reduced in India and parts of Latin America, but the campaign failed in many less developed areas, especially sub-Saharan Africa. It was halted in 1969—four years before the U.S. DDT ban—so whatever happened could not have been the result of the U.S. ban. What did happen?[43]

Malaria eradication failed in less developed nations because spraying alone didn't work. Spraying *along with* good nutrition, reduction of insect breeding grounds, education, and health care did work, which explains why malaria was eradicated in developed nations like Italy and Australia, but not in sub-Saharan Africa. Like nearly all public health initiatives, the program needed people's cooperation and understanding.

Indoor Residual Spraying—the central technique used—worked by leaving insecticide on the walls and ceilings of dwellings. This meant that people needed not to wash, paint, or replaster their walls, and many people didn't understand this, as it contradicted most other public health directives. Others just didn't like the idea, as it seemed to instruct them to have dirty homes. But the most important reason that eradication was only partially successful was that mosquitoes were developing resistance. In the United States, DDT use peaked in 1959—thirteen years before the ban—because it was already starting to fail.

BUGS AND BACTERIA offer the best evidence we have of natural selection. When an insecticide wipes out part of a population, the ones that survive pass on their genes to their offspring, and it is only a matter of time before the population adapts to the insecticide-laden environment. Insect generations last a few days to a few months, so they evolve with enormous speed—far faster than slow-breeding species like humans and most animals. So they show the effects of natural selection in a time frame that we can directly observe—sometimes in as little as a few years.

Insect resistance to DDT was first recognized in 1947, just a few years after DDT's wartime triumphs. Mosquito control workers in Fort Lauderdale reported that "the normal application of a 5 percent DDT solution had no discernible effect on salt marsh mosquitoes . . . the miraculous 'magic dust' had lost its efficacy against the hordes of salt marsh mosquitoes along Florida's east coast."[44] Resistance increased rapidly during the 1950s, and soon many pest control districts were abandoning DDT for other alternatives.

Sadly, most of the resistance that insects developed to DDT came from agricultural use, not from disease control. There *is* a tragedy in this story, but it is not the one that the Competitive Enterprise Institute thinks it is. It is that the attempt to grow food cheaply, especially in the United States, was largely responsible for the development of insect resistance. The failure of DDT in disease control is in part the result of its excess use in agriculture. Here's why.

The most efficient way to use pesticides against disease is through application to the insides of buildings—the Indoor Residual Spraying on which the World Health Organization largely relied. DDT is particularly potent in this use, as an application can last up to a year. Most important, it doesn't produce resistance very quickly, because most insects don't wind up in buildings and therefore aren't subjected to the poison. Indoor Residual Spraying just affects the small percentage of the population that make it indoors, where they are likely to bite people and transmit disease, so the selection pressure on the insect population isn't very high. It's a very sensible strategy.

However, when pesticides are sprayed over large agricultural areas, they kill a large fraction of the total insect population, ensuring that the hardy survivors breed only with other hardy survivors; the very next generation may display resistance. The more extensive the agricultural use, the more likely bugs are to evolve resistance rapidly, and the less effective the pesticide is likely to be when you need it for disease control.

We now know that agricultural spraying produced insect immunity in only seven to ten years. This isn't merely hindsight: Rachel Carson discussed insect resistance in *Silent Spring*.[45] DDT was also widely used for agriculture in countries where it was being used for disease control, so it became ineffective for disease control much sooner than it might otherwise have. In the 1950s, we already knew that insects evolved very rapidly, but our political institutions evolved much more slowly than the bugs did.

Events proved that DDT alone was not sufficient to eradicate malaria, but was DDT necessary? Was it essential in the regions where malaria was controlled? The answer here is no, too. Most people have forgotten that in the nineteenth century malaria was endemic in the United States—and a major anxiety for settlers in places like Arkansas, Alabama, and Mississippi.[46] Even California struggled with malaria.

By the 1930s mosquito control districts throughout the nation had largely brought malaria under control by drainage, removal of breeding sites, and pesticides other than DDT.[47] Malaria infection in Florida, for example, declined every year after 1935, even though DDT was yet to be introduced.[48] Urbanization played a role, too, as more Americans lived away from mosquito breeding grounds. After World War II, DDT became an additional tool in the arsenal, helping to eradicate the remaining cases—by then few and far between.

Another case is worth mentioning: the Panama Canal. Led by Ferdinand de Lesseps (who had also led the construction of the Suez Canal) the canal project was started by a French company in 1882, but faltered in part because of the impact of yellow fever and malaria. By 1889, more than twenty-two thousand workers had been felled by these two diseases, and the construction effort collapsed.

In 1904, the U.S. government took over and the new American leadership appointed a medical officer to the post of chief sanitation officer, William Crawford Gorgas. Gorgas believed what was then a radical hypothesis: that these diseases were carried by insects. He drained swamps and wetlands, removed standing pools of water from around buildings, and sent teams of men to destroy mosquito larvae with oil and to fumigate the buildings. He also equipped the buildings, especially the workers' dormitories, with screens. Between 1906 and the completion of the canal in 1914, there was only a single case of yellow fever, and the death rate in the population declined from 16.21 per thousand in 1906 to 2.58 per thousand in December 1909.[49] Yellow fever was completely eradicated—thirty-one years before Müller's discovery of DDT's insecticidal properties. While malaria proved more recalcitrant, it too was controlled in many regions by similar techniques. The lesson of history is clear: DDT alone did not eradicate insect-borne diseases, and those diseases have been controlled in places with little or no use of DDT.[50]

WHEN THE UNITED STATES took action against DDT in 1971, EPA administrator William Ruckelshaus made clear that the new ban would not apply outside the United States. (How could it? EPA had no authority over other countries.) Ruckelshaus stressed that U.S. manufacturers were free to continue to manufacture and sell the product for disease control overseas, and that his agency would "not presume to regulate the felt necessities of other countries."[51] Whatever subsequently happened in Africa, it was hardly Rachel Carson's fault—or William Ruckelshaus's.

As for Baldwin—the scientist whom John Tierney claims got the science right—the work that Tierney quotes wasn't a piece of scientific research at all, it was a *book review*: a review of *Silent Spring*. Baldwin acknowledged that *Silent Spring* was "superbly written and beautifully illustrated," and constituted "an exhaustive study of the facts bearing on the problem."[52] He also allowed that Carson's approach "will undoubtedly result in wider recognition of the fact that [pesticides] are poisons and in a more careful and rigorous control of every step in the pathway that pesticides must travel . . . There are serious hazards involved in the use of pesticides."[53]

So what was Baldwin's complaint? That the book was impassioned, rather than balanced, and read as if written by a prosecutor. That was true: Carson *was* trying to make a case. But above all, Baldwin complained that Carson had written the wrong book. He wanted to read a progress story about how the development of chemicals—pesticides included—constituted a "chemical revolution . . . that has most intimately affected every aspect of our daily life." He wanted a book that recounted how technology had made life better, emphasizing that "the span of our life has been greatly extended; our clothes are composed of fibers unknown 20 years ago; our machinery and household utensils are made of new and strange materials."[54] He wanted to be told about the benefits that science and technology had brought us, not their frightening unintended consequences. Perhaps John Tierney felt the same way.

Like virtually all of Carson's critics who followed, Baldwin insisted that pesticides were the key to the productivity of modern agriculture, and that greater use of pesticides was the key to wiping out world hunger (although most social scientists disagree, pointing out that there is plenty of food in the world; the problem we face is one of unequal distribution). Rather than answer Carson's points and address her evidence, Baldwin changed the subject: focusing on the good that modern technology has brought, and refusing to address her central argument about ecosystem harms. Contrary to Tierney's claim, Baldwin conceded the science. Like virtually all of Carson's critics—including Tierney—his faith in technology and anthropocentrism caused him to miss Carson's most important point.

In 1962, evidence of human deaths from DDT was scant. Carson acknowledged this. While she suggested that DDT was likely to cause cancer, she never claimed that large numbers of people had been killed by it. What she emphasized was the overwhelming evidence of harm to *ecosystems*, harm that she believed would sooner or later reach us. Carson's argument was that any war on nature was one that we were bound to lose. Fish and birds were killed, while fast-evolving insects came back stronger than ever. Finally—and perhaps above all—it was a mistake to assume that the only harms that counted were *physical*. Even if DDT caused not one human death, humans would be affected: our world would be impoverished if spring came and no birds sang.

If DDT's defenders have exaggerated its benefits, have its detractors exaggerated the harms? If DDT rarely harms people and sometimes helps, why not reintroduce it? Isn't Bjørn Lomborg right at least that DDT saved more lives than it cost?

The argument is a red herring. DDT was not banned on the basis of harm to humans; it was banned on the basis of harms to the environment. The scientific evidence of those harms was not only affirmed by PSAC and the EPA; it has been reaffirmed by numerous studies in areas where DDT and its metabolite, DDE, persist.[55] DDT kills birds, fish, and beneficial insects, and continues to do so long after spraying has stopped. Even today, birds in the Catalina Islands show signs of DDT poisoning, probably from eating fish that have ingested materials from the sea floor laced with residual DDT, left over from its manufacture in California decades ago.[56]

What about humans? Tierney argues that when DDT was banned "there wasn't evidence that it was carcinogenic." This is true. But since then we have learned a great deal about the risks of pesticides, and there is now strong scientific evidence that many pesticides carry serious risks to humans. (Recall that *Silent Spring* was not just about DDT; it was about pesticides in general.) Since 1971, the cancer-causing properties of diverse pesticides have been demonstrated by numerous peer-reviewed scientific studies, both in animal models and exposed humans.[57] We have also learned much more about the manner in which DDT does, in fact, harm humans.

A recent review in the *Lancet*—the world's leading medical journal—concluded that when used at levels required for mosquito control, DDT causes significant human impacts, particularly on reproductive health. (This is not surprising, given that some of the earliest evidence against DDT was that it interfered with reproduction in birds and rats.) Abundant scientific evidence reveals DDT's impact on child development, including preterm birth, low birth weight, and possible birth defects. High concentrations of DDT in breast milk are correlated with shortened duration of lactation and early weaning—itself highly correlated with infant and childhood mortality. The *Lancet* authors conclude that any saving of lives from malaria might well be abrogated by infant and early childhood mortality caused by DDT.[58] Some lives might have been saved by continued use of DDT, but others would have been lost.

And what about cancer? A few years ago, medical researchers realized that there was a shocking flaw in previous studies that investigated DDT exposure and breast cancer. Most of them were done *after* DDT use was already on the decline, or even after the ban, so the women being studied had probably been exposed only to low levels (if at all), and exposed later in life when the body is less vulnerable. To really know whether or not DDT had an effect, you'd need to study women who'd been exposed to DDT early in life, at a time when environmental exposures were high.

In a remarkable piece of medical detective work, Dr. Barbara A. Cohn and her colleagues identified women who had been part of medical study of pregnant women in the 1960s, and therefore might have been exposed as children or teenagers when DDT use was widespread in the 1940s and '50s. These women had given blood samples at the time, samples that could now be reanalyzed for DDT and its metabolites. In 2000–2001, they measured DDT-related compounds in these samples and compared them with breast cancer rates. The average age at the time of the original study was twenty-six; these women were now in their fifties and sixties—an age by which breast cancer might reasonably be expected to appear. The results showed a *fivefold increase* in breast cancer risk among women with high levels of serum DDT or its metabolites.[59] DDT *does* cause cancer, it *does* affect human health, and it *does* cost human lives. Rachel Carson was not wrong.

Admittedly, some public health experts think that DDT could play a useful role in malaria control in some places in the world today, but it never was the miracle cure that Lomborg, Sowell, Cohen, and Tierney have made it out to be. There is no scientific evidence to support the claim that millions of lives have been needlessly lost, and there is substantial scientific evidence that a good deal of harm—both to humans and the other species we share this planet with—has been avoided.

So what is going on here? Are these folks just confused? Misinformed? Ignorant? Even hysterical? Would that it were so.

We've seen how some people have fought the facts about the hazards of tobacco, acid rain, ozone depletion, secondhand smoke, and global warming. Their denials seemed plausible, at least to some, because they involved matters that were still under scientific investigation, where many of the details were uncertain even if the big picture was becoming clear. But the construction of a revisionist history of DDT gives the game away, because it came so long after the science was settled, far too long to argue that scientists had not come to agreement, that there was still a real scientific debate. The game here, as before, was to defend an extreme free market ideology. But in this case, they didn't just deny the facts of science. They denied the facts of history.

Denial as Political Strategy

Each of the stories we've told so far involved a handful of actors attempting to prevent regulation of specific products. But the twenty-first-century attack on Carson had nothing to do with preventing regulation; the regulation was long established. Nor was it an effort to overturn that regulation. It was well understood in American science, government, and agriculture that DDT was no longer effective in the United States. So why does DDT matter? Why attack a woman who has been dead for nearly half a century?

We saw in chapter 3 that as the acid rain story was emerging in the 1960s, the American environmental movement was changing its orientation away from an aesthetic environmentalism toward legal regulation. Carson's voice was fundamental to that reorientation. After all, what was the value of a national park if no birds sang in it? If Carson was wrong, then the shift in orientation might have been wrong, too. The contemporary environmental movement could be shown to have been based on a fallacy, and the need for government intervention in the marketplace would be refuted.

We see this narrative first emerging from someone we have already met: Dixy Lee Ray. In *Trashing the Planet*, Ray sang the praises of DDT and constructed a set of "facts" that have circulated ever since. She told a story of how DDT was wrongly abandoned in Sri Lanka, where "public health statistics . . . testify to the effectiveness of the spraying program." It began like this:

> In 1948, before the use of DDT there were 2.8 million cases of malaria [in Sri Lanka]. By 1963, there were only 17. Low levels of infection continued until the late 1960s, when the attacks on DDT in the U.S. convinced officials to suspend spraying. In 1968 there were one million cases of malaria. In 1969, the number reached 2.5 million, back to the pre-DDT levels. Moreover by 1972, the largely unsubstantiated charges against DDT in the United States had a worldwide effect.[60]

Is this account true? Partly—the part up to 1963. Between 1948 and 1963, DDT worked, and malaria cases dropped dramatically. Although resistance was seen as early as 1958, eradication appeared to be working overall. In 1963, the small handful of new cases should have made it controllable; indeed, malaria should have been on the path to eradication in Sri Lanka. But then Ray started to leave out key facts.

In 1968, malaria flared up again, and DDT couldn't control it. Still, the Sri Lankans persisted, using even more DDT over larger areas at more frequent intervals. Still, it didn't work. In its 1976 study of pesticide resistance, the World Health Organization's Expert Committee reported:

> In Sri Lanka a revised programme started in March 1975 that had been planned in the light of the limited financial resources available . . . The use of DDT at $1 g/m^2$ at 4 monthly intervals with particular attention to improved coverage did not result in any significant difference in malaria prevalence as compared with an area with normal (lesser) coverage, and no improvement was obtained either by using DDT at the rate of $2 g/m^2$ at 4 monthly intervals.[61]

Finally they switched to malathion, a more expensive agent, but one that the region's insect population hadn't yet adapted to. This brought the malaria rate down again, although not to the extremely low levels seen in 1963.[62]

So Sri Lanka didn't stop using DDT because of what the United States did, or for any other reason. DDT stopped working, but they kept using it anyway. We can surmise why: since DDT had appeared to work at first, officials were reluctant to give it up, even as malaria became resurgent. It took a long time for people to admit defeat—to accept that tiny mosquitoes were in their own way stronger than us. As a WHO committee concluded in 1976, "It is finally becoming acknowledged that resistance is probably the biggest single obstacle in the struggle against vector-borne disease and is mainly responsible for preventing successful malaria eradication in many countries."[63]

Resistance is never mentioned in Ray's account, an especially notable omission given that she was a zoologist. In a particularly egregious example of the pot calling the kettle black, Ray accused both environmentalists and William Ruckelshaus of giving credibility to pseudoscience, by creating "an atmosphere in which scientific evidence can be pushed aside by emotion, hysteria, and political pressure."[64] But it was she, not Ruckelshaus, who was spreading hysteria.

Ray had not accused Rachel Carson of mass murder, but others soon did. We met Steve Milloy in chapter 5, as he founded The Advancement of Sound Science Coalition on behalf of Philip Morris in 1993 to defend a product that really *had* caused millions of deaths. Soon thereafter, he began to spread the "millions of deaths" claim about DDT. According to his 1997 annual report, he began working with J. Gordon Edwards, an entomologist at San Jose State University, to help him publish an account of the DDT controversy.[65] Edwards's account finally appeared in 2004 in the *Journal of American Physicians and Surgeons*, published by the Association of American Physicians and Surgeons. This is a Libertarian political group that shares a board member with the Oregon Institute of Science and Medicine—a group that had also promoted skepticism about global warming. Edwards contended that "the worldwide effect of the U.S. ban has been millions of preventable deaths."[66] While suggesting that "the term genocide is used in other contexts to describe such numbers of casualties," he never mentioned the fact of pesticide resistance—a striking omission for an entomologist.

Milloy continued the antiscientific crusade in his post-TASSC career, and continues it to this day. "It might be easy for some to dismiss the past 43 years of eco-hysteria over DDT with a simple 'never mind,'" Milloy asserted recently, "except for the blood of millions of people dripping from the hands of the WWF, Greenpeace, Rachel Carson, Environmental Defense Fund, and other junk science–fueled opponents of DDT."[67] Milloy is well-known for his attacks on science related to all kinds of environmental issues, including global warming (which he calls a "swindle"), acid rain (which he notes helps slow global warming—although he doesn't believe in global warming anyway), and the ozone hole (which he considers to be of no real significance).[68] Milloy's current project is junkscience.com, but, as we saw in chapter 5 "junk science" was a term invented by the tobacco industry to discredit science it didn't like. Junkscience.com was originally established in a partnership with the Cato Institute, which, after Milloy's continued tobacco funding came to light, severed its ties.[69]

The disinformation campaign continues on the Web, supported by organizations and institutes that are by now familiar. After Rush Limbaugh parroted the "Rachel was wrong" attack, the Competitive Enterprise Institute promoted him for the Nobel Peace Prize.[70] The Competitive Enterprise Institute shares philosophical ground with the American Enterprise Institute, which promoted the work of the late fiction writer Michael Crichton. His 2004 novel, *State of Fear*, portrayed global warming as a liberal hoax meant to bring down Western capitalism.[71] Crichton also took on the DDT issue, as one character in the novel insists, "Banning DDT killed more people than Hitler . . . It was so safe you could eat it."[72]

The "Rachel was wrong" chorus is echoed particularly loudly at the Heartland Institute, a group dedicated to "free-market solutions to social and economic problems."[73] Their Web site insists that "some one million African, Asian, and Latin American lives could be saved annually" had DDT not been banned by the U.S. Environmental Protection Agency.[74]

The Heartland Institute is known among climate scientists for persistent questioning of climate science, for its promotion of "experts" who have done little, if any, peer-reviewed climate research, and for its sponsorship of a conference in New York City in 2008 alleging that the scientific community's work on global warming is a fake.[75] But Heartland's activities are far more extensive, and reach back into the 1990s when they, too, were working with Philip Morris.

In 1993, Richard C. Rue, a project director for the Heartland Institute, wrote to Roy E. Marden, manager of Industrial Affairs for Philip Morris Management, to solicit continued support. Rue enclosed a copy of an op-ed piece, evidently an excerpt from a forthcoming book, written by Joseph Bast, the Institute president and CEO.[76] He recounted other recent Institute activities, boasting of distributing almost nine thousand copies of a special publication of the Chemical Manufacturers Association, of which eight thousand were sent to "state legislators and constitutional officers and other public opinion leaders."[77]

Philip Morris also used Heartland to distribute reports that they (Philip Morris) had commissioned. In April 1997, Roy Marden wrote to Thomas Borelli (who we met in chapter 5) discussing a task force report they had prepared in conjunction with the Association for Private Enterprise Education. Marden wrote:

> . . . the Heartland Institute, an Illinois-based policy group with whom we work, [will] publish a 24-page summary of the report/paper as a policy study. This will be released late next week, with a distribution of at least 3000 (half journalists, the remainder to state Constitutional officers and business types). Heartland would be willing to do a full run of 10,000 (which would include every state legislator and Member of Congress) if they can get the

funding for the 7000 differential. I am getting faxed later what this will cost . . . and I think we should consider this.[78]

Heartland Institute officials also met with members of Congress on behalf of the tobacco industry, organized "off-the record" briefings, wrote and placed op-ed pieces, and organized radio interviews and letters to editors.[79]

In 1997, Philip Morris paid $50,000 to the Heartland Institute to support its activities, but this was just the tip of the iceberg of a network of support to supposedly independent and nonpartisan think tanks. The stunning extent of Philip Morris's reach is encapsulated in a ten-page document from 1997 listing policy payments that were made to various organizations. Besides the $50,000 to the Heartland Institute, there was $200,000 for TASSC, $125,000 for the Competitive Enterprise Institute, $100,000 for the American Enterprise Institute, and scores more.[80] Payments were for as little as $1,000 or as much as $300,000, and many went to groups with no evident interest in the tobacco issue, such as the Ludwig von Mises Institute or Americans for Affordable Electricity. Numerous other documents attest to activities designed to undermine the Clinton health care reform plan.[81] Often financial contributions were referred to in company documents as "philanthropy," and because these organizations were all nonprofit and nonpartisan, the donations were all tax deductible.[82]

The following image is the first page of this ten-page document listing the "policy" organizations to which the Philip Morris Corporation contributed. Note how nearly all of these were described as having a focus in either "Individual Liberties," "Regulatory Issues," or both, and how the Cato Institute, the American Enterprise Institute, and the Competitive Enterprise Institute—all of whom have questioned the scientific evidence of global warming—each received six-figure contributions. Note also the funding to the American Civil Liberties Union. Additional pages document contributions to the Frontiers of Freedom Institute, the Acton Institute, the Alexis de Tocqueville Institute, and the Independent Institute; to seemingly grass-roots organizations—the Citizens Against Government Waste, the Independent Women's Forum, and the Institute for Youth Development—and to university groups such as the George Mason Law and Economics Center and the University of Kansas Law and Organizational Economics Center.

1997 Policy Payments for Slavit

Primary Focus	Secondary Focus	Multiple Focus	Organization	Amount Paid in '97	Amt. Charged to BAC	Amt. Charge to Pol/Civ	Amt. Charged to USA BAC	Amt. Charged to Admin
Commercial Speech	Individual Liberties		Libertad, Inc.	100,000.00	-	-	-	100,000.00
Commercial Speech	Regulatory Issues		Freedom of Expression Foundation	20,000.00	20,000.00	-	20,000.00	-
Individual Liberties	Regulatory Issues	X	American Enterprise Institute	100,000.00	-	-	-	100,000.00
Individual Liberties	Regulatory Issues	X	Cato Institute	175,000.00	150,000.00	100,000.00	50,000.00	25,000.00
Individual Liberties	Regulatory Issues	X	Competitive Enterprise Institute	125,000.00	125,000.00	100,000.00	25,000.00	-
Individual Liberties	Regulatory Issues	X	ETV of South Carolina	300,000.00	150,000.00	150,000.00	-	150,000.00
Individual Liberties	Regulatory Issues	X	Free Congress Foundation	300,000.00	300,000.00	300,000.00	-	-
Individual Liberties	Regulatory Issues	X	American Civil Liberties Union Foundation	75,000.00	75,000.00	37,500.00	37,500.00	-
Privacy	Individual Liberties		Defenders of Property Rights	45,000.00	45,000.00	-	45,000.00	-
Public Opinion Research			Roper Center for Public Opinion Research	5,000.00	5,000.00	5,000.00	5,000.00	-
Regulatory Issues	Tax	X	Manhattan Institute	20,000.00	20,000.00	10,000.00	10,000.00	-
Regulatory Issues		X	Hudson Institute	25,000.00	25,000.00	25,000.00	-	-
Sound Science	Regulatory Issues		TASCC	200,000.00	-	-	-	200,000.00
Sound Science	Regulatory Issues		Reason Foundation	20,000.00	20,000.00	10,000.00	10,000.00	-
Tax	Divestment		Municipal Treasurers Association of the US and Canada	2,500.00	2,500.00	-	2,500.00	-
Tax	Regulatory Issues	X	Heartland Institute	50,000.00	50,000.00	25,000.00	25,000.00	-

The Orwellian Problem

The network of right-wing foundations, the corporations that fund them, and the journalists who echo their claims have created a tremendous problem for American science. A recent academic study found that of the fifty-six "environmentally skeptical" books published in the 1990s, 92 percent were linked to these right-wing foundations (only thirteen were published in the 1980s, and 100 percent were linked to the foundations).[83] Scientists have faced an ongoing misrepresentation of scientific evidence and historical facts that brands them as public enemies—even mass murderers—on the basis of phony facts.

There is a deep irony here. One of the great heroes of the anti-Communist political right wing—indeed one of the clearest, most reasoned voices against the risks of oppressive government, in general—was George Orwell, whose famous 1984 portrayed a government that manufactured fake histories to support its political program.[84] Orwell coined the term "memory hole" to denote a system that destroyed inconvenient facts, and "Newspeak" for a language designed to constrain thought within politically acceptable bounds.

All of us who were children in the Cold War learned in school how the Soviet Union routinely engaged in historical cleansing, erasing real events and real people from their official histories and even official photographs. The right-wing defenders of American liberty have now done the same. The painstaking work of scientists, the reasoned deliberations of the President's Science Advisory Committee, and the bipartisan American agreement to ban DDT have been flushed down the memory hole, along with the well-documented and easily found (but extremely inconvenient) fact that the most important reason that DDT failed to eliminate malaria was because insects *evolved*. That is the truth—a truth that those with blind faith in free markets and blind trust in technology simply refuse to see.

The rhetoric of "sound science" is similarly Orwellian. Real science—done by scientists and published in scientific journals—is dismissed as "junk," while misrepresentations and inventions are offered in its place. Orwell's Newspeak contained no science at all, as the very concept of science had been erased from his dystopia. And not surprisingly, for if science is about studying the world as it actually is—rather than as we wish it to be—then science will always have the potential to unsettle the status quo. As an independent source of authority and knowledge, science has always had the capacity to challenge ruling powers' ability to control people by controlling their beliefs. Indeed, it has the power to challenge anyone who wishes to preserve, protect, or defend the status quo.

Lately science has shown us that contemporary industrial civilization is not sustainable. Maintaining our standard of living *will* require finding new ways to produce our energy and less ecologically damaging ways to produce our food. Science has shown us that Rachel Carson was not wrong.

This is the crux of the issue, the crux of our story. For the shift in the American environmental movement from aesthetic environmentalism to regulatory environmentalism wasn't just a change in political strategy. It was the manifestation of a crucial realization: that unrestricted commercial activity was doing damage—real, lasting, pervasive damage. It was the realization that pollution was global, not just local, and the solution to pollution was *not* dilution. This shift began with the understanding that DDT remained in the environment long after its purpose was served. And it grew as acid rain and the ozone hole demonstrated that pollution traveled hundreds or even thousands of kilometers from its source, doing damage to people who did not benefit from the economic activity that produced it. It reached a crescendo when global warming showed that even the most seemingly innocuous by-product of industrial civilization—CO_2, the stuff on which plants depend—could produce a very different planet.

To acknowledge this was to acknowledge the soft underbelly of free market capitalism: that free enterprise can bring real costs—profound costs—that the free market does not reflect. Economists have a term for these costs—a less reassuring one than Friedman's "neighborhood effects." They are "negative externalities": negative because they aren't beneficial and external because they fall outside the market system. Those who find this hard to accept attack the messenger, which is science.

We all expect to pay for the things we buy—to pay a fair cost for goods and services from which we expect to reap benefits—but external costs are unhinged from benefits, often imposed on people who did not choose the good or service, and did not benefit from their use. They are imposed on people who did not benefit from the economic activity that produced them. DDT imposed enormous external costs through the destruction of ecosystems; acid rain, secondhand smoke, the ozone hole, and global warming did the same. This is the common thread that ties these diverse issues together: they were all market failures. They are instances where serious damage was done and the free market seemed unable to account for it, much less prevent it. Government intervention was required. This is why free market ideologues and old Cold Warriors joined together to fight them. Accepting that by-products of industrial civilization were irreparably damaging the global environment was to accept the reality of market failure. It was to acknowledge the limits of free market capitalism.

ORWELL UNDERSTOOD THAT those in power will always seek to control history, because whoever controls the past controls the present. So our Cold Warriors—Fred Seitz and Fred Singer, Robert Jastrow and Bill Nierenberg, and later Dixy Lee Ray, too, who had dedicated their lives to fighting Soviet Communism, joined forces with the self-appointed defenders of the free market to blame the messenger, to undermine science, to deny the truth, and to market doubt. People who began their careers as fact finders ended them as fact fighters. Evidently accepting that their ends justified their means, they embraced the tactics of their enemy, the very things they had hated Soviet Communism for: its lies, its deceit, its denial of the very realities it had created.

Why would any scientist participate in such a fraud? We've seen that Steve Milloy and John Tierney, the Competitive Enterprise Institute and the Heartland Institute, were late entries in this tournament, echoing arguments that had been first constructed by scientists. Our story began in the 1950s, when the tobacco industry first enlisted scientists to aid its cause, and deepened in the 1970s when Frederick Seitz joined forces with tobacco, and then with Robert Jastrow and Bill Nierenberg to defend the Strategic Defense Initiative. It continued in the early 1980s as Fred Singer planted the idea that acid rain wasn't worth worrying about, and Nierenberg worked with the Reagan White House to adjust the Executive Summary of his Acid Rain Peer Review Panel. It continued still further, and turned more personal, in the 1990s as the Marshall Institute, with help from Singer and Ray, challenged the evidence of ozone depletion and global warming and personally attacked distinguished scientists like Sherwood Rowland and Ben Santer.

Why did this group of Cold Warriors turn against the very science to which they had previously dedicated their lives? Because they felt—as did Lt. General Daniel O. Graham (one of the original members of Team B and chief advocate of weapons in space) when he invoked the preamble to the U.S. Constitution—they were working to "secure the blessings of liberty."[85] If science was being used against those blessings—in ways that challenged the freedom of free enterprise—then they would fight it as they would fight any enemy. For indeed, science *was* starting to show that certain kinds of liberties are not sustainable—like the liberty to pollute. Science was showing that Isaiah Berlin was right: liberty for wolves does indeed mean death to lambs.

Of Free Speech and Free Markets

Our Founding Fathers placed freedom of the press in the first amendment of the U.S. Constitution, because democracy requires it. Citizens need information to make decisions, and a free press is crucial to its flow. Two centuries later the Fairness Doctrine was established in law, and although the legal doctrine was dismantled in the Reagan years, the notion of "equal time" remains enshrined in Americans' sense of justice and fair play.

But not every "side" is right or true; opinions sometimes express ill-informed beliefs, not reliable knowledge. As we've seen throughout this book, some "sides" represent deliberate disinformation spread by well-organized and well-funded vested interests, or ideologically driven denial of the facts. Even honest people with good intentions may be confused or mistaken about an issue. When every voice is given equal time—and equal weight—the result does not necessarily serve us well. Writing in *Democracy in America* long ago, Alexis de Tocqueville lamented the cacophony that passed for serious debate in the young republic: "A confused clamor rises on every side, and a thousand voices are heard at once."[1]

That was two hundred years ago; today the problem is much worse. With the rise of radio, television, and now the Internet, it sometimes seems that anyone can have their opinion heard, quoted, and repeated, whether it is true or false, sensible or ridiculous, fair-minded or malicious. The Internet has created an information hall of mirrors, where any claim, no matter how preposterous, can be multiplied indefinitely. And on the Internet, disinformation never dies. "Electronic barbarism" one commentator has called it—an environment that is all sail and no anchor.[2] Pluralism run amok.

The result is plain to see. A third of all Americans think that Saddam Hussein was behind the attacks on September 11.[3] Nearly a quarter still think that there's no solid evidence that smoking kills.[4] And as recently as 2007, 40 percent of Americans believed that scientific experts were still arguing about the reality of global warming.[5] Who can blame us? Everywhere we turn someone is questioning something, and many of the important issues of our day are reduced to he said/she said/who knows? Any person could be forgiven for being confused.

This cacophony of conflicting claims is particularly unhelpful when it comes to sorting out matters related to science, because science depends on evidence, and not all positions are equally grounded in it. Indeed, we've seen throughout this book how a small group of men with scientific bona fides and deep political connections deliberately distorted public debate, running effective campaigns to mislead the public and deny well-established scientific knowledge over four decades. And we've seen how many skeptical claims are based on *ignoring* evidence. This presents a real difficulty, one that is not easily resolved, for how can you prove that someone has ignored something? One can often show what something is; it is far harder to demonstrate what it is not. Clearly, people have a right to speak; the question is, to whom should we be listening?

For half a century the tobacco industry, the defenders of SDI, and the skeptics about acid rain, the ozone hole, and global warming strove to "maintain the controversy" and "keep the debate alive" by fostering claims that were contrary to the mainstream of scientific evidence and expert judgment. They promoted claims that had already been refuted in the scientific literature, and the media became complicit as they reported these claims as if they were part of an ongoing scientific debate. Often the media did so without informing readers, viewers, and listeners that the "experts" being quoted had links to the tobacco industry, were affiliated with ideologically motivated think tanks that received money from the tobacco industry (or in later years the fossil fuel industry), or were simply habitual contrarians, who perhaps enjoyed the attention they got promoting outlier views. Perhaps correspondents felt that adding this information would be editorializing. Or perhaps they did not know.

Many journalists we have spoken with have been surprised at our revelations, and in some cases even skeptical, until we showed them the documents. The degree of research we have done for this book cannot be done in time for a daily or weekly deadline, so it is understandable that most journalists would not know what we have discovered in five years of research. But the pressures on contemporary journalism cannot be the whole story, because we have seen how, at least in the early stages of this story, media leaders were openly courted by the tobacco industry. Arthur Hays Sulzburger, Edward R. Murrow, and William Randolph Hearst Jr. were hardly unsophisticated people, yet they evidently accepted the argument that the tobacco industry's view of the harms tobacco generates merited equal consideration as the scientific community's view. That is rather hard to explain, except to suppose that journalists, like the rest of us, are reluctant to accept information we'd rather was not true. Edward R. Murrow no doubt hoped that tobacco smoking wouldn't kill him. And who among us wouldn't prefer a world where acid rain was no big deal, the ozone hole didn't exist, and global warming didn't matter? Such a world would be far more comforting than the one we actually live in. Faced with challenging situations, we welcome reassurance that everything is going to be all right. We may even prefer comforting lies to sobering facts. And the facts denied by our protagonists were more than sobering. They were downright dreadful.

Whatever the explanation, it is clear that the media *did* present the scientific debate over tobacco as unsettled long after scientists had concluded otherwise. In 1999, researchers Gail Kennedy and Lisa Bero at the University of California, San Francisco, examined newspaper and magazine coverage of research on passive smoking and found that 62 percent of all articles published between 1992 and 1994 concluded that the research was "controversial."[6] Yet, as we saw in chapter 5, the scientific community had by that point reached consensus, and the tobacco industry had known the degree of danger even before that.

A similar phenomenon developed with acid rain in the 1990s, as the media attended to the idea that its cause was still not established—more than a decade after that was no longer true—or the claim that it would cost more to fix than it was worth, which was unsupported by evidence.[7] The press continued to report well into the 1990s that the ozone hole was perhaps caused by volcanoes.[8] Until recently the mass media presented global warming as a raging debate—twelve years after President George H. W. Bush had signed the U.N. Framework Convention on Climate Change, and *twenty-five years* after the U.S. National Academy of Sciences first announced that there was no reason to doubt that global warming would occur from man's use of fossil fuels. "Balance" had become a form of bias, whereby the media coverage was biased in favor of minority—in some cases extreme minority—views.[9]

In principle, the media could act as gatekeepers, ignoring the charlatans and snake oil salesmen, but if they have tried, our story shows that at least where it comes to science they have failed. As we have seen, it wasn't just obviously right-wing outlets that reported false claims about tobacco and these other subjects; it was the "prestige press"—indeed, the allegedly liberal press—as well.

Maybe now the tide is starting to turn. In April 2008, the *New York Times* reported that many of the retired generals featured on broadcast and cable news networks, speaking as independent experts on the state of the Iraq War, were not independent at all. Many were paid employees of military contractors with a stake in the armaments and other systems being used in the war, while others were working for the White House. Several of these generals had been groomed by the Pentagon—some met personally with Defense Secretary Donald Rumsfeld—to spread the word that the war was going well, despite considerable evidence to the contrary. "Message force multipliers" the Pentagon called them, as if they were some kind of weapon. And they were: they were weapons in a propaganda war, a campaign to mislead the American people. The media either didn't know about these behind-the-scenes machinations, or didn't care.[10]

Perhaps the media were taken in because these men really *were* generals. After all, they had relevant credentials and presumably knew a lot about the prosecution of war, in general, even if not about the details of this particular one. The problem was that they were not independent—they represented a particular side with a vested interest—and they were retired, so whatever they knew about war in general was not necessarily pertinent to the prosecution of this one.

The generals' campaign to convince us that the war in Iraq was going well is just one example; there are many others. The media—and all of us—have been the repeated victims of misinformation campaigns in which "experts" are used either to sell "facts" that aren't or to fight facts that are. But the generals' campaign offers a particularly useful parallel, because Jastrow, Seitz, Nierenberg, and Singer were all retired physicists—the four-star generals of American science in the Cold War—and like the generals, they knew how to make their claims sound credible. In this case, making their claims credible meant making them look like science.

A Scientific Potemkin Village

A key strategy in the campaigns to market doubt was to create the appearance that the claims being promoted were scientific.[11] The tobacco industry created the Tobacco Institute to foster research, but its primary purpose was to develop a cadre of experts who could be called upon in time of need. The industry also sponsored conferences and workshops whose papers could be cited on the industry's behalf, and they created additional institutes to address ostensibly independent topics, such as the Center for Indoor Air Research, designed to deflect attention away from tobacco onto other causes of lung troubles.[12] Historian Robert Proctor has recently documented the creation of newsletters, magazines, and journals—including journals with ostensible peer review—in which the results of industry-sponsored research could be reported, published, and then cited, as if they were independent. These included *Tobacco and Health*, *Science Fortnightly*, and the *Indoor Air Journal*.[13] It was a simulacrum of science, but not science itself.

When the George C. Marshall Institute began to challenge the claims of the scientific community on the ozone hole and global warming, they didn't create their own journal, but they did produce reports with the trappings of scientific argumentation—graphs, charts, references, and the like. At least one of these reports was read and taken seriously at the White House. Yet they were not subject to independent peer review—the most basic requirement of any truly scientific work. Had they been, it's likely they would have failed, because at least one of them seriously misrepresented the science, presenting only one portion of a key graph, which if presented in its entirety would have refuted their argument.

The scientists in our story also turned to techniques that were clearly outside the realm of normal scientific behavior. Scientists debate each other's findings in the halls of science—universities, laboratories, government agencies, conferences, and workshops. They do not normally organize petitions, particularly public ones whose signatories may or may not circulated information soliciting signatures on a petition "refuting" global warming.[14] He did this in concert with a chemist named Arthur Robinson, who composed a lengthy piece challenging mainstream climate science, formatted to look like a reprint from the *Proceedings of the National Academy of Sciences*. The "article"—never published in a scientific journal, but summarized in the *Wall Street Journal*—repeated a wide range of debunked claims, including the assertion that there was no warming at all.[15] It was mailed to thousands of American scientists, with a cover letter signed by Seitz inviting the recipients to sign a petition against the Kyoto Protocol.[16]

Seitz's letter emphasized his connection with the National Academy of Sciences, giving the impression that the whole thing—the letter, the article, and the petition—was sanctioned by the Academy. Between his mail-in card and a Web site, he gained about fifteen thousand signatures, although since there was no verification process there was no way to determine if these signatures were real, or if real, whether they were actually from scientists.[17] In a highly unusual move, the National Academy held a press conference to disclaim the mailing and distance itself from its former president.[18] Still, many media outlets reported on the petition as if it were evidence of genuine disagreement in the scientific community, reinforced, perhaps, by Fred Singer's celebration of it in the *Washington Times* the very same day the Academy rejected it.[19]

The "Petition Project" continues today. Fred Seitz is dead, but his letter is alive and well on the Internet, and the project's Web site claims that its signatories have reached thirty thousand.[20]

Many skeptical claims about global warming have been published in the *Journal of Physicians and Surgeons*, which is associated with the Oregon Institute of Science and Medicine, who sponsored the anti–global warming petition.[21] The journal, previously known as the *Medical Sentinel*, is the outlet of the Association of American Physicians and Surgeons, which among other things filed a suit on behalf of Rush Limbaugh when his medical records were seized as part of his prosecution on drug charges.[22] The *Sentinel* published articles questioning the link between HIV and AIDS, including a commentary by Michael Fumento, the journalist we met in chapter 5 who was defending pesticides while accepting money from the Monsanto chemical corporation.[23] The journal also published the work of J. Gordon Edwards, whom we met in chapter 7 when he was working with Steve Milloy to spread the erroneous claim that banning DDT cost millions of lives. (Neither the Web of Science nor MEDLINE/PubMed lists the journal among its peer-reviewed scientific sources.)

We could go on citing examples, but the point is clear. In creating the appearance of science, the merchants of doubt sold a plausible story about scientific debate. They erected a Potemkin village populated, in only a few cases, with actual scientists. A reasonable journalist, not to mention an ordinary citizen, could be forgiven for having been fooled by it.

But if this wasn't a scientific debate, then what was it?

Free Speech and Free Markets

In January 1973, Emil Mrak, the retired chancellor of the University of California, Davis, gave a presentation entitled "Some Experience Related to Food Safety."[24] In a long, thoughtful, and nuanced speech, he recounted his experiences in Washington, D.C., serving on a major government panel. He expressed confidence in the new EPA administrator, William Ruckelshaus, as well as in the regulatory process overall, but he was concerned that toxicology and oncology were rife with uncertainties, that analytical techniques were inadequate, and that there was a tendency in Washington to overreact to incomplete evidence of possible harms. In short, Mrak was thinking hard about what was right and what was wrong in regulatory procedures. He was, honestly and sincerely, presenting both sides—not to the public, however, but to the Philip Morris tobacco company.

In the same year that Mrak spoke to Philip Morris, Richard Nixon dissolved the President's Science Advisory Committee. That was a shame, because when the nation had to face serious questions about acid rain, the ozone hole, secondhand smoke, and global warming, we didn't have PSAC to set the scientific record straight as they had for DDT. But perhaps it didn't matter, because the campaigns that were carried out to undermine the relevant science were so extensive, so sophisticated, and so well funded that PSAC would scarcely have been a match for them.

We have seen throughout this story how the merchandising of doubt was aided and abetted by ideologically motivated think tanks that promoted and spread the message. We've documented that several of these think tanks had links to the tobacco industry. Journalists Chris Mooney, Ross Gelbspan, and Bill McKibben have documented how these think tanks were in turn funded by conservative foundations including Scaife, Olin, and Adolph Coors, and giant corporations such as Exxon Mobil.[25] In 2005, for example, Chris Mooney documented how in just a few years Exxon Mobil had channeled more than $8 million to forty different organizations that challenged the scientific evidence of global warming. The organizations did not just include probusiness and conservative think tanks, but also "quasi-journalistic outlets like TechCentralStation.com (a website providing 'news, analysis, research, and commentary' that received $95,000 from ExxonMobil in 2003), a *FoxNews.com* columnist, and even religious and civil rights groups."[26] Mooney also noted how former ExxonMobil chairman and CEO Lee Raymond served as vice-chairman of the board of trustees for the American Enterprise Institute, which received $960,000 in funding from ExxonMobil, and how in 2002, ExxonMobil explicitly earmarked $60,000 for "legal activities" by the Competitive Enterprise Institute.

Mooney described what happened when scientists released the comprehensive *Arctic Climate Impact Assessment*, which concluded that the Arctic was warming at twice the rate of the rest of the world—much as the Jason scientists predicted it would back in 1979. The report was blasted in a column by Steve Milloy, now working as a columnist for *FoxNews.com* and serving as an adjunct scholar at the Cato Institute, which received $75,000 from ExxonMobil. The *Washington Times* reprinted Milloy's column, and neither *Fox News* nor the *Washington Times* disclosed that Milloy had received money from ExxonMobil: $40,000 to The Advancement of Sound Science Center and $50,000 to the Free Enterprise Action Institute—both of which are registered to Milloy's home address.[27]

ExxonMobil's support for doubt-mongering and disinformation is disturbing but hardly surprising. What is surprising is to discover how extensive, organized, and interconnected these efforts have been, and for how long.

We met the Heartland Institute in chapter 7, vilifying Rachel Carson and promoting the myth of the mistake of banning DDT. We also noted that Heartland has been active in challenging the evidence of global warming. In September 2000, Bill Nierenberg had started organizing a "competitive review" of the Intergovernmental Panel on Climate Change's upcoming Third Assessment Report (which, of course, was itself a review) with meteorologist Richard Lindzen; his death that month left the idea in limbo.[28] But only temporarily. Fred Singer picked up the idea in 2007, and carried out a "Nongovernmental International Panel on Climate Change" review sponsored by Heartland.[29] Heartland also sponsored a conference in 2008 insisting that global warming is not and never has been a serious problem.[30] But Heartland's activities are far more extensive, and reach back into the 1990s when they, too, were working with Philip Morris.

We saw in chapter 7 how the Heartland Institute distributed reports, sent faxes, and met with members of Congress on behalf of Philip Morris. Heartland also sponsored the National Journalism Center, "developed to train budding journalists in free market political and economic principles."[31] Philip Morris's point of view, of course, was that people should not be discouraged from smoking, but they made common cause with various groups and individuals committed to "free market political and economic principles."

Perhaps this is why among the scores of think tanks and organizations that Philip Morris supported, we find the seemingly obscure Ludwig von Mises Institute. Ludwig von Mises, an Austrian aristocrat, was one of the founders of modern laissez-faire economics.[32] And this brings us to the crux of our story, the pivot around which these diverse actors came together. The link that unites the tobacco industry, conservative think tanks, and the scientists in our story is the defense of the free market.

Throughout our story, the people involved demanded the right to be heard, insisting that we—the public—had the right to hear both sides and that the media had an obligation to present it. They insisted that this was only fair and democratic. But were they attempting to preserve democracy? No. The issue was not free speech; it was free markets. It was the appropriate role of government in monitoring the marketplace. It was *regulation*. So we must consider the ideology that drove the merchants of doubt—the ideology of laissez-faire economics—before we finally turn to the question of how to make sure that we don't get fooled again.

Market Fundamentalism and the Cold War Legacy

During the second half of the twentieth century, American foreign policy was dominated by the Cold War and American domestic politics was dominated by anti-Communism. Our protagonists—Fred Seitz, Fred Singer, Bill Nierenberg, and Robert Jastrow—were fiercely anti-Communist, and viewed science as crucial in helping to contain its spread.

In the early stages of their careers, they helped to build the weapons and rocketry programs that played a key role in American nuclear defense; in later years they used their positions of expertise and authority to defend the maintenance and expansion of the nuclear state, providing "scientific" credibility to arguments against détente and for continuous rearmament. As we saw in chapter 2, Jastrow, Nierenberg, and Seitz created the "scientific" pro–Star Wars lobby, which gave them tremendous credibility in hawkish conservative political circles.

When the Cold War ended, these men looked for a new great threat. They found it in environmentalism. Environmentalists, they implied, were "watermelons": green on the outside, red on the inside. Each of the environmental threats we've discussed in this book was a market failure, a domain in which the free market had created serious "neighborhood" effects. But despite the friendly sound of this term, these effects were potentially deadly—and global in reach. To address them, governments would have to step in with regulations, in some cases very significant ones, to remedy the market failure. And this was precisely what these men most feared and loathed, for they viewed regulation as the slippery slope to Socialism, a form of creeping Communism.

Fred Singer gave his game away when he denied the reality of the ozone hole, suggesting that people involved in the issue "probably [have] . . . hidden agendas of their own—not just to 'save the environment' but to change our economic system . . . Some of these 'coercive utopians' are socialists, some are technology-hating Luddites; most have a great desire to regulate—on as large a scale as possible."[33] He revealed a similar anxiety in his defense of secondhand smoke: "If we do not carefully delineate the government's role in regulating [danger] . . . there is essentially no limit to how much government can ultimately control our lives."[34] Today tobacco, tomorrow the Bill of Rights. Milton Friedman said much the same in *Capitalism and Freedom*: that economic freedom is as important as civic freedom, because if you lose one, it is only a matter of time before you lose the other.[35] And so one must defend free markets with the same vigor and vigilance as free speech, free religion, and free assembly.

The billionaire investor George Soros has coined a term to describe this perspective: "free market fundamentalism." It is the belief not simply that free markets are the best way to run an economic system, but that free markets are the *only* way that will not ultimately destroy our other freedoms. "The doctrine of laissez-faire capitalism holds that the common good is best served by the uninhibited pursuit of self-interest,"[36] Soros wrote. Like its bête noire, Marxism, laissez-faire economics claimed to be scientific, based upon immutable laws of nature, and also like Marxism, it has not stood the test of experience. If it were a scientific theory, it would have long ago been rejected.[37] Free-market fundamentalism is an article of faith.

"Scientific Socialism" wasn't scientific because when evidence suggested that some of its central claims might be wrong, its advocates refused to accept that; for the same reason, free market fundamentalism isn't scientific, either. The basic tenet of laissez-faire, that "free and competitive markets bring supply and demand into equilibrium and thereby ensure the best allocation of resources," is an axiom that turns out not to be true.[38] Prices can be displaced from their "equilibrium ideal" for long periods of time, as any American impacted by the ongoing housing market collapse can attest.

Even Milton Friedman acknowledged that there may be external costs that markets fail to account for—and pollution is the clearest example. Regulation is needed to address external costs, either by preventing them or by compensating those who are saddled with them.

Friedman was a true believer in the market—he thought that external costs were rarely high enough to justify government intervention. But most of us want our governments to protect us from harm in many, diverse ways. We want police and firefighters to protect our homes; we want to make sure that our food supply is not contaminated and the water that comes out of our tap is clean; we want to know that drugs we buy at the pharmacy won't kill us. And in recent months, we've come to see the consequences of insufficient regulation of financial markets.

Moreover, the idea that free markets produce optimum allocation of resources depends on participants having perfect information. But one of several ironies of our story is that our protagonists did everything in their power to ensure that the American people did not have good (much less perfect) information on crucial issues. Our protagonists, while ostensibly defending free markets, distorted the marketplace of ideas in the service of political goals and commercial interests. The American belief in fairness and the importance of hearing "both sides" was used and abused by people who didn't want to admit the truth about the impacts of industrial capitalism.

Free market fundamentalists can perhaps hold to their views because often they have very little direct experience in commerce or industry. The men in our story all made their careers in programs and institutions that were either directly created by the federal government or largely funded by it. Robert Jastrow spent the lion's share of his career at the Goddard Institute for Space Studies—part of NASA. Frederick Seitz and Bill Nierenberg launched their careers in the atomic weapons programs, and expanded them at universities whose research activities were almost entirely funded by the federal government at taxpayer expense. Fred Singer worked directly for the government, first at the National Weather Satellite Service, later in the Department of Transportation. If government is bad and free markets are good, why did they not reject government support for their own research and professional positions and work in the private sector?

Many honest people who actually run businesses welcome reasonable government regulation with rules that prevent bad behavior—like unfair business practices or polluting the environment—so long as the rules are clear and fair, and create a stable, level playing field. After all, a corporation that invests in pollution control wants to know that it won't suffer in the marketplace for doing the right thing.[39] But the most serious critique of the central tenet of free market fundamentalism is simply that it is wrong, factually.

History shows that markets do fail, sometimes spectacularly. During the Great Depression, capitalism was in crisis, and citizens of widely varying political and moral persuasions accepted the New Deal as necessary to save it. The alternative, it appeared to almost everyone, was a complete collapse that could indeed lead to Communism, or some other form of totalitarianism.[40] At the same time, the phrase "free enterprise" was invented and marketed by the business community, along with the notion of "the American Way," to articulate the anxiety that something important might be lost if the New Deal went too far.[41] The exigencies of the Depression and World War II, however, made arguments for the "invisible hand" seem quaint, and the New Deal concentrated power and authority in the federal government in a manner undreamed of by our Founding Fathers.

The Cold War revived these arguments—Soviet abuses of power became increasingly clear, even to former leftists—and they have driven American conservative ideology ever since. Ronald Reagan is credited with challenging the New Deal, with its presumptions of the necessity and beneficence of big government, but the ideals he instantiated had already been articulated by Friedman in 1962—the year of the Cuban Missile Crisis—the coldest moment of the Cold War. Indeed, Friedman later argued that Reagan's positions were the same as Barry Goldwater's; it had just taken twenty years for their wisdom to be recognized.[42] Bill Nierenberg agreed. On the occasion of the twenty-fifth anniversary of the University of California, San Diego, Roger Revelle was asked about Bill Nierenberg's politics, and he replied that they were more than just conservative: "Bill Nierenberg . . . thinks the whole New Deal was a mistake, no kidding."[43]

Since the Cold War was responsible for both the general resurgence of free market ideology in the United States and the specific professional success of the scientists in our story, it's not entirely surprising that these men would demonize their latter-day agonists as enemies of freedom. As we've seen already, Robert Jastrow, Fred Singer, and Dixy Lee Ray—along with political propagandists such as George Will and Rush Limbaugh—routinely accused environmentalists (and sometimes scientists whose work contributes to environmental goals) of being Communists, Socialists, or fellow travelers. We noted earlier how George Will asserted that environmentalism was a "green tree with red roots."[44] But he was hardly the only one.

When Dixy Lee Ray addressed the Progress Foundation Economic Conference in 1992 on the subject of "Global Warming and Other Environmental Myths," she began by declaring, "I believe in freedom. I believe in liberty." (As if climate scientists didn't!) The story of the twentieth century was a progress tale, she explained, except that environmentalists insisted that progress must now stop. Sustainability was replacing progress as the leitmotif of the century, and this was a problem because liberty depended on progress.[45] Without economic progress there would be no economic growth, and without growth, governments would be forced to control resources. And to control resources, governments would have to control people.

The specter of expanded government control was often linked to the threat of global governance. This theme emerged strongly in the run-up to the Earth Summit at Rio, as Ray and others feared that a global treaty on climate change would decrease national sovereignty. They also feared that this would happen not of *necessity*, but by *design*. Ray concluded her speech to the Progress Foundation by frankly insisting that the agenda of the Earth Summit was Socialist, its objective to "bring about a change in the present system of independent nations . . . [a] World Government with central planning by the United Nations. Fear of environmental crises, whether such crises are real or contrived, is expected to lead to total compliance."[46]

Ray recapitulated this argument in an interview with the Acton Institute for the Study of Religion and Liberty, whose opening question to her was this: "With the world-wide decline of socialism, many individuals think that the environmental movement may be the next great threat to freedom. Do you agree?" Ray replied, "Yes, I do . . . The International Socialist Party, which is intent upon continuing to press countries into socialism, is now headed up by people within the United Nations. They are the ones in the UN environmental program, and they were the ones sponsoring the so-called Earth Summit." When asked, "Do you see a big influence by the radical environmentalists there?" again she replied, "Oh yes. No question about that, the radicals are in charge."[47] And who accompanied Ray to Rio? Fred Smith, the founder and head of the Competitive Enterprise Institute.[48]

Ray was not the only one to strike the theme that the Earth Summit was a Socialist front. Fred Singer similarly argued in the *Wall Street Journal* that the Earth Summit would "shackle the planet."[49] Patrick Michaels argued that "we're about to centrally plan the world's energy economy based on the threat of global warming."[50] Steve Milloy repeatedly attacked *Consumer Reports* for what one commentator has described as "socialism, sensationalism, and scaring consumers away from products."[51] More recently, Patrick Michaels, a longtime critic of climate science and policy scholar at the Cato Institute, criticized plans for a cap and trade system to control greenhouse gases as "Obamunism."[52]

Perhaps the best example of the thinking behind our story comes from Richard Darman, head of the Office of Management and Budget in the administration of George H. W. Bush. In 1990 Darman gave a speech in which he attacked environmentalists as having lost faith in America and accepting the inevitability of American decline. Darman's bête noire, the *New York Times* reported, was green (perhaps they should have said vert), as he accused environmentalists of being closet Socialists: "Americans did not fight and win the wars of the 20th century to make the world safe for green vegetables."[53]

All this had real impact. After George H. W. Bush signed the U.N. Framework Convention on Climate Change, formulated at Rio, the Republican Party turned against it and led the charge against its follow-up, the Kyoto Protocol, which would have put teeth into the general principles established at Rio. President Bush's pledge to take concrete action to protect the planet had vanished along with his promise of no new taxes. The world would not be made safe for green vegetables. It would not even be safe for polar bears. Or people living on Pacific Islands.

Political scientist Peter Jacques, with sociologist colleagues Riley Dunlap and Mark Freeman, has shown that books skeptical of the reality of environmental issues increased fivefold in the 1990s over the preceding decade (even as the scientific consensus about them was coalescing), and the Republican turn against environmentalism occurred even as popular support for U.S. environmentalism was rising.[54] This observation brings us back to the early stages of our story and the debate over SDI and nuclear winter.

With the collapse of the Soviet Union, Cold Warriors looked for another great threat. They found it in environmentalism, which just at that very moment had identified a crucial global issue that required global response. In the early 1990s, global warming changed from a prediction about the future to a fact about the present. Global warming became the most charged

of all environmental debates, because it *is* global, and it implicates everything and everyone. If the rules of economic activity are the central concern
of contemporary conservatives, then global warming has to be central, too,
because it stems from how we produce and use energy, and energy is involved in all economic activity. Nicholas Stern, formerly chief economist
and senior vice president of the World Bank from 2000 to 2003, and principal author of the *Stern Review of the Economics of Climate Change* (commissioned by U.K. prime minister Gordon Brown), has called climate change
"the greatest and widest-ranging market failure ever seen."[55] No wonder the
defenders of free market capitalism are worried.

The "reds dressed in green" refrain continues today. In December 2009,
as world leaders tried yet again to craft an agreement to control greenhouse
gases—seventeen years after the U.N. Framework Convention on Climate
Change committed them to do just that—Charles Krauthammer declaimed
in the *Washington Post* that environmentalism was socialism by other
means, a brazen attempt to transfer wealth from rich to poor. "With socialism dead, the gigantic heist is now proposed as a sacred service of the
newest religion: environmentalism . . . the Left was adrift until it struck
upon a brilliant gambit: metamorphosis from red to green." Whether an
agreement was achieved in Copenhagen or not, Krauthammer went on,
Americans needed to beware of the enemy within: the EPA. "Since we op-
erate an overwhelmingly carbon-based economy, the EPA will [soon] be
regulating practically everything . . . Not since the creation of the Internal
Revenue Service has a federal agency been given more intrusive power over
every aspect of economic life . . . Big Brother isn't lurking in CIA cloak.
He's knocking on your door, smiling under an EPA cap."[56]

Some environmentalists no doubt *are* Socialists, but in our experience
very few climate scientists are. Moreover, even if all environmentalists
were socialists, it does not follow that global warming is a myth. One can
believe in the superiority of the capitalist system and advocate for market-
based solutions to pollution—as many people do—but it does not follow
that one should doubt the science that demonstrates the need for such
solutions. Acid rain, secondhand smoke, the destruction of stratospheric
ozone, and global warming are all real problems; the real question is how
to address them. Denying their truth does not make them go away. On the
contrary, the longer we delay, the worse these problems get, increasing the
odds that governments will have to take the draconian actions that conservatives most fear.

Which leads to the second great irony of our story. Men like Bill Nierenberg were proud of the role they had played in defending liberty during the Cold War and understood their latter-day activities as an extension of that role. They feared that overreaction to environmental problems would provide the justification for heavy-handed government intervention in the marketplace and intrusion in our personal lives. That was not an unreasonable anxiety, but by denying the scientific evidence—and contributing to a strategy of delay—these men helped to create the very situation they most dreaded.

Consider the case of Gus Speth.

We met Gus Speth in chapter 3 as a member of President Jimmy Carter's Council on Environmental Quality, and an advocate for action against acid rain. Speth is no rock-throwing radical. Born in South Carolina, he is the consummate Southern gentleman: well-spoken, well educated, well regarded. As an undergraduate he attended Yale, went to Oxford as a Rhodes Scholar, and returned to Yale for law school. During his long career he taught at Yale and Georgetown, served as an advisor to President Carter, worked for the United Nations, and in 1999 returned to Yale once again as the dean of the School of Forestry and Environmental Studies. *Time* magazine once called him the "ultimate insider."[57]

But after forty years as an "inside" environmentalist, Speth has become radicalized by the world's failure to act on problems we have known about for a long time. He now concludes that radical change is needed. "The global economy is crashing against the Earth," he warns in his recent book, *The Bridge at the Edge of the World*. Environmental deterioration is driven by economic activity, so we must consider if there is a fundamental flaw in our economic system. His conclusion "after much searching and considerable reluctance, is that most environmental deterioration is a result of systemic failures of the capitalism that we have today and that long-term solutions must seek transformative change in the key features of this contemporary capitalism."[58]

The merchants of doubt have produced just the effect they most dreaded. Southern gentlemen are now preparing to dismantle capitalism.

Can't Technology Save Us?

Back in the 1980s, the Reagan administration made clear to the National Academy of Sciences their view that "technology will ultimately be the answer to the problems of providing energy and protecting the environment."[59]

Many liberals and academics agree that without change in our energy technologies, there will be no solution to global warming. The question is not whether we turn to technology for help; the question is whether we can *assume* that free markets will produce those technologies freely, of their own accord. The question is also whether they will do so *in time*—so we can relax in the comforting knowledge that they will—or whether we need to get out of our chairs and *do* something.

The belief that technology can solve society's problems is central to the school of thought known as Cornucopianism, promoted by the economist Julian Simon. Cornucopians see themselves as responding to the philosophy of Thomas Malthus, who famously argued that the poor were poor because they had too many children, and the Enlightenment belief in the continued improvement of mankind was erroneous, because unchecked population growth would eventually outstrip resources. This would stop progress in its tracks, leading ultimately to a good deal of human suffering. So humans had to stop population growth, or else suffer the consequences. (Malthus was no liberal; he thought the best way to curtail population was to cut off aid to the poor.)

Cornucopians—along with historians of science and technology— recognize that Malthus's dire predictions did not come true in large part because he failed to appreciate that technological innovation can make it possible to do more with the same resources (or in some cases even less). Even though the world is far more populous today than it was when Malthus wrote, we manage to feed many (if not all) of these mouths in large part to the technological innovations of the green revolution.[60] However, Cornucopians go one step further than historians, arguing that this will *always* be the case, so long as human creativity and innovation are not circumscribed. The best way to do this, they submit, is to keep markets free, so innovators can innovate and reap the benefits of their inventions. (Contra Malthus, Cornucopians also oppose population control, feeling that human ingenuity is the ultimate renewable resource.) They insist that optimism is the correct order of the day, because history is a story of steady progress, at least in those times and places where people and their markets have been free.[61]

Simon's admirers like to call him the "doomslayer" for his insistence that gloom and doom scenarios are wrong and the future remains bright.[62] In his 1984 book, *The Resourceful Earth* (coedited with Herman Kahn), Simon insisted that, contrary to the view expressed by the members of the President's Council on Environmental Quality in their *Global*

2000 *Report to the President* (published in 1980), the future world would be "less crowded . . . less polluted, more economically stable, and less vulnerable to resource-supply disruption" than the world today.[63] In his 1995 follow-up, *State of Humanity* (which billed itself as a "balanced" look at the question), Simon began with the Panglossian assertion that the outlook for the future was "even more happy than before" because conditions had improved in nearly all areas they had studied, and more and more areas were coming under such study.[64]

Simon's philosophy can be seen ramifying throughout the stories we have followed. Dixy Lee Ray presented herself as "unmasking" the "doom-crying opponents of all progress."[65] Bill Nierenberg responded to the Reagan administration admonition not to present a "wolf-crying" scenario by painting a Cornucopian idyll of future technological solutions to global warming. And Fred Singer had been vexed by doomsday predictions since 1970, when he wrote a guest editorial for *Science* suggesting that scientists might be overreacting to concerns like global warming.

In 1970, few scientists had even thought about global warming, so Singer was clearly ahead of the curve. In hindsight we might also say that he got ahead of himself in thinking that the problem was already resolved. "After many years of speculation and discussion, the effects of fossil-fuel burning on climate seem to be reasonably clear," Singer wrote in 1970. "While there has been an actual increase in the CO_2 content, the 'greenhouse effect' of climate warming has been small, and even negative, because of the overwhelming effects of atmospheric dust which tends to cool the atmosphere."[66] Singer may have been right in his analysis of why no warming had been seen at that time: many scientists now believe that an early warming signal was not seen because of the countervailing effects of atmospheric aerosols, predominantly sulfates produced by burning coal (the same pollutants that contribute to acid rain). Singer, however, dismissed the possibility that as CO_2 continued to mount, the warming effect would eventually emerge (as it has). Rather, he assumed that the cooling effects of atmospheric aerosols would continue to counterbalance the warming effect of CO_2, and all would be well with the world. Thus, he concluded his editorial with an admonition for scientists "not to cry 'wolf' needlessly or too often."[67]

In the 1980s Singer recognized that Cornucopian arguments rest on the presumption of an adequate supply of affordable energy, but he later dropped this concern and moved solidly into the Cornucopian camp.[68] His 1999 book, *Hot Talk, Cold Science: Global Warming's Unfinished Debate* (with a forward by Frederick Seitz), was published by the Independent Institute, for which Julian Simon served on its board of advisors and was an influential guiding spirit.[69] Singer also published a chapter in Simon's *State of Humanity* book, along with Patrick Michaels, whom we met in chapter 6 denying global warming, and Laurence Kulp, whom we met in chapter 3 defending the Reagan administration's position on acid rain.

The Cornucopian argument was recently given new life by Danish political scientist Bjørn Lomborg. Lomborg's work has appeared in major newspapers around the globe, including the *Wall Street Journal*, the *New York Times*, the *Economist*, the *Los Angeles Times*, and the *Boston Globe*. He has appeared many times on television in the United States and Europe, including *60 Minutes*, *Larry King Live*, *20/20*, and the BBC.[70] His most well-known book, *The Skeptical Environmentalist: Measuring the Real State of the World*, closely follows the Cornucopian line that the world is getting steadily better, and that environmental claims are at best exaggerations, if not outright distortions and falsehoods. Indeed, he begins *The Skeptical Environmentalist* with a quote from Julian Simon:

> This is my long-run forecast in brief: The material conditions of life will continue to get better for most people, in most countries, most of the time, indefinitely. Within a century or two, all nations, and most of humanity will be at or above today's Western living standards. I also speculate, however, that many people will continue to think and say that the conditions of life are getting worse.[71]

In *The Skeptical Environmentalist*, Lomborg repeats what are by now familiar claims: that Rachel Carson was wrong about DDT, that global warming isn't a serious problem, that our forests are doing fine. Life, in general, is much better for just about everyone, and there's "no need to worry about the future."[72] So what are the environmentalists fussing for?[73]

Lomborg's book has been criticized as a textbook example of the misuse of statistics.[74] In 2002, four leading scientists enumerated in *Scientific American* the ways in which Lomborg's math was "misleading." In Denmark, a struggle erupted over the book, and charges of scientific dishonesty were leveled against Lomborg.[75] Ultimately, the Danish Ministry of Science, Technology, and Innovation ruled that Lomborg couldn't be guilty of scientific dishonesty, because it had not been shown that *The Skeptical Environmentalist* was a work of science![76]

However you characterize *The Skeptical Environmentalist*, its arguments contain two fatal flaws. In opposing brisk action against global warming, Lomborg insists that other problems, such as world hunger, are more pressing. This is the classic false dichotomy, because there is no intrinsic reason why humans cannot address both, and unchecked climate change will almost surely increase hunger, as poor nations struggle to respond to strained circumstances.[77] Moreover, as we have pointed out elsewhere, world hunger persists for many reasons, but *not* because the Western world has been busy doing anything about global warming.[78] A second flaw in Lomborg's reasoning is that his statistics are almost entirely based on impacts on humans (life expectancy, calories eaten, etc.). He freely admits that he writes from the perspective of human needs and wants: his statistics largely address the quantity of years of human life lived and the numbers of individual human lives saved by various improvements and innovation.[79] (He also counts the lives allegedly lost to the banning of DDT.) Such measures say nothing about the impact of human activities on nonhuman species, or the condition of the world that our children will inherit. It is quite possible to live in a manner that is better for us, but leaves the world impoverished for our descendants. Lomborg's arguments also say nothing about the quality of our lives, yet that is precisely what was at stake in traditional arguments for conservation, and remains a central element of many environmental concerns today.

Rachel Carson was not indifferent to humans—a good deal of *Silent Spring* was about bioaccumulation and its potential long-term impact on humans—but even if DDT had been proven harmless to people, her argument would have stood: that DDT was doing serious harm to nature. Carson's concern—shared with many contemporary environmentalists—was with the ethics of eradicating whole species (whether or not they were of use to us) and leaving our children a world that was ecologically and aesthetically impoverished. A rare flower may be beautiful even if its contribution to atmospheric oxygen is negligible; a venus flytrap may thrill us even if it does little to protect us from malaria-carrying mosquitoes. As we have put it elsewhere, Lomborg and his followers make the philosophical error of thinking that things that can't be counted don't count.[80]

Lomborg has been defended in the *Financial Times*, the *Wall Street Journal*, the *Economist*, and by many advocates of laissez-faire economics, such as the Center for the Defense of Free Enterprise.[81] He also has links to many of the ideologically driven think tanks we have already encountered, including the Competitive Enterprise Institute, the Hoover Institution, and the Heartland Institute.[82] This is not a surprise, for the Cornucopian philosophy is linked to free market fundamentalism in its conviction that the state is the problem, not the solution. Among its various activities, the Independent Institute (publisher of Singer's book *Hot Talk, Cold Science*) sponsors the Sir John M. Templeton essay contest for college students and junior faculty. The essay prompt for 2010 is:

> Everyone wants to live at the expense of the state. They forget that the state wants to live at the expense of everyone.
>
> —Frederic Bastiat (1801–1850)

> Assuming Bastiat is correct, what ideas or reforms could be developed that would make people better aware that government wants to live at their expense?[83]

Of course, the Cornucopians are not entirely wrong. Some governments do grow at the expense of their people, and many aspects of modern life are better (at least for many people) than they were in previous centuries. The problem with their view is twofold.

The first problem is their presumption that these advances will *necessarily* continue. If we have indeed reached a tipping point, as many leading scientists fear, then the past may not be a guide to the future. Past environmental changes were mostly local and reversible. Today, human activities have a global reach. We are changing our planet in radical ways, and we may not have the wherewithal to respond to the challenges ahead, at least not without enduring a good deal of discomfort and dislocation. Moreover, some of these changes—like sea level rise and the melting of Arctic ice—are almost certainly irreversible.

The second problem with Cornucopianism is its assertion that past advances have been the result—and could *only* have been the result—of free market systems. This assertion is demonstrably false.

Technofideism

The history of technology does not support the Cornucopian view of the relation between technological innovation and free markets. Many technologies crucial to the advance of civilization were invented *before* the advent of capitalism. Moreover, the Soviet Union, for all its failures, was a technologically innovative society. Most famously, they launched an artificial satellite into space—*Sputnik*—before the United States did. The problem with the Soviet Union was not that they lacked technological innovation. The problem was that the benefits did not accrue to their people. Cornucopians hold to a blind faith in technology that isn't borne out by the historical evidence. We call it "technofideism."

Why do they hold this belief when history shows it to be untrue? Again we turn to Milton Friedman's *Capitalism and Freedom*, where he claimed that "the great advances of civilization, in industry or agriculture, have never come from centralized government."[84] To historians of technology, this would be laughable had it not been written (five years after *Sputnik*) by one of the most influential economists of the second half of the twentieth century. The most important technology of the industrial age was the ability to produce parts that were perfectly identical and interchangeable. Blacksmiths and carpenters couldn't do it; in fact, humans can't do it routinely in any profession. Only machines can. It was the U.S. Army's Ordnance Department that developed this ability to have machines make parts for other machines, spending nearly fifty years on this effort—an inconceivable period of research for a private corporation in the nineteenth century.[85] Army Ordnance wanted guns that could be repaired easily on or near a battlefield by switching out the parts. Once the basic technology to do this—machine tools, as we know them today—was invented, it spread rapidly through the American economy. Despite efforts to prevent it, it soon spread to Europe and Japan, as well. Markets spread the technology of machine tools throughout the world, but markets did not create it. Centralized government, in the form of the U.S. Army, was the inventor of the modern machine age.

Machine tools are not the exception that proves the rule; there are many other cases of government-financed technology that were commercialized and redounded to the benefit of society. Even while Friedman was writing his soon-to-be-famous book, digital computers were beginning to find uses beyond the U.S. government's weapons systems, for which they were originally developed. Private enterprise transformed that technology into something that could be used and afforded by the masses, but the U.S. government made it possible in the first place. The U.S. government also played a major role in the development of Silicon Valley.[86] In recent years, something we now all depend on—the Internet, originally ARPANET—was developed as a complex collaboration of universities, government agencies, and industry, funded largely by the Department of Defense's Advanced

Research Projects Agency. It was expanded and developed into the Internet by the government support provided by the High Performance Computing and Communication Act of 1991, promoted by then-senator Al Gore.[87]

In other cases, new technologies were invented by individual or corporate entrepreneurs, but it was government action or support that transformed them into commercially viable technologies; airplanes and transistors come to mind.[88] (Transistors were explicitly promoted by the U.S. government when they realized that Minutemen missiles needed onboard rather than remote controls, and vacuum tubes would not suffice.)[89] Still other technologies were invented by individuals but were spread through government policy. Electricity was extended beyond the major cities by a federal loan-guarantee program during the Great Depression.[90] The U.S. interstate highway system, which arguably created postwar America as we know it, was the brainchild of President Dwight Eisenhower, who recognized the role it could play both in the U.S. economy and in national defense; it became the model for similar highway systems around the globe. And nuclear power, which may help us out of the global warming conundrum, was a by-product of the technology that launched the Cold War: the atomic bomb. The relationship between technology, innovation, and economic and political systems is varied and complex. It cannot be reduced to a simple article of faith about the virtues of a free market.

What this all adds up to—to return to our story—is that the doubt-mongering campaigns we have followed were not about science. They were about the proper role of government, particularly in redressing market failures. Because the results of scientific investigation seem to suggest that government really did need to intervene in the marketplace if pollution and public health were to be effectively addressed, the defenders of the free market refused to accept those results. The enemies of government regulation of the marketplace became the enemies of science.

Why Didn't Scientists Stand Up?

If the skeptical arguments pursued by our protagonists were not about science—if they were politics camouflaged as science—then why didn't scientists recognize this, and say something? Why did the scientific community stand by while this was happening?

With the notable exception of the atmospheric science community's defense of Ben Santer, scientists fighting back have been conspicuously scarce. We would have liked to have told heroic stories of how scientists set the record straight—and in a few cases we have. Gene Likens and his colleagues sought help from the National Academy of Sciences when the White House interfered with their acid rain peer review report. F. Sherwood Rowland tried to correct Fred Singer's misrepresentations of the ozone debate. Climate modeler Stephen Schneider, who appeared in chapter 3 in the nuclear winter debate, has for many years actively spoken out against misrepresentations of climate science, including the problem of false balance in the media.[91] But these voices are, unfortunately, scant. Clearly, scientists knew that many contrarian claims were false. Why didn't they do more to refute them? Where were the real scientific voices speaking out against the Potemkin village?

One reason for scientists' silence involves the complex dance that takes place in science between the individual and the group. Scientists are strongly motivated by the accolades and prestige that accrue from making a major discovery. Yet, at the same time they are often reluctant to attract the limelight for themselves. The reason is twofold. First, nearly all modern science is the result of teamwork—a point to which we shall shortly return—and second, knowledge *counts* as science when it reflects the consensus of expert opinion, even if it originated in the genius or creativity of one person. In the modern world, any scientific breakthrough is likely to be the result of the collective effort of several dozen, scores, or hundreds of researchers. The IPCC today attempts to summarize the work of *thousands*. A scientist who steps out to speak on behalf of his colleagues risks censure, lest colleagues think he is trying to take all the credit for himself.

The scientific societies have tried to address this by developing formal statements on climate change that reflect the collective wisdom of their members, but these statements tend to be dry at best, and often nearly impossible for a normal person to decipher. Who among us has read the IPCC *Summary for Policymakers*, much less the thousands of pages of actual reports? Indeed, who on the planet has read all this stuff? What average citizen knows that the American Meteorological Society even exists, much less knows to visit its home page to look for its climate-change statement?[92]

Clearly, it's ridiculous to imagine that anyone would, so someone has to summarize and communicate it. Then another difficulty arises. Scientists are finely honed specialists trained to create new knowledge, but they have little training in how to communicate to broad audiences, even less in how to defend scientific work against determined and well-financed contrarians.

They often have little talent or taste for it, either. Until recently, most scientists have not been particularly anxious to take the time to communicate broadly. They consider their "real" work to be the production of knowledge, not its dissemination, and they often view these two activities as mutually exclusive. Some even sneer at colleagues who communicate to broader audiences, dismissing them as "popularizers."

Scientists' commitment to expertise and objectivity also places them in a delicate position when it comes to refuting false claims. If a scientist jumps into the fray on a politically contested issue, he may be accused of "politicizing" the science and compromising his objectivity—as Carl Sagan was when he tried to call public attention to the dangers of nuclear winter. This places scientists in a double bind: the demands of objectivity suggest that they should keep aloof from contested issues, but if they don't get involved, no one will know what an objective view of the matter looks like.[93]

Scientists have also been afraid to get involved because they have seen what happens when they do. Ben Santer's experience is unfortunately not unique. In 2005, Penn State researcher Michael Mann was the subject of an angry attack by Congressman Joe Barton of Texas, who demanded that Dr. Mann hand over detailed information about the sources of his research support, the location of all his data, and much more—even though the scientific results under scrutiny had been published in peer-reviewed journals and there was no evidence that Mann had done anything wrong, anything that is, except provide compelling evidence that the Earth was rapidly warming.[94] In the course of writing this book, we have been attacked, too, including by Senator James Inhofe of Oklahoma.[95] Nearly fifteen years after Ben Santer was first attacked by Fred Seitz, Fred Singer, and Bill Nierenberg, he continues to be harassed. Most recently Climate Audit, run by Steve McIntyre (a Canadian geologist with links to the mining industry, who was previously involved in the attacks on Michael Mann), has used the Freedom of Information Act (FOIA) to demand details about his research. FOIA, of course, was designed to enable citizens to know what their own government was up to, not to help foreigners harass our own scientists.[96] In any case, the whole point of the Lawrence Livermore Model Intercomparison Project, which Santer leads, is to make the relevant data and models available to any scientist who wants to use them. Anyone wanting to replicate Santer's work is free to do so. There is no need for FOIA.

These attacks have had a chilling effect. At a recent conference, a colleague told one of us that in IPCC discussions, some scientists have been reluctant to make strong claims about the scientific evidence, lest contrarians "attack us."[97] Another said that she'd rather err on the side of conservatism in her estimates, because then she feels more "secure."[98] Biologist Kåre Fog has described how many Danish scientists gave up trying to correct the many false claims propagated by Bjørn Lomborg because they did not wish to be subject to misrepresentations of their work and victims of vicious personal attacks.[99] Intimidation works.

Perhaps the most forgivable reason why scientists have not gotten more involved is because they love science, and believe that truth wins out in the end. It is their job—their singular job—to figure out what that truth is. Someone else can best popularize it. Someone else can better communicate it. And if there's garbage being promoted somewhere, someone else can deal with it. Indeed, it would be wrong for them to take time away from their work to deal with quotidian matters. As we noted, one leading scientist said about the 1983 report, Changing Climate, "We knew it was garbage, so we just ignored it."[100] Unfortunately, garbage doesn't just go away. Someone has to deal with it, and that someone is all of us: journalists who report scientific findings, specialist professional bodies who represent the scientific fields, and all of us as citizens.

Recently, the distinguished economist Robert Samuelson repeated an argument in the pages of the *Washington Post*—and again in *Newsweek*—that Bill Nierenberg made twenty-five years ago: Global warming can't really be solved—so we just have to get used to it.[101] But there *are* solutions. Global warming is a big problem, and to solve it we have to stop listening to disinformation. We have to pay attention to our science and harness the power of our engineering. Rome may not be burning, but Greenland is melting, and we are still fiddling. We *all* need a better understanding of what science really is, how to recognize real science when we see it, and how to separate it from the garbage.

A New View of Science

IMAGINE A GIGANTIC BANQUET. Hundreds of millions of people come to eat. They eat and drink to their hearts' content—eating food that is better and more abundant than at the finest tables in ancient Athens or Rome, or even in the palaces of medieval Europe. Then, one day, a man arrives, wearing a white dinner jacket. He says he is holding the bill. Not surprisingly, the diners are in shock. Some begin to deny that this is *their* bill. Others deny that there even *is* a bill. Still others deny that they partook of the meal. One diner suggests that the man is not really a waiter, but is only trying to get attention for himself or to raise money for his own projects. Finally, the group concludes that if they simply ignore the waiter, he will go away.

This is where we stand today on the subject of global warming. For the past 150 years, industrial civilization has been dining on the energy stored in fossil fuels, and the bill has come due. Yet, we have sat around the dinner table denying that it is our bill, and doubting the credibility of the man who delivered it. The great economist John Maynard Keynes famously summarized all of economic theory in a single phrase: "There is no such thing as a free lunch." And he was right. We have experienced prosperity unmatched in human history. We have feasted to our hearts' content. But the lunch was not free.

It's not surprising that many of us are in denial. After all, we didn't *know* it was a banquet, and we didn't know that there would be a bill. Now we do know. The bill includes acid rain and the ozone hole and the damage produced by DDT. These are the environmental costs of living the way citizens of the wealthy, developed nations have lived since the Industrial Revolution. Now we either have to pay the price, change the way we do business, or both. No wonder the merchants of doubt have been successful. They've permitted us to think that we could ignore the waiter while we haggled about the bill.

The failure of the United States to act on global warming and the long delays between when the science was settled and when we acted on tobacco, acid rain, and the ozone hole are prima facie empirical evidence that doubt-mongering worked. Decision theory explains why. In their textbook, *Understanding Scientific Reasoning*, Ronald Giere, John Bickle, and Robert Mauldin show that the outcome of a rational decision-theory analysis is that if your knowledge is uncertain, then your best option is generally to do nothing. Doing something has costs—financial, temporal, or opportunity costs—and if you aren't confident those costs will be repaid in future benefits, you're best off leaving things alone. Moreover, acting to prevent future harm generally means giving up benefits in the present: certain benefits, to be weighed against uncertain gains. If we didn't know that smoking was dangerous, but we did know that it gave us pleasure, we would surely decide to smoke, as millions of Americans did before the 1960s. Uncertainty favors the status quo. As Giere and his colleagues put it, "Is it any wonder that those who benefit the most from continuing to do nothing emphasize the controversy among scientists and the need for continued research?"[1]

To change the way the problem of global warming looks, Giere and his colleagues conclude, you'd need "undeniable evidence both that doing nothing will lead to warming and that doing something could prevent it."[2] But as we have seen, *any* evidence can be denied by parties sufficiently determined, and you can never prove anything about the future; you just have to wait and see. So the question becomes, Why do we expect "undeniable" evidence in the first place?

The protagonists of our story merchandised doubt because they realized—with or without the help of academic decision theory—that doubt works. And it works in part because we have an erroneous view of science.

We think that science provides certainty, so if we lack certainty, we think the science must be faulty or incomplete. This view—that science could provide certainty—is an old one, but it was most clearly articulated by the late-nineteenth-century positivists, who held out a dream of "positive" knowledge—in the familiar sense of absolutely, positively true. But if we have learned anything since then, it is that the positivist dream was exactly that: a dream. History shows us clearly that science does not provide certainty. It does not provide proof. It only provides the consensus of experts, based on the organized accumulation and scrutiny of evidence.

Hearing "both sides" of an issue makes sense when debating politics in a two-party system, but there's a problem when that framework is applied to science. When a scientific question is unanswered, there may be three, four, or a dozen competing hypotheses, which are then investigated through research. Or there may be just one generally accepted working hypothesis, but with several important variations or differences in emphasis. When geologists were debating continental drift in the 1940s, Harvard professor Marlin Billings taught his students no less than nineteen different possible explanations for the phenomena that drift theory—later plate tectonics—was intended to explain.

Research produces evidence, which in time may settle the question (as it did as continental drift evolved into plate tectonics, which became established geological theory in the early 1970s). After that point, there are no "sides." There is simply accepted scientific knowledge. There may still be questions that remain unanswered—to which scientists then turn their attention—but for the question that has been answered, there is simply the consensus of expert opinion on that particular matter. That is what scientific knowledge *is*.

Most people don't understand this. If we read an article in the newspaper presenting two opposing viewpoints, we assume both have validity, and we think it would be wrong to shut one side down. But often one side is represented only by a single "expert"—or as we saw in our story—one or two. When it came to global warming, we saw how the views of Seitz, Singer, Nierenberg, and a handful of others were juxtaposed against the collective wisdom of the entire IPCC, an organization that encompasses the views and work of thousands of climate scientists around the globe—men and women of diverse nationality, temperament, and political persuasion. This leads to another important point: that modern science is a collective enterprise.

For many of us, the word "science" does not actually conjure visions of science; it conjures visions of *scientists*. We think of the great men of science—Galileo, Newton, Darwin, Einstein—and imagine them as heroic individuals, often misunderstood, who had to fight against conventional wisdom or institutions to gain appreciation for their radical new ideas. To be sure, brilliant individuals are an important part of the history of science;

men like Newton and Darwin deserve the place in history that they hold. But if you asked a historian of science, When did modern science *begin*? She would not cite the birth of Galileo or Copernicus. Most likely, she would discuss the origins of scientific *institutions*.

From its earliest, days, science has been associated with institutions—the Accademia dei Lincei, founded in 1609, the Royal Society in Britain, founded in 1660, the Académie des Sciences in France, founded in 1666—because scholars (savants and natural philosophers as they were variously called before the nineteenth-century invention of the word "scientist") understood that to create new knowledge they needed a means to test each other's claims. Medieval learning had largely focused on study of ancient texts—the preservation of ancient wisdom and the appreciation of texts of revelation—but later scholars began to feel that the world needed something more. One needed to make room for *new* knowledge.

Once one opened the door to the idea of new knowledge, however, there was no limit to the claims that might be put forth, so one needed a mechanism to vet them. These were the origins of the institutional structures that we now take for granted in contemporary science: journals, conferences, and peer review, so that claims could be reported clearly and subject to rigorous scrutiny.

Science has grown more than exponentially since the 1600s, but the basic idea has remained the same: scientific ideas must be supported by evidence, and subject to acceptance or rejected. The evidence could be experimental or observational; it could be a logical argument or a theoretical proof. But whatever the body of evidence is, both the idea and the evidence used to support it must be judged by a jury of one's scientific peers. Until a claim passes that judgment—that *peer review*—it is only that, just a claim. What counts as knowledge are the ideas that are accepted by the fellowship of experts (which is why members of these societies are often called "fellows"). Conversely, if the claim is rejected, the honest scientist is expected to accept that judgment, and move on to other things. In science, you don't get to keep harping on a subject until your opponents just give up in exhaustion.

The he said/she said framework of modern journalism ignores this reality. We think that if someone disagrees, we should give that someone due consideration. We think it's only fair. What we don't understand is that in many cases, that person has already received due consideration in the halls of science. When Robert Jastrow and his colleagues first took their claims to the halls of public opinion, rather than to the halls of science, they were stepping outside the institutional protocols that for four hundred years have tested the veracity of scientific claims.

Many of the claims of our contrarians had already been vetted in the halls of science and failed to pass the test of peer review. At that point, their claims could not really be considered scientific, and our protagonists should have moved on to other things. In a sense they were poor losers. The umpires had made their call, but our contrarians refused to accept it.

Moreover, in many cases these contrarians did not even attempt to have their claims vetted. In fact, many of them had stopped doing scientific research. Our story began in the 1970s, when Fred Seitz was already retired from the Rockefeller University and began defending tobacco, although he was a solid-state physicist, not a biologist, oncologist, or physician. The story continued in the 1980s, when Seitz joined forces with Robert Jastrow and William Nierenberg. How much original research on SDI or acid rain or the ozone hole or secondhand smoke or global warming did any of them do? The answer is nearly none. A search of the Web of Science—an index of peer-reviewed scientific publications maintained by the Institute for Scientific Information—shows that Frederick Seitz stopped doing original scientific research around 1970. After that he continued to publish here and there, but mostly book reviews, editorials and letters to editors, and a few works on great men in the history of science. Bill Nierenberg and Robert Jastrow similarly published little in the peer-reviewed journals during this period.

Fred Singer has perhaps the most credible claim to have been a working scientist during the course of our story. In the 1950s and '60s he published a substantial number of articles on physics and geophysics, many in leading journals such as *Nature*, *Physical Review*, and the *Journal of Geophysical Research*. But around 1970, he too shifted, from then on writing a large number of letters and editorials.[3] Web of Science lists some of these as articles, but it is at least debatable as to whether most of these constitute original scientific research, such as Singer's 1992 piece, "Warming Theories Need Warning Labels," published in the *Bulletin of Atomic Scientists* (which, not incidentally, contains an illustration of the domino effect— shades of his anti-Communism).[4] In the 1980s, Singer did a series of articles for the *Wall Street Journal* on oil resources, yet he was not a geologist, a petroleum engineer, or a resource economist, and had done little or no peer-reviewed research on the topic.[5]

The fact is that these men were never really experts on the diverse issues to which they turned their attention in their golden years. They were physicists, not epidemiologists, ecologists, atmospheric chemists, or climate modelers. To have been truly expert on all the different topics on which they commented, they would have to have been all of these things: epidemiologist and ecologist, atmospheric chemist and climate modeler. No one in the modern world is all of those things. Modern science is far too specialized for that. It requires a degree of focus and dedication that makes it a daunting task to be an expert in any area of modern science, much less in several of them at once. If nothing else, this should have clued observers in that these men simply could not have been real experts. An all-purpose expert is an oxymoron.

Journalists were fooled by these men's stature, and we are all fooled by the assumption that a smart person is smart about everything: physicists have been consulted on everything from bee colony collapse to spelling reform and the prospects for world peace.[6] And, of course, smoking and cancer. But asking a physicist to comment on smoking and cancer is like asking an Air Force captain to comment on the design of a submarine. He might know something about it; then again, he might not. In any case, he's not an expert.

So what do we do?

We all have to make decisions every day, and we do so in the face of uncertainty. When we buy a car, when we buy a house, when we choose health insurance or save for retirement, we make decisions and we don't allow uncertainty to paralyze us. But we may rely on people who we think can help us.

Normally, we try to make decisions based on the best information that we can get about the question. Let's say you need to buy a car. No doubt you will take some test drives, but you'll also talk to friends, especially those who know something about cars, and maybe read some magazines that evaluate cars, like *Consumer Reports* or *Car and Driver*. While you know that magazines make mistakes, and that prices and availability of options can vary, you assume that the information you find is reasonably accurate and realistic. Call it car and driver realism.

The metaphor isn't quite apt for our discussion though, because in the end buying a car is highly subjective, based to a large extent on questions of taste. I can decide what I think is the right car for *me*, but there are no experiments I can do or observations I can make that will settle the question for *others*. There is, in the end, no truth of the matter.

So consider a different example.

One of the largest financial decisions most of us make in our lives is the decision to buy a home. When we do, we consider numerous factors: the size and location; access to work, shopping, and recreation; safety and security; the quality of local schools; and of course the price. The process of deciding to make an offer can be wrenching, involving, like the car, a host of subjective factors, but with far more at stake. Once we've made the decision to make an offer, however, we need to do one more thing—something to which most of us give really rather little thought, considering how much is at stake.

We do a title search. Or rather, we *hire* someone to do a title search. We need to know that the title on the property actually belongs to the person who is selling it, and there are no outstanding claims or liens to stand in our way of ownership. If the person we hire to do the search is incompetent or dishonest, we could end up in a financial disaster. Yet we *do* trust the title search. Why? The short answer is because we don't have much choice. Someone has to do the title search, and we do not have the expertise to do it ourselves. We trust someone who is trained, licensed, and experienced to do it for us.

The sociologist Michael Smithson has pointed out that all social relations are trust relations. We trust other people to do things for us that we can't or don't want to do ourselves.[7] Even legal contracts involve a degree of trust, because the person involved could always flee to Venezuela. If we don't trust others or don't want to relinquish control, we can often do things for ourselves. We can cook our own food, clean our own homes, do our own taxes, wash our own cars, even school our own children. But we cannot do our own science.

So it comes to this: we must trust our scientific experts on matters of science, because there isn't a workable alternative.[8] And because scientists are not (in most cases) licensed, we need to pay attention to who the experts actually are—by asking questions about their credentials, their past and current research, the venues in which they are subjecting their claims to scrutiny, and the sources of financial support they are receiving.

If the scientific community has been asked to judge a matter—(as the National Academy of Sciences routinely is)—or if they have self-organized to do so (as in the Ozone Trends Panel or the IPCC), then it makes sense to take the results of their investigations very seriously. These *are* the title searches of modern science and public policy. It does not make sense to dismiss them just because some person, somewhere, doesn't agree. And it especially does not make sense to dismiss the consensus of experts if the dissenter is superannuated, disgruntled, a habitual contrarian, or in the pay of a group with an obvious ideological agenda or vested political or economic interest. Or in some cases, all of the above.

Sensible decision making involves acting on the information we have, even while accepting that it may well be imperfect and our decisions may need to be revisited and revised in light of new information. For even if modern science does not give us certainty, it does have a robust track record. We have sent men to the moon, cured diseases, figured out the internal composition of the Earth, invented new materials, and built machines to do much of our work for us—all on the basis of modern scientific knowledge. While these practical accomplishments do not prove that our scientific knowledge is true, they do suggest that modern science gives us a pretty decent basis for *action*.

In the early 1960s, one of the world's leading epidemiologists, initially skeptical of the idea that tobacco was deadly, came around to accepting that the weight of evidence showed that it was. In response to those who still doubted it and insisted that more data were needed, he replied,

> All scientific work is incomplete—whether it be observational or experimental. All scientific work is liable to be upset or modified by advancing knowledge. That does not confer upon us a freedom to ignore the knowledge we already have, to postpone action that it appears to demand at a given time. Who knows, asks Robert Browning, but the world may end tonight? True, but on available evidence most of us make ready to commute on the 8:30 next day.[9]

Don't get us wrong. Scientists have no special purchase on moral or ethical decisions; a climate scientist is no more qualified to comment on health care reform than a physicist is to judge the causes of bee colony collapse. The very features that create expertise in a specialized domain lead to ignorance in many others. In some cases lay people—farmers, fishermen, patients, indigenous peoples—may have relevant experiences that scientists can learn from. Indeed, in recent years, scientists have begun to recognize this: the Arctic Climate Impact Assessment includes observations gathered from local indigenous groups.[10] So our trust needs to be circumscribed, and focused. It needs to be very *particular*. Blind trust will get us into as least as much trouble as no trust at all. But without some degree of trust in our designated experts—the men and women who have dedicated their lives to sorting out tough questions about the natural world we live in—we are paralyzed, in effect not knowing whether to make ready for the morning commute or not. We are left, as de Tocqueville recognized two hundred years ago, with nothing but confused clamor. Or as Shakespeare suggested centuries ago, life is reduced to "a tale told by an idiot, full of sound and fury, signifying nothing."[11]

C. P. Snow once argued that foolish faith in authority is the enemy of truth. But so is a foolish cynicism.

In writing this book, we have plowed through hundreds of thousands of pages of documents. As historians during the course of our careers we have plowed through millions more. Often we find that, in the end, it is best to let the witnesses to events speak for themselves. So we close with the comments of S. J. Green, director of research for British American Tobacco, who decided, finally, that what his industry had done was wrong, not just morally, but also intellectually: "A demand for scientific proof is always a formula for inaction and delay, and usually the first reaction of the guilty. The proper basis for such decisions is, of course, quite simply that which is reasonable in the circumstances."[12]

Or as Bill Nierenberg put it in a candid moment, "You just know in your heart that you can't throw 25 million tons a year of sulfates into the Northeast and not expect some . . . consequences."[13]

We agree.

Acknowledgments

A very large number of people gave generously of their time and expertise to help us research, write, and fact-check this book. We are enormously grateful to the various people who lived through parts or all of this history, spoke to us about it, and read parts of the manuscript: Henry Abarbanel, Richard Ayres, Chris Bernabo, Edward Frieman, Stanton Glantz, Justin Lancaster, James Hansen, Donald Kennedy, Gene Likens, John Perry, Sherwood Rowland, Ben Santer, Anthony Socci, and Richard Somerville; and to our colleagues in science studies who have helped us develop our understanding both of this story and of science in general: Keith Benson, Richard Creath, Max Boykoff, Nancy Cartwright, Riley Dunlap, James Fleming, Dale Jamieson, Myanna Lahsen, Mary Morgan, Jane Maienschein, Minakshi Menon, Robert Proctor, Narendra Subramanian, Paul Thacker, Spencer Weart, Peter Westwick, and Zuoyue Wang.

This project would truly not have been possible without the resource of the Legacy Tobacco Documents Library, and the work of Professor Stanton Glantz and his colleagues at the University of California, San Francisco. We are also grateful to Walter Munk for helping us to obtain the early Jason reports on climate change; to John Mashey for his enthusiasm for this project, for defending our honor on the Internet, and for reading the entire manuscript at a crucially helpful moment; and to Mott Greene, who helped us to develop the concept of Title Search Realism.

No historian can do his or her work without the existence of archives and the support of archivists. We thank the dedicated staff of the archives of the Scripps Institution of Oceanography, particularly Deborah Day and Carolyn Rainey; Janice Goldblum at the National Academy of Sciences; the archives of the Massachusetts Institute of Technology; the Lyndon B. Johnson Presidential Library; and the George H.W. Bush Presidential Library.

We were fortunate to be aided by several outstanding research assistants, including Benjamin Wang, Afsoon Foorohar, Karin Matchett, Matthew Crawford, Matthew Shindell, Krystal Tribbett, and above all, the indefatigable Charlotte Goor, without whom this book truly would not have been finished, because we would have long ago given up in frustration.

We also thank Holly Hodder, formerly of Westview Press, who first realized that this project would be a trade book, our agent Ayesha Pande, who from the beginning totally got the project, and our outstanding editors at Bloomsbury Press, Peter Ginna and Pete Beatty, whose advice has been consistently clear, intelligent, helpful, and kind.

Erik Conway thanks Blaine Baggett and Stephen Kulczycki of the Jet Propulsion Laboratory in Pasadena, California, for approving the leave necessary to complete his contributions to this work, and the management of JPL and the California Institute of Technology for maintaining a policy enabling staff to pursue their interests on their own time. Conway is also grateful to Eric Fetzer, Frederick W. Irion, John T. Schofield, and David Kass for many lunchtime conversations about the atmospheric sciences.

Naomi Oreskes thanks Christopher Patti for helpful legal advice, Lynn Russell, Richard Somerville and Larry Armi for unending moral support, Tony Heymet for his commitment to academic integrity, the University of California, San Diego, for its steadfast support of academic freedom, and Shannon Sloan, for her vigilant protection against the menace of daily (academic) life. Last but by no means least, Oreskes thanks her beloved husband, Kenneth Belitz, who, as he likes to say, did not have to read the book, because he lived it, and her children, Hannah and Clara Belitz, who have gamely tolerated many months of maternal neglect.

Permissions

Portions of chapter 4 are drawn from Erik M. Conway, *Atmospheric Science at NASA: A History* (Baltimore, Md.: The Johns Hopkins University Press, 2008).

Portions of chapter 6 are drawn from Naomi Oreskes, Erik M. Conway, and Matthew Shindell, "From Chicken Little to Dr. Pangloss: William Nierenberg, Global Warming, and the Social Deconstruction of Scientific Knowledge," *Historical Studies in the Natural Sciences* 38, no. 1: 109–52. Used by permission.

Portions of chapter 7 are drawn from Naomi Oreskes, "Science and public policy: What's proof got to do with it?" *Environmental Science and Policy* 7, no. 5: 369–83.

Portions of the conclusion are drawn from Naomi Oreskes, "The scientific consensus on climate change: How do we know we're not wrong?" *Climate Change: What It Means for Us, Our Children, and Our Grandchildren*, edited by Joseph F. C. DiMento and Pamela Doughman (Cambridge, Mass.: MIT Press, 2007), 65–99. Used by permission of MIT Press.

Notes

NOTE ON SOURCES

This book draws heavily on the documents unearthed by the tobacco litigation of the 1990s. These documents are housed at the University of California, San Francisco and have been placed online at the Legacy Tobacco Documents Library: http://legacy.library.ucsf.edu.

The online database of more than fifty million pages is full-text searchable, and every document has a unique identification number called the Bates Number (BN). Because the library contains many documents with very similar names, in the notes that follow we have chosen to emphasize the Bates Number as the simplest way to retrieve individual documents. We have also omitted direct URLs, as these take up space and may change in the future.

INTRODUCTION

1. *Massachusetts et al. vs. Environmental Protection Agency et al.*, no. 05-1120 (Washington, D.C., November 29, 2006), http://www.supremecourtus.gov/oral_arguments/argument_transcripts/05-1120.pdf; David Rosner and Gerald Markowitz, "You say Troposphere, I say Stratosphere," The Pump Handle Crowd Blog, posted January 8, 2007, http://thepumphandle.wordpress.com/2007/01/08/you-say-troposphere-i-say-stratosphere/.
2. "The IPCC Controversy," Science and Environment Policy Project, http://www.sepp.org/Archive/controv/ipcccont/ipcccont.html.
3. Frederick Seitz, "A Major Deception on 'Global Warming,'" *Wall Street Journal*, June 12, 1996, Op-Ed, Eastern edition, A16.
4. Benjamin Santer et al., letter to the editor, *Wall Street Journal*, June 25, 1996, Eastern edition, A15.
5. Susan K. Avery et al., "Special insert: An open letter to Ben Santer," UCAR—University Corporation for Atmospheric Research, *Communications Quarterly* (Summer 1996), http://www.ucar.edu/communications/quarterly/summer96/insert.html.
6. Ibid.
7. Paul N. Edwards and Stephen H. Schneider, "The 1995 IPCC Report: Broad Consensus or Scientific Cleansing," *Ecofable/Ecoscience* 1, no. 1 (1997): 3–9.
8. S. Fred Singer, letter to the editor, *Wall Street Journal*, July 11, 1996, Eastern edition, A15. For a detailed account of the entire event, see Edwards and Schneider, "The 1995 IPCC Report," 3–9.

9. Jonathan DuHamel, "The Assumed Authority—The IPCC Examined," Climate Realists Blog (formerly CO$_2$ sceptics), posted May 29, 2008, http://climaterealists.com/index.php?id=1368; "The IPCC Controversy," Science and Environment Policy Project.

10. "IPCC Global Warming Report," American Liberty Publishers, http://www.amlibpub.com/essays/ipcc-global-warming-report.html.

11. John Schwartz, "Philip Morris Sought Experts to Cloud Issue, Memo Details," *Washington Post*, May 9, 1997, A02, http://www.washingtonpost.com/wp-srv/national/longterm/tobacco/stories/second.htm.

12. Richard Leroy Chapman, "A Case Study of the U.S. Weather Satellite Program: The Interaction of Science and Politics" (Ph.D. thesis, Syracuse University, 1967).

13. S. Fred Singer and Kent Jeffreys, *The EPA and the Science of Environmental Tobacco Smoke*, Alexis de Tocqueville Institution, University of Virginia, 1994, BN: TICT0002555 and BN: TI31749030, Legacy Tobacco Documents Library.

14. *Bad Science: A Resource Book*, 26 March 1993, BN: 2074143969, Legacy Tobacco Documents Library.

15. Frederick Seitz, Robert Jastrow, and William A. Nierenberg, eds., *Global Warming: What Does the Science Tell Us?* (Washington, D.C.: The George C. Marshall Institute, 1989). The Marshall Institute republished this as *Scientific Perspectives on the Greenhouse Problem* in 1989, 1990, and 1991. See Robert Jastrow, William Nierenberg, and Frederick Seitz, *Scientific Perspectives on the Greenhouse Problem* (Washington, D.C.: George C. Marshall Institute, 1989).

16. Leslie Roberts, "Global Warming: Blaming the Sun," *Science* 246, no. 4933 (November 24, 1989): 992–93.

17. *S. Fred Singer v. Justin Lancaster*, Mass. Civil Action 93-2219 (August 2, 1993); Gary Taubes, "The Ozone Backlash," *Science* 260 (June 11, 1993): 1580–83; F. Sherwood Rowland, "President's Lecture: The Need for Scientific Communication with the Public," *Science* 260 (June 11, 1993): 1573–76; S. Fred Singer letter to the editor and Rowland response, *Science* 261 (August 27, 1993), 1101–3.

18. Myanna Lahsen, "Experiences of Modernity in the Greenhouse: A Cultural Analysis of a Physicist 'Trio' Supporting the Backlash against Global Warming," *Global Environmental Change* 18 (2008): 204–19.

19. As recently as 2007, when the IPCC Fourth Assessment declared that warming was "unequivocal," the *New York Times* still quoted Fred Singer as disagreeing. See Cornelia Dean, "Even Before Its Release, World Climate Report Is Criticized as Too Optimistic," *New York Times*, February 2, 2007, http://www.nytimes.com/2007/02/02/science/02oceans.html?scp=1&sq=Fred+Singer&st=nyt.

CHAPTER 1

1. *Executive Summary*, 1987, Bates Number (BN): 507720494, Legacy Tobacco Documents Library; *A Discussion of Tobacco Industry and R. J. Reynolds Industries' Support of Biomedical Research*, BN: 504480429, Legacy Tobacco Documents Library.

2. Frederick Seitz to H. C. Roemer, 1 May 1978, BN: 504480670, Legacy Tobacco Documents Library.

3. Executive Summary, 1987, BN: 507720494; see also William D. Hobbs to J. Paul Sticht,

RE: Corporate Support of Biomedical Research, 29 May 1980, BN: 504480340, Legacy Tobacco Documents Library; and *A Discussion of Tobacco Industry and R. J. Reynolds Industries' Support of Biomedical Research*, BN: 504480429, Legacy Tobacco Documents Library.

4.   "Atabrine and New Pharmacology," The AMINCO-Bowman SPF: Special Spotlights,
      http://history.nih.gov/exhibits/bowman/SSatabrine.htm.

5.   *Executive Summary*, BN: 507720494, Legacy Tobacco Documents Library.

6.   Ibid.

7.   These experiments later got him into trouble; see Nicholas Wade, "Gene Therapy
      Caught in More Entanglements," *Science* 212, no. 4490 (April 23, 1981): 24–25, and dis-
      cussion later in this chapter.

8.   Frederick Seitz, *On the Frontier: My Life in Science* (New York: American Institute of
      Physics Press, 1994).

9.   See, for example, Colin Stokes, *Chairman of R. J. Reynolds Industries to G. Barry Pierce,
      MD*, 2 November 1978, BN: 503036338, Legacy Tobacco Documents Library.

10.   First two quotes in William D. Hobbs to J. Paul Sticht, BN: 504480340, Legacy Tobacco
      Documents Library; last quote in Colin Stokes, *RJR's Support of Biomedical Research,
      International Advisory Board, Draft II: Presentation Prepared by RJR Managerial Employee
      for Review and Approval by RJR In-house Legal Counsel*, May 1979, BN: 50480518,
      Legacy Tobacco Documents Library; see also the nearly identical document, Colin
      Stokes, *RJR's Support of Biomedical Research, International Advisory Board*, May 1979,
      BN: 504697359, Legacy Tobacco Documents Library.

11.   William D. Hobbs to J. Paul Sticht, BN: 504480340, Legacy Tobacco Documents Library.

12.   Ibid.

13.   Colin Stokes, *RJR's Support of Biomedical Research*, BN: 504480518, Legacy Tobacco
      Documents Library.

14.   Ibid.

15.   Ibid.

16.   Ibid.

17.   Robert N. Proctor, *The Nazi War on Cancer* (Princeton, N.J.: Princeton University Press,
      1999).

18.   Frederick Seitz to John L. Bacon, Director–Corporate Contributions, 24 October 1984, BN:
      508455343, Legacy Tobacco Documents Library. On at least one occasion, Prusiner had
      dinner with tobacco executives; see *Memorandum to J. Tylee Wilson*, BN: 505628702G,
      Legacy Tobacco Documents Library.

19.   Ernest L. Wynder et al., "Experimental Production of Carcinoma with Cigarette Tar,"
      Cancer Research 13 (December 1953): 855–64.

20.   Stanton A. Glantz et al., *The Cigarette Papers* (Berkeley: University of California Press,
      1996), 25.

21.   Proctor, *Nazi War*; Robert N. Proctor, *Cancer Wars: How Politics Shapes What We Know
      and Don't Know about Cancer* (New York: Oxford University Press, 2001); Devra Lee
      Davis, *The Secret History of the War on Cancer* (New York: Basic Books, 2007).

22.   This chapter draws on a number of accounts of the tobacco industry's efforts to challenge
      scientific evidence, as well as our own original research in the tobacco legacy documents.
      Our sources include: Glantz et al., *The Cigarette Papers*; Proctor, *Nazi War* and *Cancer
      Wars*; Allan M. Brandt, *The Cigarette Century: The Rise, Fall, and Deadly Persistence of the*

Product that Defined America (New York: Basic Books, 2007); Davis, *War on Cancer*; David Michaels, *Doubt Is Their Product: How Industry's Assault on Science Threatens Your Health* (New York: Oxford University Press, 2008); and Sheldon Rampton and John Stauber, *Trust Us, We're Experts! How Industry Manipulates Science and Gambles with Your Future* (New York: Tarcher, 2000).

23. *Background Material on the Cigarette Industry Client*, 15 December 1953, BN: 280706554, Legacy Tobacco Documents Library.

24. U.S. Department of Justice, Civil Division, "Litigation Against Tobacco Companies," http://www.usdoj.gov/civil/cases/tobacco2/index.htm; U.S. complaint against the tobacco industry, filed September 22, 1999—*United States of America v. Philip Morris, R. J. Reynolds, et al.*, http://www.usdoj.gov/civil/cases/tobacco2/complain.pdf; *United States of America v. Philip Morris USA Inc. et al.*, Closing Arguments, vol. 115, CA99-02496, (Washington, D.C., June 9, 2005), BN: DOJDCS060905, Legacy Tobacco Documents Library.

25. Only Liggett and Myers—manufacturer of Larks and Chesterfields—declined to participate, thinking it better just to ignore the whole thing. What was that whole thing? The "health crisis" produced by the mounting scientific evidence that tobacco use killed. For this discussion see "Background Material on the Cigarette Industry Client," December 15, 1953, Exhibit 2, in http://www.tobacco.neu.edu/box/BOEKENBox/hkwaxman.html, "A Frank Statement to Cigarette Smokers," January 1, 1964, Legacy Tobacco Documents Library, BN: TINY0001786.

The "Boeken Box" Web site presents key documents in the 2001 case of *Boeken v. Philip Morris*, which resulted in a $3 billion punitive damages award against the defendant, compiled by lawyers involved in the case (e-mail communication Ray Goldstein to Charlotte Goor, December 3, 2009). For interested readers, the "exhibits" in the Boeken Box correspond to the following Bates numbers in the Legacy Tobacco Documents Library: Exhibit 1, BN: TINY0001772; Exhibit 3, BN: TINY0001776; Exhibit 4, BN: TINY0001786; Exhibit 5, BN: TINY0001788; Exhibit 6, BN: TINY0001792; Exhibit 7, BN: TINY0001800; Exhibit 8, BN: TINY0001805; Exhibit 9, BN: TINY0001828; Exhibit 10, BN: TINY0001836; Exhibit 11, BN: TINY0001841; Exhibit 12, BN: TINY0001848; Exhibit 13, BN: TINY0001852; Exhibit 14, BN: TINY0001859. (Exhibit 2 is not a Legacy Tobacco document, it is MN A.G. Trial Exhibit 18905).

The Boeken documents were also referenced in Congressional hearings led by U.S. Representative Henry Waxman in 1994, leading to the report *The Hill and Knowlton Documents, Waxman Report: How the Tobacco Industry Launched its Disinformation Campaign, A Staff Report*, Majority Staff, Subcommitte on Health and the Environment, May 26, 1994, U.S. House of Representatives, Committee on Energy and Commerce, 13 pp. http://www.tobacco.neu.edu/box/BOEKENBox/Waxman%201994%20Hill-Knowlton/HK%20Conspiracy%20Waxman%20Report.pdf; BN: TINY0001756-1770, Legacy Tobacco Documents Library.

26. "Background Material on the Cigarette Industry Client."

27. *United States of America v. Philip Morris, R.J. Reynolds, et al.*, 1999, p. 3.

28. "Background Material on the Cigarette Industry Client"; see also *United States of America v. Philip Morris USA Inc., et al.*, Final Opinion, CA-02496 (August 17, 2006), http://www.usdoj.gov/civil/cases/tobacco2/ORDER_FINAL.pdf, and Amended Final Opinion,

CA99-02496, filed September 8, 2006, http://www.usdoj.gov/civil/cases/tobacco2/amended%20opinion.pdf.

29. Ibid.

30. *United States of America v. Philip Morris, R. J. Reynolds, et al.*, 1999, and documents cited therein.

31. House Committee on Energy and Commerce, Subcommittee on Health and Environment, *The Hill and Knowlton Documents, Waxman Report: How the Tobacco Industry Launched Its Disinformation Campaign*, Majority Staff Report, May 26, 1994, http://www.tobacco.neu.edu/box/BOEKENBox/Waxman%201994%20Hill-Knowlton/HK%20Conspiracy%20Waxman%20Report.pdf.

32. Waxman Report, *How the Tobacco Industry Launched Its Disinformation Campaign*, 8, and 12. See also "Conferences with LIFE and Reader's Digest, July 17–18, 1956," Exhibit 14, http://www.tobacco.neu.edu/box/BOEKENBox/Waxman%201994%20Hill-Knowlton/Waxman14.pdf, BN: TINY001859, Legacy Tobacco Documents Library.

33. Glantz et al., *The Cigarette Papers*, chap. 4; Brandt, *The Cigarette Century*, 247–48, 275–76, 364, 379.

34. George D. Snell, "Clarence Cook Little," *Biographical Memoirs v. 46* (Washington, D.C.: National Academies Press, 1974), 240–63, http://www.nap.edu/openbook.php?record_id=569&page=241. Little ran the Tobacco Industry Research Committee from 1954 to 1969.

35. Waxman Report, *How the Tobacco Industry Launched Its Disinformation Campaign*, 7.

36. Mark Parascandola, "Public Health Then and Now: Cigarettes and the US Public Health Service in the 1950s," *American Journal of Public Health* 91, no. 2 (February 2001): 196–205, on p. 198.

37. Tobacco Institute Research Council, *A Scientific Perspective on the Cigarette Controversy*, 1954, BN: 946078469, Legacy Tobacco Documents Library.

38. Hill and Knowlton, "Public Relations Report to the Tobacco Industry Research Committee," 14 February 1956, exhibit 13, BN: TINY0001852, Legacy Tobacco Documents Library.

39. Hill and Knowlton, "Preliminary Recommendations for Cigarette Manufacturers," December 24, 1953, Exhibit 3, on pp. 5–6, http://www.tobacco.neu.edu/box/BOEKENBox/Waxman%201994%20Hill-Knowlton/Waxman03.pdf, BN: TINY0001775, Legacy Tobacco Documents Library, quotes on pp. 5–6.

40. Hill and Knowlton, "Preliminary Recommendations," 5–6.

41. Stanton A. Glantz et al., "Tobacco Industry Sociological Programs to Influence Public Beliefs About Smoking," *Social Sciences and Medicine* 66, no. 4 (February 2008): 970–81.

42. David A. Kessler, *A Question of Intent: A Great American Battle with a Deadly Industry* (New York: Public Affairs, 2002); Brandt, *The Cigarette Century*; Rampton and Stauber, *Trust Us, We're Experts!*; Thomas O. McGarity and Wendy Wagner, *Bending Science: How Special Interests Corrupt Public Health Research* (Cambridge, Mass.: Harvard University Press, 2008).

43. Museum of Broadcast Communications, "Fairness Doctrine: U.S. Broadcasting Policy," http://www.museum.tv/archives/etv/F/htmlF/fairnessdoct/fairnessdoct.htm.

44. Leonard M. Schuman, "The Origins of the Report of the Advisory Committee on Smoking and Health to the Surgeon General," *Journal of Public Health Policy* 2, no. 1 (March 1981): 19–27; Brandt, *The Cigarette Century*; Glantz et al., *The Cigarette Papers*.

45. Waxman Report, *How the Tobacco Industry Launched Its Disinformation Campaign*, 7, 1994, 79.

46. David Halberstam, *The Powers That Be* (Urbana-Champaign: University of Illinois Press, 2000), 38.

47. Hill and Knowlton, "Report on TIRC Booklet 'A Scientific Perspective on the Cigarette Controversy,'" May 3, 1954, Exhibit 6, on p. 1, http://www.tobacco.neu.edu/box/ BOEKENBox/Waxman%201994%20Hill-Knowlton/Waxman06.pdf, BN: TINY0001792, Legacy Tobacco Documents Library.

48. *Tobacco Industry Research Committee: Council for Tobacco Research*, BN: 2015002362, Legacy Tobacco Documents Library.

49. Ibid.

50. *Tobacco Industry Research Committee*, BN: 2015002362, Legacy Tobacco Documents Library.

51. Ibid.; on the origins of the 1964 committee see Schuman, "Origins of the Report."

52. Schuman, "Origins of the Report," 19–27.

53. Stanton Glantz notes that the industry was terrified of what health groups would do with the report, but these groups did not respond as strongly as the industry expected. See Glantz et al., *The Cigarette Papers*, 50–52.

54. Brandt, *The Cigarette Century*, 220, 228–30.

55. Parascandola, "Public Health Then and Now," 196–205.

56. Mark Parascandola, "Two Approaches to Etiology: The Debate over Smoking and Lung Cancer in the 1950s," *Endeavour* 28, no. 2 (June 2008): 81–86, on p. 85.

57. Dean F. Davies, "A Statement on Lung Cancer," *CA: A Cancer Journal for Clinicians* 9, no. 6 (1959): 207–8.

58. Glantz et al., *The Cigarette Papers*, 15, 32; also see Kessler, *A Question of Intent*, for a discussion of his efforts to regulate nicotine as a drug, and industry opposition in the 1990s.

59. Glantz et al., *The Cigarette Papers*, 29.

60. Ibid., first quote on 15, second quote on 18.

61. Schuman, "Origins of the Report," 19–27.

62. Parascandola, "Public Health Then and Now," 196–205; Alan Blum et al., "The Surgeon General's Report on Smoking and Health 40 Years Later: Still Wandering in the Desert," *Lancet* 363, no. 9403 (January 10, 2004): 97–98; National Library of Medicine, "The Reports of the Surgeon General: The 1964 Report on Smoking and Health," Profiles in Science, http://profiles.nlm.nih.gov/NN/Views/Exhibit/narrative/smoking.html. See also Centers for Disease Control and Prevention, "History of the Surgeon General's Report on Smoking and Health," http://www.cdc.gov/Tobacco/data_statistics/sgr/history/index .htm, and Tobacco.org, "1964: The First Surgeon General's Report," http://www.tobacco .org/resources/history/1964_01_11_1st_sgr.html.

63. Centers for Disease Control and Prevention, "History of Surgeon General's Report on Smoking and Health."

64. Glantz et al., *The Cigarette Papers*, 50.

65. *Tobacco Industry Research Committee: Council for Tobacco Research*, BN: 2015002362, Legacy Tobacco Documents Library.

66. Glantz et al., *The Cigarette Papers*, 51.

67. Ibid., 53–54.

68. Ibid., 45.

69. *Press Release: From the Council for Tobacco Research—USA*, 11 March 1964, BN: 961009573, Legacy Tobacco Documents Library.

70. *Tobacco News Summary no. 31, Condensed from Public Sources by Hill and Knowlton, Inc.*, 31 March 1965, BN: 680280682, Legacy Tobacco Documents Library.

71. U.S. Department of Health, Education and Welfare, *The Health Consequences of Smoking A Public Health Service Review: 1967* (Washington, D.C.: U.S. Government Printing Office, 1967), http://profiles.nlm.nih.gov/NN/B/B/K/M/_/nnbbkm.pdf.

72. Ibid.

73. *Company Statement on Smoking and Health*, 12 May 1967, BN: 282001858, Legacy Tobacco Documents Library.

74. *The Original Ed Gibbs Newsletter of the Beer, Wine and Liquor Industries*, 7 February 1969, BN: TI55842608, Legacy Tobacco Documents Library.

75. On the industry's view of the impact of the Fairness Doctrine, see Glantz et al., *The Cigarette Papers*, 262. On the industry not opposing the advertising ban, and its benefits to the industry, see idem., 256, 258.

76. Glantz et al., *The Cigarette Papers*, 256.

77. *Newsletter of the Beer, Wine and Liquor Industries*, BN: TI55842608, Legacy Tobacco Documents Library.

78. U.S. Department of Health, Education and Welfare, *Smoking and Health: Report of the Advisory Committee to the Surgeon General of the Public Health Service* (Washington, D.C.: U.S. Government Printing Office, 1964), http://profiles.nlm.nih.gov/NN/B/B/M/Q/_/ nnbbmq.pdf. For statistics on the number of doctors who smoked, see Jonathan Foulds, "How Many Medical Doctors Smoke?" Healthline, http://www.healthline.com/blogs/ smoking_cessation/labels/medical.html; and Derek R. Smith and Peter A. Leggat, "An International Review of Tobacco Smoking in the Medical Profession: 1974–2004," *BMC Public Health* 7, no. 115 (June 20, 2007), http://www.biomedcentral.com/1471-2458/7/115; on the *New York Times* quoting tobacco industry representatives, see Clark Hoyt, "The Doctors Are In. The Jury Is Out," *New York Times*, February 17, 2008, http:// www.nytimes.com/2008/02/17/opinion/17pubed.html.

79. *Annual Report, 1969, R. J. Reynolds Tobacco Company*, BN: 500435078, Legacy Tobacco Documents Library.

80. Glantz et al., *The Cigarette Papers*, chap. 7.

81. Altria, "History of Tobacco Litigation: 1954–1978," http://www.altria.com/media/03_06_01_02_01_1954-1978.asp. One industry document suggests that in 1979 the industry paid $794 million in cigarette excise taxes. Colin Stokes, *RJR's Support of Biomedical Research, Draft III*, May 1979, BN: 504480518, Legacy Tobacco Documents Library.

82. Glantz et al., *The Cigarette Papers*, 19.

83. Ibid., chap. 7, esp. 241–47.

84. *Summary, Re: Research Funding*, BN: 502370120, Legacy Tobacco Documents Library.

85. Stokes, *RJR's Support of Biomedical Research*, BN: 504480518, Legacy Tobacco Documents Library.

86. *Summary, Re: Research Funding*, Legacy Tobacco Documents Library.

87. Stokes, *RJR's Support of Biomedical Research*, BN: 504480518, Legacy Tobacco Documents Library.

88. *Biography of Frederick Seitz*, November 1985, BN: 87697430, Legacy Tobacco Documents Library.

89. Frederick Seitz, *Draft of Presentation to the International Advisory Committee—R. J. Reynolds Industries*, 9 May 1979, BN: 503955384, Legacy Tobacco Documents Library. See also Stokes, *RJR's Support of Biomedical Research, Draft III*, BN: 504480518, Legacy Tobacco Documents Library.

90. H. C. Roemer et al., *Proposal for Profit, Expenditure, or Policy Change: Financial Support of Research Efforts of Rockefeller University*, 11 September 1975, BN: 502521448, Legacy Tobacco Documents Library.

91. Seitz, *On the Frontier*.

92. Murph Goldberger, personal communication with Naomi Oreskes, January 24, 2008; see also Zuoyue Wang, *In Sputnik's Shadow: The President's Science Advisory Committee and Cold War America* (New Brunswick, N.J.: Rutgers University Press, 2008).

93. Frederick Seitz et al., "Eugene Paul Wigner," *Biographical Memoirs v. 74* (Washington, D.C.: National Academies Press, 1998), 364–88, http://books.nap.edu/openbook.php?record_id=6201&page=365; "Wigner, Eugene P.," *New World Encyclopedia*, http://www.newworldencyclopedia.org/entry/Eugene_P._Wigner.

94. Seitz, *On the Frontier*.

95. Seitz, *Draft of Presentation to the International Advisory Committee*, BN: 503955384, Legacy Tobacco Documents Library.

96. Frederick Seitz, *Presentation to Operating Committee R. J. Reynolds Industries, Inc.*, 8 August 1979, BN: 504779244, Legacy Tobacco Documents Library.

97. Ibid.

98. Ibid.

99. Michel Ter-Pogossian et al., "Radioactive Oxygen 15 in Studies of Kinetics of Oxygen of Respiration," *American Journal of Physiology* 201 (1961): 582–86, http://ajplegacy.physiology.org/cgi/content/abstract/201/3/582.

100. Seitz, *On the Frontier*, 380.

101. His memoir includes numerous comments about the disgraceful state of popular culture, particularly the movies emanating from Hollywood. He also discusses how William Schockley's elitism—later racism—had its roots, he felt, in Hollywood influences.

102. Seitz, *On the Frontier*, 37.

103. "Emphysema," Aetna InteliHealth: Health A to Z, http://www.intelihealth.com/IH/ihtIH/WSIHW000/9339/10600.html.

104. Karen Bartholomew, "Century at Stanford," *Stanford Magazine* (July/August 2003), http://www.stanfordalumni.org/news/magazine/2003/julaug/dept/century.html.

105. U.S. Department of Energy, "Chronic Beryllium Disease," http://www.energy.gov/safetyhealth/cbd.htm; see also Natural Resources Defense Council, "Settlement of the Contempt Action against the Department of Energy: Joint Stipulation and [Proposed] Order," http://www.nrdc.org/nuclear/9812doe.asp. The U.S. government developed a rule to prevent beryllium disease, see "Chronic Beryllium Disease Prevention" (10 CFR 850); and David Michaels and Celeste Monforton, "Beryllium's Public Relations Problem: Protecting Workers When There Is No Safe Exposure Level," *Public Health Reports* 123

(January–February 2008): 79–88, http://www.defendingscience.org/upload/Berylliums_PR_Problem.pdf.

106. Seitz, *On the Frontier*, 108.

107. Murph Goldberger, personal communication with Naomi Oreskes, January 24, 2008.

108. Frederick Seitz to Colin Stokes, *Meetings of the R. J. Reynolds Advisory Committee in Bermuda*, 7 November 1979, BN: 502742718, Legacy Tobacco Documents Library; see also Seitz's comments in *Vanity Fair*: Mark Hertsgaard, "While Washington Slept," *Vanity Fair*, May 2006, 5, http://www.vanityfair.com/politics/features/2006/05/warming200605.

109. For a specific breakdown of project funding, see Colin Stokes, *RJR's Support of Biomedical Research, International Advisory Board*, May 1979, BN: 504697359; *R. J. Reynolds Industries Support of Biomedical Research*, 12 September 1979, BN: 515449696; *Executive Summary of the RJR Nabisco, Inc. Biomedical Research Grants Program for 1987*, BN: 508265593; William D. Hobbs to J. Paul Sticht, *Corporate Support for Biomedical Research*, 29 May 1980, BN: 504480340; see also, Seitz to H. C. Roemer, 19 June 1979, BN: 503137602; Frank G. Colby to Alan Rodgman, *Information for Dr. Laurene's Weekly Meeting with Mr. Hobbs*, 17 October 1979, BN: 502443876, all Legacy Tobacco Documents Library.

110. Hobbs to Sticht, *Corporate Support for Biomedical Research*, BN: 504480340, Legacy Tobacco Documents Library.

111. David Dickson, "NIH Censure for Dr. Martin Cline," *Nature* 291, no. 4 (June 1981): 369; Wade, "Gene Therapy Caught in More Entanglements," 24–25.

112. Dickson, "NIH Censure," 369.

113. Ibid.

114. *Deposition Transcript of James J. Morgan*, 17 April 1994, BN: 2063670882, Legacy Tobacco Documents Library; *Deposition Transcript of Martin J. Cline*, 20 May 1997, BN: 516969762, Legacy Tobacco Documents Library, on p. 55 for discussion of Cline's lost research grant.

115. "*Norma R. Broin et al. v. Philip Morris Incorporated*—Further Readings," http://law.jrank.org/pages/12908/Broin-et-al-v-Philip-Morris-Incorporated-et-al.html; see also *Deposition Transcript of James J. Morgan*, BN: 2063670882, Legacy Tobacco Documents Library.

116. Colin Stokes, Chairman of R. J. Reynolds Industries, to G. Barry Pierce, M.D., BN: 503036338, Legacy Tobacco Documents Library.

117. *Deposition Transcript of Martin J. Cline*, BN: 516969762, Legacy Tobacco Documents Library, 19–21.

118. Ibid., 20.

119. Ibid., first quote on 23–24; second quote on 46.

120. *Joint Defendants' Initial List of Fact Witnesses*, U.S. v. Philip Morris Inc. et al., CA99-CV-02496 (GK), 18 January 2002, BN: 94690287, Legacy Tobacco Documents Library. The witness list starts on p. 3, Prusiner is mentioned on p. 12. See also an earlier document, from 1998, mentioning his work: *Combined Exhibit List—Additions*, August 1998, BN: 2084317019, Legacy Tobacco Documents Library, and *United States' Final Proposed Finding of Facts*, U.S. v. Philip Morris Incorporated et al., CA99-CV-02496 (GK), July 1, 2004, http://www.library.ucsf.edu/sites/all/files/ucsf_assets/uspm.pdf. On the tobacco industry settlements with the state governments, see "Tobacco Settlement Agreements," Government Relations, PhilipMorrisUSA, http://www.philipmorrisusa.com/en/cms/Responsibility/Government_Relations/TSA/MSA_10yrs_Later/default.aspx.

121. U.S. Department of Justice, Civil Division, *Litigation Against Tobacco Companies*; U.S. complaint against the tobacco industry, filed September 22, 1999—*United States of America v. Philip Morris, R. J. Reynolds, et al.*; *United States of America v. Philip Morris USA et al.*, Closing Arguments, vol. 115, CA99-02496, BN: DOJDCS060905, Legacy Tobacco Documents Library.

122. *Amended Final Opinion*, CA99-02496, filed September 8, 2006.

123. Glantz et al., *The Cigarette Papers*, 289–339.

124. *A Discussion of Tobacco Industry and R. J. Reynolds Industries' Support of Biomedical Research*, BN: 504480429, Legacy Tobacco Documents Library. Other companies developed this strategy, too: see Glantz et al., *The Cigarette Papers*, 44. The euphemism "special projects" frequently referred to efforts "designed to find scientists and medical doctors who might serve as industry witnesses in lawsuits or in a legislative forum," *Memo from Ernest Peoples to J. E. Edens, Chairman and CEO of Brown & Williamson*, 4 April 1978, quoted in Glantz et al., *The Cigarette Papers*, 44.

125. Kessler, *A Question of Intent*; Alicia Mundy and Lauren Etter, "Senate Passes FDA Tobacco Bill," June 12, 2009, *Wall Street Journal*, http://online.wsj.com/article/SB12447 4789599707175.html; U.S. Department of Health and Human Services, "FDA Seeks Public Input on Tobacco Regulation," FDA: News and Events, June 30, 2009, http://www.fda.gov/NewsEvents/Newsroom/PressAnnouncements/ucm169853.htm.

126. "Smoking—The Health Effects," *BBC News Online*, February 8, 2003, http://news.bbc.co.uk/1/hi/health/medical_notes/473673.stm.

127. Ross C. Brownson et al., "Demographic and Socioeconomic Differences in Beliefs About the Health Effects of Smoking," *American Journal of Public Health* 82, no. 1 (January 1992): 99–103, http://www.pubmedcentral.nih.gov/picrender.fcgi?artid=1694417 &blobtype=pdf.

128. Stuart Berg Flexner and Eugene F. Shewmaker, eds., *The Random House Dictionary of the English Language* (New York: Random House, 1973), 235.

129. *Smoking and Health Proposal*, 1969, BN: 680561778, Legacy Tobacco Documents Library; see also "Doubt of Tobacco Hazard, Ads Goal," *Indianapolis Star*, July 7, 1981, BN: 690834753, Legacy Tobacco Documents Library; *Press Query*, 2 July 1981, BN: 170012852, Legacy Tobacco Documents Library, and discussions in Michaels, *Doubt Is Their Product*.

130. Seitz, *On the Frontier*, 382.

Chapter 2

1. Harry Rubin, "Walter M. Elsasser," *Biographical Memoirs v. 68* (Washington, D.C.: National Academies Press, 1995), 103–66, http://books.nap.edu/openbook.php?record_id=4990&page=103.

2. Alexander Holtzman to Bill Murray, *Subject: Fred Seitz*, 31 August 1989, BN: 2023266534, Legacy Tobacco Documents Library.

3. Philip Mirowski and Dieter Plehwe, *The Road from Mont Pèlerin: The Making of the Neoliberal Thought Collective* (Cambridge, Mass.: Harvard University Press, 2009).

4. "Remembering Frederick Seitz," March 4, 2008, George C. Marshall Institute, http://www.marshall.org/article.php?id=579.

5. Quoted in Anne Hessing Cahn, *Killing Détente: The Right Attacks the CIA* (College Station: Pennsylvania State University, 1998), 125.

6. Ibid.

7. Teller's interpretation is given in Hessing Cahn, *Killing Détente*, 125.

8. Central Intelligence Agency (CIA), *National Intelligence Estimate NIE 11-3/8-75: Soviet Forces for Intercontinental Conflict, Through the mid-1980s, Volume 1, Key Judgments and Summary, Advance Dissemination*, created November 17, 1975, CIA, 40, http://www.foia.cia.gov/browse_docs.asp?doc_no=0000268110.

9. The intelligence official quoted here is George Carver. See Hessing Cahn, *Killing Détente*, 130.

10. Hessing Cahn, *Killing Détente*, 132–35.

11. Ibid., 126–27.

12. On Graham see "Meet the Staff," High Frontier, http://www.highfrontier.org/Highfrontier/main/Contact/Meet%20th%20Staff.htm; on Wolfowitz see Hessing Cahn, *Killing Détente*, 147–52.

13. Central Intelligence Agency (CIA), *Intelligence Community Experiment in Competitive Analysis, TCS 889140-76, Soviet Strategic Objectives: An Alternative View, Report of Team "B,"* December 1976, 5, http://www.foia.cia.gov/browse_docs.asp?doc_no=0000278531.

14. Ibid., 13–14.

15. Ibid., 45–46, emphasis in original.

16. Ibid., 47.

17. Ibid., 32.

18. C. S. Lewis, *The Four Loves* (San Diego: Harcourt Brace, 1960), 60.

19. Ibid., 61.

20. CIA, *Soviet Strategic Objectives*, 34.

21. Hessing Cahn, *Killing Détente*, 176–79.

22. Ibid., 189.

23. Quoted in Frances FitzGerald, *Way Out There in the Blue: Reagan, Star Wars, and the End of the Cold War* (New York: Simon and Schuster, 2000), 207.

24. Ibid., 179–82.

25. Rebecca Slayton, "Discursive Choices: Boycotting Star Wars between Science and Politics," *Social Studies of Science* 37, no. 1 (2007): 27–66.

26. The eight thousand warhead figure is drawn from figure 3, Stephen I. Schwartz, ed., *Atomic Audit: The Costs and Consequences of U.S. Nuclear Weapons Since 1940* (Washington, D.C.: Brookings Institution Press, 1998), 23.

27. Keay Davidson, *Carl Sagan: A Life* (New York: John Wiley and Sons, 1999), 358–59.

28. Slayton, "Discursive Choices," 27–66.

29. Peter Galison and Barton Bernstein, "In Any Light: Scientists and the Decision to Build the Superbomb, 1942–1954," *Historical Studies in the Physical and Biological Sciences* 19, no. 2 (1989): 267–347.

30. Slayton, "Discursive Choices," 27–66.

31. John Schwartz, "Robert Jastrow, Who Made Space Understandable, Dies at 82," *New York Times*, February 12, 2008, http://www.nytimes.com/2008/02/12/science/space/12jastrow.html.

32. Henry Abarbanel, personal communication with Naomi Oreskes, September 17, 2009.

33. Daniel Patrick Moynihan, "Reflections: The Salt Process," *New Yorker*, November 19, 1979, 136.

34. Ibid., 159.

35. This was the MX missile.

36. Moynihan, "Reflections," 162.

37. Robert Jastrow, "Why Strategic Superiority Matters," *Commentary* 75, no. 3 (March 1983): 27–32.

38. Ibid., 32.

39. Hessing Cahn, *Killing Détente*, 167.

40. William Burr and Svetlana Savranskaya, eds., "Previously Classified Interviews with Soviet Officials Reveal U.S. Strategic Intelligence Failure over Decades," Nuclear Vault, National Security Archive, released online September 11, 2009, http://www.gwu.edu/~nsarchiv/nukevault/ebb285/index.htm#1.

41. FitzGerald, *Way Out There in the Blue*, 498.

42. Luis W. Alvarez, *Alvarez: Adventures of a Physicist* (New York: Basic Books, 1987), 252–58.

43. R. P. Turco et al., "Nuclear Winter: Global Consequences of Multiple Nuclear Explosions," *Science* 222, no. 4630 (December 23, 1983), 1283–92; see also Paul R. Ehrlich et al., *The Cold and the Dark: The World after Nuclear War* (New York: W. W. Norton and Company, 1984), 83–85.

44. Turco et al., 1292.

45. Ehrlich et al., *The Cold and the Dark*, xiii–xvii.

46. Nuclear Winter had been scheduled for an earlier outing, December 9, 1982, via a paper at the annual Geophysical Union meeting. The Ames Research Center's management blocked it temporarily, because it had not been properly cleared via internal review. This led to charges that NASA was trying to suppress the science, but Ames set up a special review for the paper and provided additional funding to the group to improve it. See Lawrence Badash, "Nuclear Winter: Scientists in the Political Arena," *Physics in Perspective* 3, no. 1 (March 2001), 85.

47. Ehrlich et al., *The Cold and the Dark*, xiii–xvii.

48. Carl Sagan, "The Nuclear Winter," *Parade Magazine*, October 30, 1983, 4–6; Lawrence Badash, *A Nuclear Winter's Tale: Science and Politics in the 1980s* (Cambridge, Mass.: MIT Press, 2009), 63–76. Badash agrees that Sagan had "jumped the gun with his high impact article in *Parade*," but he does not see it as a violation of scientific norms.

49. Carl Sagan, "Nuclear War and Climate Catastrophe: Some Policy Implications," *Foreign Affairs* 62, no. 2 (Winter 1983/1984): 257–92.

50. William D. Carey, "A Run Worth Making," *Science* 222, no. 4630 (December 23, 1983): 1281.

51. See, for example, Lawrence Badash, *Scientists and the Development of Nuclear Weapons: From Fission to the Limited Test Ban Treaty* (Atlantic Highlands, N.J.: Humanities Press, 1995).

52. Curt Covey et al., "Global Atmospheric Effects of Massive Smoke Injections from a Nuclear War: Results from General Circulation Model Simulations," *Nature* 308 (March 1, 1984): 21–25; Starley L. Thompson and Stephen H. Schneider, "Nuclear Winter Reappraised," *Foreign Affairs* 64 (Summer 1986): 981–1005. See also Stephen H. Schneider, *Science as a Contact Sport: Inside the Battle to Save Earth's Climate* (Washington, D.C.: National Geographic, 2009), 95–108.

53. John Maddox, "What Happened to Nuclear Winter?" *Nature* 333 (May 19, 1988): 203; Starley L. Thompson and Stephen Schneider, "Simulating the Climatic Effects of Nuclear War," *Nature* 333 (May 19, 1988): 221–27; Maddox attacked "environmental alarmism" in a popular book, John Maddox, *The Doomsday Syndrome* (New York: McGraw-Hill, 1972).

54. R. P. Turco et al., "Climate and Smoke: An Appraisal of Nuclear Winter," *Science* 248, no. 4939 (January 12, 1990): 166–76; see also the earlier review article by Schneider and Thompson, "Simulating the Climatic Effects," 221–27.

55. Peter Westwick, personal communication with Erik Conway, October 1, 2009.

56. K. A. Emanuel, "Towards a Scientific Exercise," *Nature* 319, no. 6051 (January 23, 1986): 259.

57. Covey et al., "Global Atmospheric Effects," 21–25.

58. Curt Covey, "Nuclear Winter Debate," *Science* 235, no. 4791 (February 20, 1987): 831.

59. On Sagan's activism, see Badash, "Nuclear Winter: Scientists in the Political Arena," 76–105.

60. *Draft Proposal for the George C. Marshall Institute*, 12 December 1984, William A. Nierenberg (WAN) papers, MC13, 75: 6, Scripps Institution of Oceanography (SIO) Archives.

61. Union of Concerned Scientists, "Founding Document: 1968 MIT Faculty Statement," http://www.ucsusa.org/about/founding-document-1968.html.

62. Richard L. Garwin and Hans A. Bethe, "Anti-Ballistic Missile Systems," *Scientific American* 218, no. 3 (March 1968): 21–31.

63. On Safeguard, John Mashey, who worked at Bell Labs at the time (and helped develop the UNIX operating system) recalls, "Safeguard was engineered by Bell Labs, and when I joined BTL in 1973, some very good software people were coming off Safeguard to do other things, including working in the department I'd joined. Uniformly, they all said: 'This was really difficult, we met the specs, we can shoot down a missile . . . and this will never be useful in a real war. We did learn a lot about software engineering . . . anyway, It wasn't just other physicists who didn't think Safeguard would really be useful, it was the people who built it." John Mashey, e-mail communication with authors, October 1, 2009.

64. Daniel J. Kevles, *The Physicists: The History of a Scientific Community in Modern Day America* (New York: Knopf, 1995), 406–7.

65. Robert Jastrow, "The War Against Star Wars," *Commentary* 78, no. 6 (December 1984): 28; FitzGerald, *Way Out There in the Blue*, 246–47.

66. Robert Jastrow to Robert Walker, 1 December, 1986, WAN papers, Accession 2001-01, 21: file label "George Marshall Institute 9/86–1/88," SIO Archives; on the role of foundations in promoting the conservative movement, see John B. Judis, *The Paradox of American Democracy* (New York: Pantheon Books, 2000), 109–36, and Sidney Blumenthal, *The Rise of the Counter-Establishment: From Conservative Ideology to Political Power* (New York: Crown, 1986), 67.

67. *Draft Proposal for the George C. Marshall Institute*, 12 December 1984, MC13, 75: 6, SIO Archives.

68. Jastrow to Walker, "George Marshall Institute," WAN papers.

69. Robert Jastrow, *How To Make Nuclear Weapons Obsolete* (Boston: Little Brown, 1985);

Ashton B. Carter, "Directed Energy Missile Defense in Space," OTA-BP-ISC_26, Washington, D.C., Office of Technology Assessment, April 1984; U.S. Congress, Office of Technology Assessment, "Ballistic Missile Defense Technologies," OTA-ISC-254, Washington, D.C., September 1985; the UCS studies are contained in John Tirman, ed., *The Fallacy of Star Wars* (New York: Vintage Books, 1984).

70. Jastrow to Walker, "George Marshall Institute," WAN papers.

71. The small number of licenses was due to the technological inability to control transmitter and receiver frequency carefully. Radio and television stations had to be kept far apart in frequency so they didn't interfere with each other, so when the radio and TV spectra were carved up in the 1930s, not many stations could be established. By the 1980s, technological improvements made much more accurate control possible, and so many more stations were possible too. The availability of cable, which could carry even more channels than over-the-air broadcast, rendered scarcity irrelevant.

72. Initially there was a fourth colleague—Karl Bendetsen—who was a member of an advisory group to Ronald Reagan that received security clearances to learn about new weapons developments such as nuclear X-ray lasers. See Kerry Richardson, "The Bohemian Grove and the Nuclear Weapons Industry: Some Connections," 1987, http://sonic.net/~kerry/bohemian/grovenukes.html. Bendetsen spent the early portion of his career in the War Department, serving among other things as assistant secretary of the army in 1950–1952, and undersecretary of the army in 1952. His most important role in history was perhaps as director of the evacuation and relocation in 1942 of persons of Japanese ancestry into internment camps during the war. Bendetsen appears in some of the early Marshall Institute materials, but then seems to have dropped out; "Oral History Interviews with Karl R. Bendetsen," Harry S. Truman Library and Museum, http://www.trumanlibrary.org/oralhist/bendet.htm#bio; Alfonso A. Narvaez, "Karl R. Bendesten, 81, Executive and High-Ranking U.S. Official," *New York Times*, June 30, 1989, http://www.nytimes.com/1989/06/30/obituaries/karl-r-bendetsen-81-executive-and-high-ranking-us-official.html.

73. Robert Jastrow, *George Marshall Institute Program Summary*, 1985–1986, WAN papers, Accession 2001-01, 21: file label "George Marshall Institute 9/86–1/88," SIO Archives.

74. James Frelk to William Nierenberg, 2 December 1986, and attached letter, Jastrow to Walker, 1 December 1986, WAN papers, Accession 2001-01, 21: file label "George Marshall Institute," WAN papers.

75. Ibid.

76. Ibid.

77. James Frelk to William Nierenberg, enclosure of an excerpt of Strategic Defense Initiative of February 19, 1986, WAN papers, Accession 2001-01, 21: file label "George Marshall Institute 9/86–1/88," SIO Archives.

78. "Grant Proposal Information," Adolph Coors Foundation, http://www.adolphcoors.org/criteria.cfm.

79. Jastrow to Walker, 1 December, 1986, WAN papers, Accession 2001-01, 21: file label "George Marshall Institute," SIO Archives.

80. Ibid.

81. Ibid.

82. On the Olin Foundation, and its support of right-wing causes, see http://media

transparency.org/funderprofile.php?funderID=7. This link is now inactive as the Olin Foundation has shut down; see "John M. Olin Foundation," SourceWatch, htpp://www.sourcewatch.org/index.php?title=John M. Olin Foundation; "John M. Olin Foundation," Right Web, http://www.rightweb.irc-online.org/profile/John M. Olin Foundation.

83. William E. Simon, *A Time for Truth* (New York: McGraw-Hill, 1978), 221.

84. Russell Seitz, letter to the editor, *Foreign Affairs* 62 (Spring 1984): 998–99; Russell Seitz, "In from the Cold: 'Nuclear Winter' Melts Down," *National Interest* 5 (Fall 1986): 3–17.

85. R. Seitz, "In from the Cold," 3.

86. Ibid., 4.

87. Ibid., 4.

88. Ibid., 5.

89. Ibid., 5.

90. Ibid., 9.

91. Ibid., 7.

92. Freeman J. Dyson, *Infinite in All Directions: Gifford Lectures Given at Aberdeen, Scotland, April–November 1985*, ed. by author (New York: Harper and Row, 1988).

93. Nicholas Dawdoff, "The Civil Heretic," *New York Times*, March 25, 2009, http://www.nytimes.com/2009/03/29/magazine/29Dyson-t.html; G. MacDonald et al., *Long Term Impact of Atmospheric Carbon Dioxide on Climate*, Jason Technical Report JSR-78-07, April 1979, xiii.

94. R. Seitz, "In from the Cold," 8.

95. Quoted in William Broad and Nicholas Wade, *Betrayers of the Truth* (New York: Simon and Schuster, 1982), 12.

96. R. Seitz, "In from the Cold," 12.

97. Broad and Wade, *Betrayers of the Truth*, 213.

98. Various tobacco industry documents track Wade's work. See for example, Victor Han to Ellen Merlo, *Subject: Burson/ETS*, Memorandum, 22 February 1993, BN: 2023920035, and Victor Han to Tom Humber, *Subject: Op Ed*, Memorandum, 23 November 1992, BN: 2021173510. The industry tracked comments that Wade made that seemed to suggest he thought the press were biased in favor of environmental groups and might be sympathetic to the industry cause. See, for example David Shaw, "Dose of Skepticism Enters Coverage on Environment, *Los Angeles Times*, September 11, 1994, BN: 500805874; David Shaw, "Nuclear Power Coverage Focused Morbidly on Risk," *Los Angeles Times*, September 13, 1994, BN: 500873034; Nicholas Wade, "The Editorial Notebook: The Titanic Lesson," *New York Times*, September 4, 1985, on p. 62 in *Environmental Tobacco Smoke and Indoor Air Quality in Modern Office Work Environments*, vol. 29, no. 1, BN: 507967179, all in Legacy Tobacco Documents Library.

99. R. Seitz, "In from the Cold," 12.

100. Ibid., 13.

101. Russell Seitz referred to this apparatus as the Office of the Presidential Science Adviser. Its actual name was President's Science Advisory Committee, or PSAC; Gregg Herken, *Cardinal Choices: Presidential Science Advising from the Atomic Bomb to SDI* (New York: Oxford University Press, 1992), 179–81.

102. R. Seitz, "In from the Cold," 14.

103. Edward Teller, "Defensive Weapons Development," *Science* 223, no. 4633 (January 20, 1984): 236.

104. Edward Teller, "Widespread After-Effects of Nuclear War," *Nature* 310 (August 23, 1984): 621–24.

105. S. Fred Singer, "On a Nuclear Winter," *Science* 227, no. 4685 (January 25, 1985): 356; Russell Seitz, "Nuclear Winter Debate," *Science* 235, no. 4791 (February 20, 1987): 832.

106. Kerry Emanuel, personal communication with Naomi Oreskes, September 16, 2009.

107. Peter Westwick shares the following, from a letter from Edward Teller to Joseph Coors. The context is that Teller has heard from George Keyworth that Bob Jastrow has been recommended to succeed Keyworth as science advisor. Teller writes to Coors:

> I have known Bob for one-third of a century. He is a good scientist, and I am very fond of him. He has done an excellent job in defending SDI. Still, I believe that his credentials are not sufficient for the job. Moreover, if he should become the President's Science Advisor, the appointment is apt to draw criticism as being a political one. Bob is qualified, but not highly qualified. I believe the appointment would give our opponents plenty of ammunition. When I discussed the situation with Jay, he strongly suggested that I make my opinions known, and I am taking this opportunity to do so [from the Teller papers at Hoover Institute, Teller to Joseph Coors, 6 January 1986 Teller, 275: Coors].

> Subsequently, Teller sent a list of recommendations to William Wilson, the Vatican envoy in order: Frederick Seitz, William Nierenberg, and Harold Agnew. (Teller to William A. Wilson [Vatican envoy], 14 February 1984, Teller, 286: Wilson.) On Nierenberg as a potential candidate for science advisor, we know from various documents in the SIO WAN papers, MC13 collection that Nierenberg was being considered in 1980. Westwick suggests that he was considered a second time later, when Teller also recommended Dixy Lee Ray. (Peter Westwick, e-mail correspondence with Naomi Oreskes, May 16, 2009—from the unorganized part of the Teller papers.)

108. Edward Reiss, *The Strategic Defense Initiative* (Cambridge University Press, 1992), 105.

109. R. Seitz, "In from the Cold," 7.

110. Russell Seitz, "An Incomplete Obituary," *Forbes* 159, no. 3 (February 10, 1997): 123; "Aliens Cause Global Warming," January 17, 2003, Speeches, Michael Crichton Official Site, http://michaelcrichton.com/speech-alienscauseglobalwarming.html.

111. Milton Friedman, *Capitalism and Freedom* (Chicago: University of Chicago Press, 1962).

112. Friedman, *Capitalism and Freedom*, esp. viii and ix.

113. Russell Seitz, "The Melting of Nuclear Winter," *Wall Street Journal*, November 5, 1986, Eastern edition, 1; also Russell Seitz, letter to the editor, *Wall Street Journal*, January 29, 1987, Eastern edition, 1. A few years later Seitz would defend secondhand smoke in Forbes; see Russell Seitz, "Making the World Safe for Cigarette Smokers," *Forbes* 160, no. 5 (September 8, 1997), 181, http://www.forbes.com/forbes/1997/0908/6005181a.html.

114. Barry Goldwater, acceptance speech for the nomination as candidate of the Republican party for President of the United States, 1963, Quotation Details, The Quotation Page, http://www.quotationpage.com/quote/34605.html; "Barry Goldwater: Extremism in the Defense of Liberty," YouTube, http://www.youtube.com/watch?v=RVNoCluoh9M.

CHAPTER 3

1. Gene E. Likens, "The Science of Nature, the Nature of Science: Long-term Ecological Studies at Hubbard Brook," *Proceedings of the American Philosophical Society* 143, no. 4 (December 1999): 558–72.

2. On "purple rain" in the Renaissance, which some observers thought might be related to pollution, see Margaret D. Garber, "Alchemical Diplomacy: Optics and Alchemy in the Philosophical Writings of Marcus Marci in Post-Rudolphine Prague, 1612–1670" (Ph.D. dissertation, University of California, San Diego, 2002), 278–314. Other sources that include discussion of unusual meteorological phenomena, for which some observers suggested a pollution/precipitation connection, include James Fleming and Roy Goodman, eds., *International Bibliography of Meteorology from the Beginning of Printing to 1889* (Upland, Penn.: Diane Publishing, 1994); Vladimir Janković, *Reading the Skies: A Cultural History of the English Weather 1650–1820* (Manchester, UK: Manchester University Press, 2000). See also Jan Golinksi, *British Weather and the Climate of Enlightenment* (Chicago: University of Chicago Press, 2007).

3. Doug Scott, *The Enduring Wilderness: Protecting Our Natural Heritage through the Wilderness Act* (Golden, Colo.: Fulcrum Publishing, 2004).

4. Gene E. Likens and F. Herbert Bormann, "Acid Rain: A Serious Regional Environmental Problem," *Science* 184, no. 4142 (June 14, 1974): 1176–79.

5. Ibid., 1176.

6. Ibid., 1177; Charles Herrick, "Predictive Modeling of Acid Rain: Obstacles to Generating Useful Information," in *Prediction: Science, Decision Making, and the Future of Nature*, ed. Daniel Sarewitz, Roger A. Pielke Jr., and Radford Byerly Jr. (Washington, D.C.: Island Press, 2000), 251–68.

7. For example, R. J. Beamish and H. H. Harvey, "Acidification of the La Cloche Mountain Lakes, Ontario, and Resulting Fish Mortalities," *Journal of the Fisheries Research Board of Canada* 29 (1972): 1131–43; and Svante Odén's work: S. Odén and T. Ahl, Sartryck, *Ur Ymer Årsbok* (1970): 103–22; Svante Odén, "The Acidity Problem—An Outline of Concepts," *Water, Air, and Soil Pollution* 6, no. 2–4 (June 1976): 137–66.

8. Linda Lear, *Rachel Carson: Witness for Nature* (New York: Houghton Mifflin Harcourt, 2009); Elizabeth Seigel Watkins, *On the Pill: A Social History of Oral Contraceptives, 1950–1970* (Baltimore, Md.: Johns Hopkins University Press, 1998).

9. Bert Bolin, ed., *Report of the Swedish Preparatory Committee for the U.S. Conference on the Human Environment* (Stockholm: Norstedt and Söner, 1971); see also Lawrence Van Gelder, "Aroused Europeans Try to Stem Industrial Pollution," *New York Times*, January 11, 1971, 41.

10. Bert Bolin et al., *Air Pollution Across National Boundaries: The Impact of the Environment of Sulfur in Air and Precipitation, Sweden's Case Study for the United Nations Conference on the Human Environment*, Royal Ministry for Foreign Affairs and Royal Ministry of Agriculture (Stockholm, Sweden, 1971).

11. Bolin et al., *Air Pollution Across National Boundaries*, quotes on p. 85. Henning Rodhe published a major paper in *Tellus*, a leading geophysical journal, analyzing the sources of sulfur in the European atmosphere. His research showed that anthropogenic sulfur sources outweighed natural ones, that this sulfur could be transported more than one

thousand kilometers before it was deposited, and that in Sweden about half the sulfur in acid rain was coming from foreign pollution; the other half was caused by Swedish emissions and natural sources. Acid rain was not caused by volcanoes, for example, or derived from sea spray, and it easily moved across international boundaries: Henning Rodhe, "A Study of the Sulfur Budget for the Atmosphere over Northern Europe," *Tellus* 24 (January 19, 1972): 128–38. Lennart Granat also published the results of sampling across Europe—the European Atmospheric Chemistry Network—over one thousand locations in Scandinavia, Britain, and northern and central Europe. The results showed that Holland and Belgium were actually the worst affected countries; while Swedish scientists had done a great deal of the work on acid rain in Europe, other countries were in fact more seriously affected: Lennart Granat, "On the Relation Between pH and the Chemical Composition in Atmospheric Precipitation," *Tellus* 24 (1972): 550–60; Lennart Granat, *Deposition of Sulfate and Acid with Precipitation over Northern Europe*, Report AC-20, Institute of Meteorology (University of Stockholm, March 1972), 30, plus appendices.

12. USDA Forest Service, *Workshop Report on Acid Precipitation and the Forest Ecosystem*, General Technical Report NE 26, 1976.

13. D. H. Matheson and F. C. Elder, eds., "Atmospheric Contribution to the Chemistry of Lake Waters, First Specialty Symposium of the International Association for Great Lakes Research, Supplement 1 to Vol. 2," *Journal of Great Lakes Research* (Longford Mills, Ontario, Canada, 1976).

14. Richard J. Beamish, "Acidification of Lakes in Canada by Acid Precipitation and the Resulting Effects on Fishes," *Water, Air, and Soil Pollution* 6, no. 2–4 (March 26, 1976): 501–14. In 1970, the Canadian Federal Department of Indian Affairs had sued the mining companies for environmental damage to lakes and forests on a nearby Indian reserve. While that particular case was settled out of court, it suggested that by 1970 indigenous peoples had been observing these effects of acidification for some time.

15. H. Leivestad and I. P. Muniz, "Fish Kill at Low pH in a Norwegian River," *Nature* 259 (February 5, 1976): 391–92.

16. Gene Likens, "Acid Precipitation," *Chemical and Engineering News* (November 22, 1976): 29–44.

17. Ibid.

18. F. Herbert Bormann, "Acid Rain and the Environmental Future," *Environmental Conservation* (November 20, 1974): 270. Bormann leaned toward the latter conclusion, in contrast to most of his colleagues.

19. For a discussion of the use of stable carbon isotopes in demonstrating that increased atmospheric carbon dioxide is caused by human activities, see "How do we know that recent CO_2 increases are due to human activities?" *Real Climate*, December 22, 2004, http://www.realclimate.org/index.php/archives/2004/12/how-do-we-know-that-recent-cosub2sub-increases-are-due-to-human-activities-updated/.

20. Jerome O. Nriagu and Robert D. Coker, "Isotopic Composition of Sulphur in Atmospheric Precipitation around Sudbury, Ontario," *Nature* 274 (August 31, 1978): 883–85. The signatures were particularly distinctive, because the nickel ores there are geologically unusual, having formed at exceptionally high temperatures in the Precambrian era.

21. N. M. Johnson, "Acid Rain: Neutralization within the Hubbard Brook Ecosystem and Regional Implications," *Science* 204, no. 4392 (May 4, 1979): 497–99; N. M. Johnson

et al., "Acid Rain, Dissolved Aluminum and Chemical Weathering at the Hubbard Brook Experimental Forest, New Hampshire," *Geochimica et Cosmochimica Acta* 45, no. 9 (September 1981): 1421–37.

22. Gene E. Likens et al., "Acid Rain," *Scientific American* 241, no. 4 (October 1979): 43–51.

23. J. N. B. Bell, "Acid Precipitation—a New Study from Norway," *Nature* 292 (July 16, 1981): 199–200, on 199.

24. U.S. Department of State, *Long-range Transboundary Air Pollution, Convention between the United States of America and Other Governments*, done at Geneva, November 13, 1979, Treaties and Other International Acts Series 10541. For a review of the implementation of this convention, see United Nations, Economic Commission for Europe, *The State of Transboundary Air Pollution, Air Pollution Studies No. 5*, Geneva, Switzerland, 1989. In a harbinger of things to come, the United States did not sign the 1985 protocol.

25. The 1985 protocol was signed by Canada, Sweden, Norway, Denmark, Finland, West Germany, Switzerland, Austria, and France.

26. Gus Speth, "The Sisyphus Syndrome: Air Pollution, Acid Rain and Public Responsibility," *Proceedings of the Action Seminar on Acid Precipitation, Nov 1st to 3rd, 1979* (Canada: A.S.A.P. Organizing Committee, 1979), 170.

27. Ibid., 164.

28. Ibid., 171.

29. Patricia M. Irving, ed., *Acidic Deposition: State of Science and Technology: Summary Report of the US National Acid Precipitation Assessment Program, 1990 Integrated Assessment Report* (Washington, D.C.: Office of the Director, National Acid Precipitation Assessment Program, 1991); see also Herrick, "Predictive Modeling"; Gene Likens, "The Role of Science in Decision Making: Is Anybody Listening? When Does Evidence-Based Science Drive Environmental Management and Policy?" 2009 Cary Conference, for submission in *Frontiers in Ecology and the Environment*, second correction sent August 12, 2009; on the value of NAPAP for policy decision making, see Christopher J. Bernabo, "Improving Integrated Assessments for Applications to Decision Making," in *Air Pollution in the 21st Century: Priority Issues and Policy, Studies in Environmental Science* 72, ed. T. Schneider (Amsterdam: Elsevier, 1998), 183–97.

30. Herrick, "Predictive Modeling," 252.

31. Royal Society of Canada, *Acid Deposition in North America: A Review of Documents Prepared under the Memorandum of Intent between Canada and the United States of America, 1980, on Transboundary Air Pollution*, II, Technical Report, May 1983.

32. Ibid., II-11.

33. Ibid., II-11.

34. Ibid., II-9.

35. Ibid., II-7.

36. Ibid., II-8.

37. Environment Canada, *The Acid Rain Story* (Ottawa: Ministry of Supply and Services, 1984).

38. Chris Bernabo, telephone conversation with Naomi Oreskes, April 17, 2009. On the release of the Interagency Task Force Report on Acid Precipitation, see Eliot Marshall, "Acid Rain Researchers Issue Joint Report," *Science* 220, no. 4604 (June 24, 1983): 1359.

39. J. Christopher Bernabo, "Global Climate Change: A Second Generation Environmental Issue," *Air and Waste Management Association*, for Presentation at the 82nd Annual Meeting and Exhibition, Anaheim, California, June 25–30, 1989; Bernabo, "Communication Among Scientists, Decision-Makers and Society: Developing Policy-Relevant Global Climate Change Research," *Climate Change Research: Evaluation and Policy Implications, Proceedings of the International Climate Change Research Conference*, ed. S. Zwerver et al. (Amsterdam: Elsevier, 1995), 103–17; Bernabo, "Improving Integrated Assessments," 183–97.

40. Chris Bernabo, telephone conversation with Naomi Oreskes, April 17, 2009; Gene Likens, telephone conversation with Naomi Oreskes, April 29, 2009.

41. Gene Likens, telephone conversation with Naomi Oreskes, April 29, 2009.

42. Richard Ayres, e-mail communication to Naomi Oreskes, April 15, 2009; Jurgen Schmandt, Judith Clarkson, and Hilliard Roderick, eds., *Acid Rain and Friendly Neighbors: The Policy Dispute Between Canada and the United States* (Charleston, N.C.: Duke University Press, 1989); Leslie R. Alm, *Crossing Borders, Crossing Boundaries: The Role of Scientists in the US Acid Rain Debate* (Westport, Conn.: Praeger, 2000); Gene E. Likens, *The Ecosystem Approach: Its Use and Abuse* (Germany: Ecology Institute, 1992).

43. David W. Schindler, chair, *Atmosphere—Biosphere Interaction: Toward a Better Understanding of the Ecological Consequences of Fossil Fuel Combustion*, Committee on the Atmosphere and Biosphere, Board on Agriculture and Renewable Resources, Commission on Natural Resources, National Research Council (Washington, D.C.: National Academy of Sciences, 1981).

44. Robert Reinhold, "Acid Rain Issue Creates Stress between Administration and Science Academy," *New York Times*, June 8, 1982, C1.

45. "Clearing the Air," editorial, *Washington Post*, June 12, 1982, A14.

46. Frank Press and William Nierenberg had a rather testy exchange over this issue; see William A. Nierenberg (WAN) papers, MC13, 140: 9, Scripps Institute of Oceanography (SIO) Archives.

47. Schindler, *Atmosphere—Biosphere Interaction*; quoted in "Dropping Acid," *Washington Post*, October 16, 1981, A26.

48. Discussed in "Acid Rain Is Caused Mostly by Pollution at Coal-Fired Midwest Plants, Study Says," *Wall Street Journal*, November 2, 1982, 2; a second Academy report in early 1983 confirmed these conclusions. See Jack Calvert, chair, *Acid Deposition: Atmospheric Processes in Eastern North America: A Review of Current Scientific Understanding*, Committee on Atmospheric Transport and Chemical Transformation in Acid Precipitation, Environmental Studies Board, Commission on Physical Sciences, Mathematics, and Resources, National Research Council (Washington, D.C.: National Academies Press, 1983).

49. In January 1981, Nierenberg received a letter from E. Pendleton James, director of presidential personnel, informing him that he was under consideration for a position in the new administration. William A. Nierenberg personal correspondence, WAN papers, MC13, 35: 8, SIO Archives.

50. William A. Nierenberg to General Alexander Haig, 17 December 1980, WAN papers, MC13, 43: 3, SIO Archives.

51. William A. Nierenberg personal correspondence, WAN papers, MC13, 35: 8, SIO Archives; see also WAN papers, MC13, 35: 13, Personal, Jan.–Dec. 1982.

52. See Naomi Oreskes et al., "From Chicken Little to Dr. Pangloss: William Nierenberg, Global Warming, and the Social Deconstruction of Scientific Knowledge," *Historical Studies in the Natural Sciences* 38, no. 1 (2007): 109–52; Naomi Oreskes, "My Facts Are Better than Your Facts: Spreading Good News about Global Warming," in *How Do Facts Travel?* ed. Mary Morgan and Peter Howlett (Cambridge: Cambridge University Press, in press); see also chapter 6 of this book.

53. Charles Townes and Walter Munk, "Obituary, William Aaron Nierenberg," *Physics Today* 54, no. 6 (June 2001), http://scitation.aip.org/journals/doc/PHTOAD-ft/vol_54/iss_6/74_1.shtml.

54. Naomi Oreskes, *Science on a Mission: American Oceanography in the Cold War and Beyond* (Chicago: University of Chicago Press, in prep.); Naomi Oreskes and Ronald Rainger, "Science and Security before the Atomic Bomb: The Loyalty Case of Harold U. Sverdrup," *Studies in the History and Philosophy of Modern Physics* 31B, (2000): 309–69; Ronald Rainger, "Constructing a Landscape for Postwar Science: Roger Revelle, the Scripps Institution and the University of California, San Diego," *Minerva* 39, no. 3 (2001): 327–52; Ronald Rainger, "Science at the Crossroads: The Navy, Bikini Atoll, and American Oceanography in the 1940s," *Historical Studies in the Physical and Biological Sciences* 30, no. 2 (2000): 349–71; Ronald Rainger, "Adaptation and the Importance of Local Culture: Creating a Research School at the Scripps Institution of Oceanography," *Journal of the History of Biology* 36 (2003): 461–500; Jacob D. Hamblin, *Oceanographers and the Cold War: Disciples of Marine Science* (Seattle: University of Washington Press, 2005); Jacob D. Hamblin, *Poison in the Well: Radioactive Waste in the Ocean at the Dawn of the Nuclear Age* (New Brunswick, N.J.: Rutgers University Press, 2008); Naomi Oreskes, "A Context of Motivation: US Navy Oceanographic Research and the Discovery of Seafloor Hydrothermal Events," *Social Studies of Science* 33 (2003): 697–742; William Nierenberg, *Oceanography: The Making of Science, People, Institutions, and Discovery,* oral history interview with Naomi Oreskes, February 10, 2000, Office of Naval Research, Oral History Project.

55. Gene Likens, e-mail comment to Naomi Oreskes on draft chapter, June 15, 2009; Sherwood Rowland, personal communication with Naomi Oreskes, September 7–8, 2005.

56. William Nierenberg, chairman, *Report of the Acid Rain Peer Review Panel, July 1984* (Washington, D.C.: White House Office of Science and Technology Policy, 1984), v.

57. Ibid., III-3 and III-7.

58. William A. Nierenberg to John Marcum, Office of Science and Technology Policy, 18 March 1982, WAN papers, MC13, 140: 4, SIO Archives, and also letter from Nierenberg to A. M. Rosenthal, 9 November 1982, MC13, 140: 2, SIO Archives: "I personally chose all the members of the Committee but one." This was published in a shortened version in the *New York Times,* letter to the editor, December 5, 1982, late city final edition, sec. 6, 174, and the comment about how many members were members of the Academy was repeated in a letter to the *Washington Post* on July 2, 1982, letters page, "Acid Rain Fallout."

59. Homer Newell, "Beyond the Atmosphere: Early Years of Space Science," *The Academy of Sciences Stakes a Claim,* chap. 5, NASA History Series, http://www.hq.nasa.gov/office/pao/History/SP-4211/ch5-1.htm. The section in question is on pp. 52–53 of chapter 5.

60. Newell, "Beyond the Atmosphere," 50–53; Walter S. Sullivan, *Assault on the Unknown: The International Geophysicist Year* (New York: McGraw-Hill, 1961), 78.

61. Erik M. Conway, *Atmospheric Science at NASA: A History* (Baltimore, Md.: John Hopkins University Press, 2008), 34.

62. His work on global warming—discussed in chapter 6—perhaps earned him the most enemies, particularly as he took on such revered figures as Roger Revelle and Jule Charney. A few years ago, Singer called one of us (Naomi Oreskes) at home one night, saying that he was a friend of a very famous scientist, a man who had been a close friend and colleague of Roger Revelle. The next morning, Oreskes called that scientist, a man who after sixty years in America still retains his Viennese accent, who replied, "Ohhhhh. I would not call Fred Singer a friend." He then proceeded to tell the story of how "Singer betrayed Roger." See chapter 6.

63. S. Fred Singer, ed., *Global Effects of Environmental Pollution: A Symposium Organized by the American Association for the Advancement of Science* (New York: Springer-Verlag, 1970), 206.

64. S. Fred Singer, ed., *Is There an Optimum Level of Population?* (New York: McGraw-Hill, 1971), 4.

65. Singer, *Is There an Optimum Level?* 157.

66. Ibid., 256.

67. In 1984, Singer wrote an essay for the Cornucopian book *The Resourceful Earth*, edited by Julian L. Simon and Herman Kahn. Simon was widely interpreted as the leading advocate of the Cornucopian school, and inspiration to Danish political scientist Bjørn Lomborg, author of *The Skeptical Environmentalist*. At the time of the book, Simon was a fellow at the Heritage Foundation—a major conservative/Libertarian think tank. Singer's position in his chapter "World Demand for Oil" revealed a full-fledged Cornucopianism: "The consumption of energy . . . does not appear to provide any kind of limit to future economic growth, nor even an important break [sic]. Contrary to popular views, the availability of oil presents no major problem," Singer, *The Resourceful Earth: A Response to Global 2000* (New York: Blackwell, 1984), 339–60, on 339.

68. S. Fred Singer, *Cost-Benefit Analysis as an Aid to Environmental Decision-Making, Report M77-106* (McLean, Va.: MITRE Corporation, 1979), 3.

69. Ibid., 126.

70. Singer, *Cost-Benefit Analysis*.

71. Robert Reinhold, "13 Experts Named to Councel [sic] Reagan's Adviser for Science," *New York Times*, February 18, 1982, http://query.nytimes.com/gst/fullpage.html?res= 980DEED8123BF93BA25751C0A964948260. On Singer's interest in serving the administration, with his attached CV, see Singer to E. E. David, 21 November 1981, WAN papers, MC13, 49: 12, file label " 'S' Misc.," SIO Archives. His CV listed numerous Republican affiliations, including Life Member of Republican National Committee, a member of the RNC Advisory Council Subcommittee on International Economics under William Casey (later CIA director), a Consultant to Treasury Secretary William Simon on International Oil and Resource Problems, and the Faculty Advisor to the University of Virginia Young Republicans; see also WAN papers, MC13, 50: 1, Misc. 1982, SIO Archives.

72. Ibid.

73. S. Fred Singer, "The World's Falling Need for Crude Oil," press clipping, no source given; S. Fred Singer, "The Price of World Oil," *Annual Review of Energy* 8 (November

1983): 451–508. Both clippings found in Nierenberg personal correspondence, 1981, WAN papers, MC13, 35: 10, SIO Archives; see also S. Fred Singer, "Oil Pricing Blunders Now Have Saudis in a Jam," *Wall Street Journal*, May 28, 1981, 24.

74. S. Fred Singer, "The Coming Revolution in World Oil Markets," *Wall Street Journal*, February 4, 1981, 26: clipping in Nierenberg personal correspondence, 1981, WAN papers, MC13, 35: 10, SIO Archives.

75. S. Fred Singer to E. E. David, 21 November 1981, WAN papers MC13, 35: 10, SIO Archives.

76. S. Fred Singer to Joseph Ryan, Personnel Office, the White House, 15 June 1981, WAN papers MC13, 35: 10, SIO Archives.

77. William Nierenberg to John Marcum, 18 March 1982, WAN papers, MC13, 140: 4, SIO Archives; see also *Biographic Data for Acid Rain Group*, 18 March 1982, WAN papers, MC13, 140: 4, SIO Archives. This typed list contains more details, and does not include Singer; *Scientific Problems of Weather Modification: A Report of the Panel on Weather and Climate Modification*, Committee on Atmospheric Sciences, NAS-NRC Publication 1236 (Washington, D.C.: National Academies Press, 1964).

78. William Nierenberg to John Marcum, 18 March 1982, WAN papers, MC13, 140: 4, SIO Archives; *Biographic Data for Acid Rain Group*, WAN papers.

79. Ibid.

80. Ibid.; see also Russell W. Peterson, "Laissez Faire Landscape," *New York Times*, October 31, 1982, SM27.

81. Ben Lieberman and Nicolas Loris, "Five Reasons the EPA Should Not Attempt to Deal with Global Warming," Heritage Foundation, Issues: Energy and Environment, April 23, 2009, http://www.heritage.org/Research/EnergyandEnvironment/wm2407.cfm.

82. Singer to Nierenberg, 19 January 1982, WAN papers, MC13, 50: 1, SIO Archives.

83. William Ackerman to John Robertson, *"Review of Subject" notes*, 4 February 1983, WAN papers, MC13, 140: 9, SIO Archives.

84. *Minutes of the January 27, 28, and 29 meeting of the Acid Rain Peer Review Panel*, WAN papers, MC13, 140: 9, stamped March 11, 1983, SIO Archives.

85. A copy of the press release, dated July 27, 1983, can be found in the William Nierenberg papers, *General Comments on Acid Rain: A Summary of the Acid Rain Peer Review Panel for the Office of Science and Technology Policy*, press release (draft), 17 June 1983, WAN papers, MC13, 140: 12. A copy of the report itself is found in WAN papers, MC13, 141: 4, SIO Archives.

86. "Reagan-appointed Panel Urges Big Cuts in Sulfur Emissions to Control Acid Rain," *Wall Street Journal*, June 28, 1983, 6.

87. *General Comments on Acid Rain*, WAN papers. MC 13, 140: 12, on p. 1, SIO Archives.

88. Ibid., discussion on pp. 1–2.

89. Ibid., 1.

90. Ibid., 4.

91. S. Fred Singer, *Overall Recommendation of the Acid Rain Peer Review Panel*, draft by Singer to panel, 4 June 1983, received by Nierenberg 10 June 1983, WAN papers, MC13, 140: 12, SIO Archives.

92. Ibid., 2.

93. Chauncey Starr to George Keyworth, 19 August 1983, and also other letters in WAN papers, MC13, 141: 1, SIO Archives.

94. Starr to Keyworth, 19 August 1983.

95. S. Fred Singer to Acid Rain Peer Review Panel, *Acid Effects on Forest Productivity?* and *Political Solutions to the Acid Rain Problem*, 1 November 1983, WAN papers, MC13, 141: 1, SIO Archives.

96. S. Fred Singer to Acid Rain Peer Review Panel, *Assessment of Crop Losses from Ozone*, 31 October 1983, WAN papers, MC13, 141: 1, SIO Archives.

97. S. Fred Singer to William A. Nierenberg, 27 July 1983, cover letter to *3rd Draft—Acid Rain Peer Review Panel*, 2 August 1983, WAN Papers MC13 141:2, SIO Archives. The draft has Singer's annotations; the cover letter makes clear that these are his changes, which he is sharing with Nierenberg.

98. Frederic Golden, "Storm over a Deadly Downpour," *Time*, December 6, 1982, 84–85.

99. Gene Likens to Tom Pestorius, 19 April 1983, WAN papers, MC13, 140: 10, SIO Archives.

100. *Critique of NRC Document on Atmosphere-Biosphere Interactions, Set 1981, Environmental Research and Technology, Inc.*, and critique by David Schindler (typescript), WAN papers, MC13, 140: 7, SIO Archives.

101. S. Fred Singer to John Robertson, letter and attachments, 15 February 1983, WAN papers, MC13, 140: 7, SIO Archives.

102. S. Fred Singer to John Robertson, *Global Environment, Resources, and Population Issues-Federal Policy in the 1980s*, 15 February 1983, WAN papers, MC13, 140: 7, SIO Archives. Quote on pp. 3 and 4.

103. Gene Likens, telephone conversation with Naomi Oreskes, April 19, 2009.

104. Nierenberg's notes show that he did accept that acid rain was a serious problem. In January 1983, for example, when the panel first convened, Nierenberg sent notes to Gene Likens and Ruth Patrick that stressed that although it was important to lay out the uncertainties, it was also important to do so in the context of a "discussion whose purpose is to avoid allowing the uncertainties and gaps to keep us from recognizing the potential seriousness of the problem and the necessity of adopting national policies that would blunt the possible grave consequences of environmental acidification . . . Even more to the point, demanding absolute economy of approach in treating problems arising from acid precipitation would differ markedly from procedures we have adopted for other major problems in our society. In dealing with other environmental problems, education, energy, welfare, and national defense, we do the best we can with the knowledge base available. Prudence dictates that we do not go too far beyond this base but it also dictates that we do not stand still either." Draft sent to Dr. Gene Likens and Dr. Ruth Patrick, 12 January 1983, WAN papers, MC13, 140: 9, SIO Archives.

105. William Nierenberg to S. Fred Singer, 26 July 1983, WAN papers, MC13, 141: 3, SIO Archives. A letter to Ruckelshaus a few months later confirms this, in a discussion of the uncertain basis for recommending specific emissions reductions: "I hope you realize that, in my trying to get this thinking straightened out, I have not yielded my position that reductions [in] emissions are required." Letter from William Nierenberg to William Ruckelshaus, 4 November 1983, WAN papers, MC13, 141: 6, SIO Archives.

106. Steve LaRue, "Early Action Urged in Fight on Acid Rain," *San Diego Union*, August 8, 1984, Local Section, B-2.

107. On Ackermann, see William J. Hall, et al., "William C. Ackermann," in *Memorial Tributes: National Academy of Engineering*, vol. 4 (Washington, D.C.: National Academies

Press, 1991), 1–8, http://books.nap.edu/openbook.php?record_id=1760&page=3. Economists were developing means to quantify the value of lakes and forests, but it was beyond the charge and expertise of this panel to try to do that.

108. S. Fred Singer to chairman of the House Committee on Science and Technology, 30 September 1983, WAN papers, MC13, 141: 10, SIO Archives.

109. *Memorandum for Acid Rain Peer Review Panel*, 3 December 1982, WAN papers, MC13, 140: 2, SIO Archives.

110. Russell W. Peterson, "Laissez Faire Landscape," *New York Times*, October 31, 1982; Gerald O. Barney, study director, *The Global 2000 Report to the President: Entering the Twenty-First Century, Prepared by the Council on Environmental Quality and Department of State* (Washington, D.C.: U.S. Government Printing Office, 1980).

111. S. Fred Singer, *Report of the Acid Rain Peer Review Panel, Final Report, July 1984*, Office of Science and Technology Policy (Washington, D.C.: U.S. Government Printing Office, 1984), appendix 5.

112. Ibid., app. 5, A5-2.

113. Ibid., app. 5, A5-8.

114. Title IV of 1990 Clean Air Act Amendments, a.k.a., Acid Deposition Control Program, including provisions for market-based incentives for controlling emission of sulfur dioxide from electricity generation—a cap and trade system—an approach that many now consider a model for how to control and reduce greenhouse gas emissions. It also led to the idea of "flexible compliance"—whereby one could trade and bank allowances—rather than a command and control approach. In 1998, post hoc assessment concluded that "quantifiable benefits could be relatively large in the areas of human health and visibility [and t]he magnitude of potential benefits from these two areas alone could exceed the costs [of compliance]"; Herrick, "Predictive Modeling," 252.

115. Milton Friedman, *Capitalism and Freedom* (Chicago: University of Chicago Press, 1962), 30–32.

116. Ibid., 32.

117. Singer, app. 5, *Report of the Acid Rain Peer Review Panel, Final Report*, A5-8.

118. Ibid., p. A5-10.

119. Quoted in Russell W. Peterson, "Laissez Faire Landscape," *New York Times*, October 31, 1982. The $6,000 estimate is found in "The 30% Club," *Time*, April 2, 1984, http://www.time.com/time/magazine/article/0,9171,954196,00.html.

120. *Press Conference with Allan MacEachen*, 11 April 1983, transcript in WAN papers, MC13, 140: 10, file label "OSTP Acid Rain Review Group 1983," SIO Archives; see also Deborah Shapely, "Acid Rain Settlement in Sight?" *Nature* 301 (January 27, 1983): 274.

121. *Press Conference with Allan MacEachen*, WAN papers.

122. "Acid Rain Report Suppressed before U.S. Vote, Group Says," *G&M*, September 13, 1984, press clipping in WAN papers, MC13, 141: 4, SIO Archives.

123. "Polluted Air and Acid Rain: A Missing Link?" *Newsweek*, September 2, 1985, 25.

124. "Impurities from Heaven," *New Republic* 190, no. 10 (March 12, 1984): 8–9.

125. Stephen Budiansky, "Acid Rain: Canada Must Act Alone," *Nature* 307, no. 5953 (February 23, 1984): 679.

126. William M. Brown, "Maybe Acid Rain Isn't the Villain," *Fortune* 109 (May 28, 1984): 170–74.

127. Alan W. Katzenstein, "Acidity Is Not the Major Factor," *Wall Street Journal*, June 28, 1984, 28.

128. John S. Eaton, letter to the editor, *Wall Street Journal*, September 5, 1984, 33. (One wonders why this letter was published so much later.)

129. *Alan W. Katzenstein*, Bates Number (BN): TI01942387, Legacy Tobacco Documents Library; see also "Alan W. Katzenstein," SourceWatch, http://www.sourcewatch.org/index.php?title=Alan_W._Katzenstein.

130. Michael Wines, "Acid Rain Must be Sharply Curbed Soon, Controversial White House Report Warns," *Los Angeles Times*, August 18, 1984, A10; Ben A. Franklin, "Legislators Say White House Suppressed Acid Rain Report," *New York Times*, August 18, 1984, p. 10.

131. Marjorie Sun, "Acid Rain Report Allegedly Suppressed," *Science* 225, no. 4668 (September 21, 1984): 1374.

132. "Acid Rain Report Suppressed," *G&M*, WAN papers.

133. William A. Nierenberg to Acid Rain Peer Review Panel members, WAN papers, MC13, 140: 11, SIO Archives.

134. Tom Pestorius to William A. Nierenberg, *Acid Rain Panel Report-Executive Summary*, 21 May 1984, WAN papers, MC13, 141:6, SIO Archives. See also Kenneth A. Rahn, *Memorandum to Members of the OSTP Acid Rain Peer Review Panel*, 13 September 1984 and enclosed copies of original and revised Executive Summaries, WAN papers, MC13, 141: 4, SIO Archives.

135. Alm, *Crossing Borders, Crossing Boundaries*, 66.

136. Rahn, *Memorandum*, WAN papers, 1.

137. John Robertson to Acid Rain Panel, 24 February 1984, WAN papers, MC13, 141: 6, SIO Archives.

138. Ibid.

139. John Robertson, *Memorandum for the Acid Rain Peer Review Panel*, 2 March 1983, WAN papers, MC13, 140:7 SIO Archives.

140. Rahn, *Memorandum*, WAN papers, on p. 2 of the letter, bottom sentence.

141. Rahn, *Memorandum*, WAN papers, 2.

142. Ibid., 3.

143. Gene E. Likens to William A. Nierenberg, 17 September 1984, WAN papers, MC13 141: 4, SIO Archives.

144. Mal Ruderman to William A. Nierenberg, 5 October 1984, WAN papers, MC13, 141: 4, SIO Archives.

145. Rahn, *Memorandum*, WAN papers, 3.

146. William A. Nierenberg to Mal Ruderman, 24 October 1984, WAN papers, MC13 141: 4, SIO Archives.

147. Tom Pestorius, OSTP, letter to Mal Ruderman, 22 October 1984, WAN papers, MC13 141: 4, SIO Archives. Moreover, the revised summary contained a line that was very distinctive. "Acid deposition belongs to a socially very important class of problems that only appear to be precisely soluble by a straightforward sum of existing technological and legislative fixes. This is deceptive. Rather, this class of problems is not permanently solved in a closed fashion, but must be treated progressively. As knowledge steadily increases, actions are taken which appear most effective, and economical." The sentence might well have been true, but it was unlike anything in the original version. It suggested that acid rain

was fundamentally insoluble; you just lived with it and did the best you could to deal with it over time. And it was very nearly the same thing Nierenberg had written the year before, in the opening line of a report on global warming: "There is a broad class of problems that have no 'solution' in the sense of an agreed course of action that would be expected to make the problem go away." Nierenberg was the *only* member common to both committees, so either these were his words, or the White House had put them in his mouth. See *Changing Climate: Report of the Carbon Dioxide Assessment Committee, Board on Atmospheric Sciences and Climate, Commission on Physical Sciences, Mathematics, and Resources* National Research Council (Washington, D.C.: National Academies Press, 1983), xiii.

148. Mal Ruderman to William A. Nierenberg, 7 November 1984, WAN papers, MC13, 141: 4, SIO Archives.

149. Gene Likens, telephone conversation with Naomi Oreskes, April 29, 2009.

150. *Changes Wanted by Keyworth*, copy of 21 May 1984 telecopy, WAN papers, MC13, 141: 5, SIO Archives. See also unmarked copy of 21 May 1984 telecopy in WAN papers, MC13, 141: 6, SIO Archives.

151. Vernon Ehlers to William A. Nierenberg, 1 July 1983, WAN papers, MC13, 36: 4.

152. William A. Nierenberg to Carol Lynch, OSTP, 8 September 1983, WAN papers, MC13, 36: 4, SIO Archives.

153. C. Stark Draper to Donald T. Regan, chief of staff to the president, 5 December 1985, WAN papers, MC13, 36: 8, SIO Archives.

154. *This Week with David Brinkley*, Sunday, 26 August 1984, transcript in WAN papers, MC13, 141: 4, SIO Archives.

155. Brown, "Maybe Acid Rain Isn't the Villain," 170–74.

156. Magda Havas et al., "Red Herrings in Acid Rain Research," *Environmental Science and Technology* 18, no. 6 (1984): 176A–86A.

157. "With a Gun at Its Head, the EPA Turns Activist," *BusinessWeek*, November 5, 1984, 35.

158. "Acid Rain: How Great a Threat?" *Consumers' Research Magazine*, March 1986, 11–15.

159. William M. Brown, "Hysteria about Acid Rain," *Fortune* 113 (April 14, 1986): 125–26.

160. Daniel Seligman, "April Fooling," *Fortune*, May 11, 1987, 153.

161. Frank W. Woods, "The Acid Rain Question: Making Decisions Today for Tomorrow," *Futurist* (January–February 1987): 37. This article does not actually question the scientific evidence, but a box next to it, drawing on William Brown, does.

162. Edward C. Krug and Charles R. Fink, "Acid Rain on Acid Soil: A New Perspective," *Science* 221, no. 4610 (August 5, 1983): 520–25.

163. James N. Galloway et al., "Acid Precipitation: Natural versus Anthropogenic Components," *Science* 226, no. 4676 (November 16, 1984): 829–31.

164. Edward C. Krug, "Fish Story: The Great Acid Rain Flimflam," *Policy Review* 52 (Spring 1990): 44–48.

165. William Anderson, "Acid Test," *Reason* (January 1992); "The EPA vs. Ed Krug," http://www.sepp.org/Archive/controv/controversies/epavskrug.html.

166. Edward C. Krug, "Save the Planet, Sacrifice the People: The Environmental Party's Bid for Power," *Imprimis* 20, no. 7 (July 1991): 1–5; The Skeptic Tank, http://www.skepticfiles.org/conspire/kr91part.htm.

167. For example, the Heartland Institute promotes a book by Joseph L. Bast et al., *Eco-Sanity: A Common-Sense Guide to Environmentalism*, Heartland Institute (Lanham,

Md.: Madison Books, 1994), Heartland.org, http://www.heartland.org/bin/media/publicpdf/23043a.pdf; see also Samuel Aldrich and Jay Lehr, "Acid Rain, Nitrogen Scares Debunked," *Environment and Climate News*, February 1, 2007, Heartland Institute, Heartland.org, http://www.heartland.org/publications/environment%20climate/article/20522/Acid_Rain_Nitrogen_Scares_Debunked.html; J. Laurence Kulp, "Acid Rain: Causes, Effects, and Control," *Regulation: The CATO Review of Business and Government* (Winter 1990): 41–50, http://www.cato.org/pubs/regulation/regv13n1/v13n1-5.pdf; Michael Sanera, "Environmental Education in Wisconsin: What the Textbooks Teach," *Wisconsin Policy Research Institute Report* 9, no. 5 (June 1996): 1–39, Competitive Enterprise Institute, http://cei.org/gencon/025,01843.cfm.

168. William L. Anderson and Jacquelynne W. McLellan, "Newspaper Ideological Bias or 'Statist Quo'? The Acid (Rain) Test," *American Journal of Economics and Sociology* 65, no. 3 (July 2006): 473–95, Business Services Industry, BNET, http://findarticles.com/p/articles/mi_m0254/is_3_65/ai_n27009297/pg_5/.

169. Daniel Seligman, "Our Government Fails an Acid Test, How to Buy Politicians, California Conspiracies, and Other Matters," *Fortune* 123 (February 11, 1991): 145–46.

170. S. Fred Singer, "Environmental Strategies with Uncertain Science," *Regulation: The Cato Review of Business and Government* 13 (Winter 1990): 65–70, http://www.cato.org/pubs/regulation/regv13n1/v13n1-8.pdf. He cites the billion-dollar claim on p. 68.

171. Likens, "Is Anybody Listening?" 8–9.

172. Charles Drake (deceased), personal communication with Naomi Oreskes, 1995–1996. See also Ronald L. Numbers, *The Creationists: From Scientific Creationism to Intelligent Design* (Cambridge, Mass.: Harvard University Press, 2006), 184–92.

173. Philip Shabecoff, "Government Acid Rain Report comes under Sharp Attack," *New York Times*, September 11, 1987, http://www.nytimes.com/1987/09/22/science/government-acid-rain-report-comes-under-sharp-attack.html?pagewanted=all.

174. U.S. Environmental Protection Agency, "Air Trends: Sulfur Dioxide," http://www.epa.gov/air/airtrends/sulfur.html. Despite later skepticism about global warming (see chapter 6), George Will encouraged conservatives to accept the scientific evidence of acid rain, noting that "acid rain falls on golf courses, too." See discussion in transcript, *This Week with David Brinkley*, Sunday, 26 August 1984, transcript in WAN papers, MC13, 141: 4, SIO Archives; Energy Information Administration, "Annual Energy Review 2008," table 8.10, "Average Retail Prices of Electricity, 1960–2008," June 26, 2009, http://www.eia.doe.gov/emeu/aer/elect.html.

175. Likens, "Is Anybody Listening?"12; Office of Management and Budget, *Informing Regulatory Decisions: 2003 Report to Congress on the Costs and Benefits of Federal Regulations and Unfunded Mandates on State, Local, and Tribal Entities*, 8, http://www.whitehouse.gov/omb/inforeg/2003_cost-ben_final_rpt.pdf.

176. Sharon Begley, "Is It All Just Hot Air?" *Newsweek*, November 20, 1989.

177. Likens, "The Science of Nature, the Nature of Science," 558–72; see also Gene E. Likens, "Some Aspects of Air Pollutant Effects on Terrestrial Ecosystems and Prospects for the Future," *Ambio* 18, no. 3 (1989): 172–78; Gene E. Likens, *The Ecosystem Approach*, 166.

178. Likens, "The Science of Nature, the Nature of Science," 567.

179. Gene E. Likens and Jerry F. Franklin, "Ecosystem Thinking in the Northern Forest—

and Beyond," *Bioscience* 59, no. 6 (2009): 511–13, on p. 512. See also Likens, "Is Anybody Listening?"

180. Likens and Franklin, "Ecosystem Thinking," 511.

181. Ibid., 512; see also Gary Randorf, "Environmental Advocacy, the Adirondacks and Air Quality," *Environmental Science and Policy* 1, no. 3 (August 1998): 175–78.

182. Likens and Franklin, "Ecosystem Thinking," 512.

183. Jeffrey Salmon, *Are We Building Environmental Literacy?* A report by the George C. Marshall Institute's Independent Commission on Environmental Education, April 15, 1997, 9; see also John H. Cushman Jr., "Critics Rise up Against Environmental Education," *New York Times*, April 22, 1997, http://www.nytimes.com/1997/04/22/us/critics-rise-up-against-environmental-education.html?pagewanted=all.

184. Margaret R. Taylor, Edward S. Rubin, and David A. Hounshell, "Regulation as the Mother of Innovation: The Case of SO_2 Control," *Law and Policy* 27, no. 2 (April 2005): 348–78.

CHAPTER 4

1. MIT, *Man's Impact on the Global Environment: Report of the Study of Critical Environmental Problems* (Cambridge, Mass.: MIT Press, 1970).

2. Ibid., 100–106.

3. Harrison Halstead, "Stratospheric Ozone with Added Water Vapor: Influence of High-Altitude Aircraft," *Science* 170, no. 3959 (November 13, 1970): 734–36.

4. McDonald cited Frederick Urbach, ed. *The Biologic Effects of Ultraviolet Radiation, with Emphasis on the Skin* (New York: Pergamon Press, 1969); Alexander Hollaender, ed., *Radiation Biology v. 2* (New York: McGraw-Hill, 1965); and H.F. Blum, *Carcinogenesis by Ultraviolet Light* (Princeton: Princeton University Press, 1959), as his sources for the UV-skin cancer link.

5. Erik M. Conway, *High-speed Dreams: NASA and the Technopolitics of Supersonic Transportation, 1945–1999* (Baltimore, Md.: Johns Hopkins University Press, 2005), 63.

6. Dotto and Schiff discuss the conference in a great deal of detail. See Lydia Dotto and Harold Schiff, *The Ozone War* (Garden City, N.Y.: Doubleday, 1978), 39–68.

7. Paul Crutzen, "The Influence of Nitrogen Oxides on the Atmospheric Ozone Content," *Quarterly Journal of the Royal Meteorological Society* 96 (1970): 320–25.

8. Two separate accounts of this sequence exist. The first is Dotto and Schiff, *The Ozone War*, 59–68. In a retrospective article published in 1992, Johnston gives a somewhat different account from his voluminous notes: Harold S. Johnston, "Atmospheric Ozone," *Annual Review of Physical Chemistry* 43 (October 1992): 1–32. We have synthesized this account from both sources, relying on Johnston where they differ.

9. Conway, *High Speed Dreams*, chap. 5.

10. Dotto and Schiff, *The Ozone War*, 68–70.

11. Ibid., 70.

12. Ibid., 83.

13. Ibid., 86. See also Thomas Donahue and Alan J. Grobecker, "The SST and Ozone Depletion," letter to editor, *Science* 187, no. 4182 (March 28, 1975): 1142 and 1145.

14. This was almost certainly prompted by the advocacy of three scientists within the

agency who believed the shuttle might have a problem: Robert Hudson of the Johnson Space Center, James King at JPL, and I. G. Poppoff at Ames. See Dotto and Schiff, *The Ozone War*, 127. Hudson managed the contract with Stolarski and Cicerone, according to Stolarski. The shuttle assessment is: R. J. Cicerone et al., *Assessment of Possible Environmental Effects of Space Shuttle Operations*, NASA CR-129003, June 3, 1973.

15. Richard Stolarski, interview with Erik Conway, April 26, 2001; Johnston, "Atmospheric Ozone," 26–27.

16. Erle Ellis, "Anthropocene," in *The Encyclopedia of Earth*, ed. Jay Gulledge (Washington, D.C.: Environmental Information Coalition, National Council for Science and the Environment, 2008), http://www.eoearth.org/article/Anthropocene.

17. Mario J. Molina and F. S. Rowland, "Stratospheric Sink for Chlorofluoromethanes: Chlorine Atom Catalyzed Destruction of Ozone," *Nature* 249 (June 28, 1974): 810–12; see also F. S. Rowland and Mario J. Molina, "Chlorofluoromethanes in the Environment," *Reviews of Geophysics and Space Physics* 13 (February 1975): 1–35.

18. F. Sherwood Rowland, e-mail communication with Erik Conway, September 17, 2009; Panel on Atmospheric Chemistry, *Halocarbons: Effects on Stratospheric Ozone* (Washington, D.C.: National Academies Press, 1976), v–vi.

19. Edward Parson, *Protecting the Ozone Layer: Science and Strategy* (New York: Oxford University Press, 2003), 32; Senate Committee on Aeronautical and Space Sciences, *Stratospheric Ozone Depletion: Hearings before the Subcommittee on the Upper Atmosphere*, 94th Congress, 1st sess., 8 September 1975.

20. Dotto and Schiff, *The Ozone War*, 149.

21. Ibid., 150.

22. Parson, *Protecting the Ozone Layer*, 35.

23. Ibid., 37.

24. Dotto and Schiff, *The Ozone War*, 201.

25. DuPont's chief scientist, Ted Cairns, wrote to Handler several times trying to get him to ensure favorable treatment in the report; Hal Johnston, who seems to have been on friendly terms with Handler, also wrote several times: e.g., T. L. Cairns to Philip Handler, 19 January 1976, *Committee on Impacts of Stratospheric Change: Panels: Atmospheric Chemistry: Report*, National Academies Archive; Harold S. Johnston to Philip Handler, 2 February 1976, *Committee on Impacts of Stratospheric Change: Panels: Atmospheric Chemistry: Report*, National Academies Archive. Schiff goes into some detail on a story that involved panelist Hans Panofsky of Penn State, who unwisely spoke to a reporter about the report and wound up the center of an investigation; Dotto and Schiff, *The Ozone War*, 266–70.

26. Quoted in Dotto and Schiff, *The Ozone War*, 156.

27. Ibid., 157.

28. Ibid., 214.

29. Ibid., 157.

30. Ibid., 218.

31. Ibid., 225.

33. Ibid., 227.

34. Parson, *Protecting the Ozone Layer*, 76; Dotto and Schiff, *The Ozone War*, 228.

35. Dotto and Schiff, *The Ozone War*, 249.

36. Parson, *Protecting the Ozone Layer*, 38.

37. Committee on Impacts of Stratospheric Change, *Halocarbons: Environmental Effects of Chlorofluoromethane Release* (Washington, D.C.: National Academy of Sciences, 1976), 7.

38. Russell Peterson, quoted in Parson, *Protecting the Ozone Layer*, 39.

39. Dotto and Schiff, *The Ozone War*, 287.

40. Western Aerosol Information Bureau quoted in Dotto and Schiff, *The Ozone War*, 280.

41. Robert T. Watson, oral history interview with Erik Conway, April 14, 2004.

42. Erik M. Conway, *Atmospheric Science at NASA: A History* (Baltimore, Md.: John Hopkins University Press, 2008), 175.

43. Sharon L. Roan, *Ozone Crisis: The 15 Year Evolution of a Sudden Global Emergency* (New York: Wiley & Sons, 1989), 131.

44. Parson, *Protecting the Ozone Layer*, 84–85; Maureen Christie, *The Ozone Layer: A Philosophy of Science Perspective* (Cambridge: Cambridge University Press, 2001), 44–45; see also Conway, *Atmospheric Science at NASA*, 173.

45. Roan, *Ozone Crisis*, 132.

46. Roan, *Ozone Crisis*, 173–79; Ellen Ruppel Shell, "Weather versus Chemicals," *Atlantic Monthly*, May 1987.

47. Ruppel Shell, "Weather versus Chemicals."

48. See, for example, the introduction to the AAOE mission definition document: NASA, *Airborne Antarctic Ozone Experiment*, Ames Research Center, MS 245-5 (July 1987), 1–2, copy from: NASA HQ History Office, file "Airborne Antarctic Ozone Experiment"; A. F. Tuck et al., "The Planning and Execution of ER-2 and DC-8 Aircraft Flights Over Antarctica, August and September 1987," *Journal of Geophysical Research* 94: D9 (August 30, 1989): 11181–11222.

49. Parson, *Protecting the Ozone Layer*, 141.

50. Ibid., 121–22.

51. Ibid., 142–44.

52. Conway, *Atmospheric Science at NASA*, 185–87.

53. Robert T. Watson, F. Sherwood Rowland, and John Gille, *Ozone Trends Panel Press Conference*, NASA Headquarters, 15 March 1988, Langley Research Center doc. CN-157273, 1988; see also *Executive Summary of the Ozone Trends Panel*, 15 March 1988, Langley Research Center doc. CN-157277, 1988.

54. Parson, *Protecting the Ozone Layer*, 156.

55. Richard Turco, Alan Plumb, and Estelle Condon, "The Airborne Arctic Stratospheric Expedition: Prologue," *Geophysical Research Letters* 17, no. 4 (March 1990, Supplement): 313–16; W. H. Brune et al., "In Situ Observations of ClO in the Arctic Stratosphere: ER-2 Aircraft Results from 59N to 80N Latitude," *Geophysical Research Letters* 17, no. 4 (March 1990, Supplement): 505–8; D. S. McKenna et al., "Calculations of Ozone Destruction during the 1988/1989 Arctic Winter," *Geophysical Research Letters* 17, no. 4 (March 1990, Supplement): 553–56; M. H. Proffitt et al., "Ozone Loss in the Arctic Polar Vortex Inferred from High-Altitude Aircraft Measurements," *Nature* 347 (September 6, 1990): 31–36.

56. Parson, *Protecting the Ozone Layer*, 206.

57. Robert E. Taylor, "Advice on Ozone May Be: 'Wear Hats and Stand in Shade,'" *Wall Street Journal*, May 29, 1987, Eastern edition, sec. A, 1.

58. John B. Judis, *The Paradox of American Democracy* (New York: Pantheon Books, 2000). On the anti-environmental aspects of late twentieth century conservatism, Samuel P. Hays, *Beauty, Health, and Permanence: Environmental Politics in the United States, 1955–1985* (Cambridge: Cambridge University Press, 1987), 491; Hal K. Rothman, *Saving the Planet: The American Responses to the Environment in the Twentieth Century* (Chicago: Ivan R. Dee, 2000), 158.

59. Judis, *The Paradox of American Democracy*, 125. See also Edwin J. Feulner Jr., interview by Adam Meyerson, "Building the New Establishment," *Policy Review* 58 (Fall 1991): 6–16.

60. Judis, *The Paradox of American Democracy*, 124–27.

61. "S. Fred Singer, Ph.D. Professional Background," http://www.sepp.org/about%20sepp/bios/singer/cvsfs.html.

62. S. Fred Singer, "Ozone Scare Generates Much Heat, Little Light," *Wall Street Journal*, April 16, 1987, 1.

63. Ibid.

64. Ibid.

65. S. Fred Singer, "Does the Antarctic Ozone Hole Have a Future?" *EOS* 69, no. 47 (November 22, 1988): 1588.

66. Ibid.

67. V. Ramanathan, "The Greenhouse Theory of Climate Change: A Test by an Inadvertent Global Experiment," *Science* 240, no. 4850 (April 15, 1988): 293–99.

68. Committee on Energy and Natural Resources, *Hearing on Greenhouse Effect and Global Climate Change*, 100th Congress, 1st sess., November 9, 1987 (Washington, D.C.: U.S. Government Printing Office, 1987), 53; also see J. Hansen et al., "Global Climate Changes as Forecast by Goddard Institute for Space Studies Three-Dimensional Model," *Journal of Geophysical Research* 93: D8 (August 20, 1988): 9341–64.

69. S. Fred Singer, "My Adventures in the Ozone Layer," *National Review* (June 30, 1989): 34–38, quoted from 36.

70. Ibid. The discussion of Dobson is on p. 37; the quote regarding CFCs is on p. 38.

71. See Roan, *Ozone Crisis*, chap. 11.

72. Christie, *The Ozone Layer*, 46–47.

73. The major consumer-level CFC replacement, HFC-134a, has "comparable cycle efficiency," and energy efficiency standards adopted for refrigerators in 1990 actually led to a large reduction in energy consumption despite the adoption of non-CFC refrigerants. See James R. Sand et al., *Energy and Global Warming Impacts of HFC Refrigerants and Emerging Technologies* (Washington, D.C.: U.S. Department of Energy, 1997), 22.

74. See "Washington Institute for Values in Public Policy," SourceWatch, http://sourcewatch.org/index.php?title=Washington_Institute_for_Values_in_Public_Policy; Singer wrote to Roger Revelle on Washington Institute letterhead during 1990: see Singer to Revelle, 2 March 1990, Revelle Papers, MC6A, 150: 10, Scripps Institute of Oceanography (SIO) Archives.

75. Wigner's link to the Unification Church is discussed in Frederick Seitz et al., *Eugene Paul Wigner, Biographical Memoirs v. 74* (Washington, D.C.: National Academy Press, 1998), 364–88, http://books.nap.edu/openbook.php?record_id=6201&page=365.

76. S. Fred Singer and Candace Crandall, "Misled by Lukewarm Data," *Washington Times*,

May 30, 1991, final edition, sec. G; S. Fred Singer, "The Science Behind Global Environmental Scares," *Consumers' Research Magazine* 74, no. 10 (October 1991): 17. The quotation is from the *Washington Times* article.

77. Dixy Lee Ray and Lou Guzzo, *Trashing the Planet: How Science Can Help Us Deal with Acid Rain, Depletion of the Ozone, and Nuclear Waste (Among Other Things)* (New York: HarperPerennial, 1990), 12; originally published by Regnery Gateway, 1990.

78. Ibid., 45.

79. Ibid., 175; see also Singer, *Global Climate Change.*

80. Rogelio Maduro, "The Ozone Layer that Won't Go Away," *21st Century Science and Technology* 2 (September/October 1989): 26; and Rogelio Maduro, "The Myth Behind the Ozone Hole Scare," *21st Century Science and Technology* 2 (July/August 1989): 11.

81. Rogelio A. Maduro and Ralf Schauerhammer, *The Holes in the Ozone Scare: The Scientific Evidence that the Sky Isn't Falling* (Washington, D.C.: 21st Century Science Associates, 1992), intro. and chap. 1. For another analysis of their "facticity," see Christie, *The Ozone Layer,* 185–202.

82. Dixy Lee Ray with Lou Guzzo, *Environmental Overkill: Whatever Happened to Common Sense?* (Washington, D.C.: Regnery Gateway, 1993), 35.

83. Gary Taubes, "The Ozone Backlash," *Science* 260 (June 11, 1993): 1580–83.

84. F. Sherwood Rowland, "President's Lecture: The Need for Scientific Communication with the Public," *Science* 260 (June 11, 1993): 1571–1576, on p. 1573.

85. Ibid., 1574.

86. David A. Johnston, "Volcanic Contribution of Chlorine to the Stratosphere: More Significant to Ozone than Previously Estimated?" *Science* 209, no. 4455 (July 25, 1980): 491–93.

87. F. Sherwood Rowland, "President's Lecture," 1574.

88. S. Fred Singer, "The Hole Truth about CFCs," *Chemistry & Industry* (March 21, 1994): 240. See also S. Fred Singer, "Bad Science Pulling the Plug on CFCs?" *Washington Times,* February 22, 1994, final edition, sec. A.

89. House Committee on Science, Subcommittee on Energy and the Environment, Hearing on *Scientific Integrity and Public Trust: The Science Behind Federal Policies and Mandates: Case Study 1—Stratospheric Ozone: Myths and Realities,* S. Fred Singer testimony, 104th Congress, 1st sess., September 20, 1995 (Washington, D.C.: U.S. Government Printing Office, 1996), 50–64, quotes on p. 50 ("misled") and p. 52 ("wrong").

90. Ibid., 54.

91. "The Nobel Prize in Chemistry 1995," Nobelprize.org, http://nobelprize.org/nobel_prizes/chemistry/laureates/1995/.

92. S. Fred Singer, "Ozone Politics with a Nobel Imprimatur," *Washington Times,* November 1, 1995, final edition, sec. A15.

93. William K. Stevens, "G.O.P. Bills Aim to Delay Ban on Chemical in Ozone Dispute," *New York Times,* September 21, 1995, A20, http://www.nytimes.com/1995/09/21/us/gop-bills-aim-to-delay-ban-on-chemical-in-ozone-dispute.html.

94. Singer, "My Adventures in the Ozone Layer," 36.

95. Data from Science and Environmental Policy Project IRS Form 990 for 2007 (lines 8d and 21), dated 15 May 2008.

96. Singer, "My Adventures in the Ozone Layer," 36–37.

97. S. Fred Singer, "Global Warming: Do We Know Enough to Act?" *Environmental Protec-*
     tion: Regulating for Results, ed. Kenneth Chilton and Melinda Warren (Boulder, Colo.:
     Westview, 1991), 45.

98. Singer, "Global Warming: Do We Know Enough to Act?" 45–46.

99. Quoted in Andrew Revkin, "Let's Be Sensible on Global Warming," *Christian Science
     Monitor*, June 30, 1992; George Will, "Chicken Little: The Persistence of Eco-Pessimism,"
     Washington Post, May 31, 1992, C7.

100. Frederick Seitz, *Global Warming and Ozone Hole Controversies: A Challenge to Scientific
     Judgment* (Washington, D.C.: George C. Marshall Institute, 1994), 16–17, 22. His state-
     ment is "this would appear to remain an open question . . . A balloon filled with Freon
     gas is much heavier than air." Our copy came from the database of the Legacy Tobacco
     Documents Library, BN: 2025479245.

101. Patrick J. Michaels, "Perils Up in the Air," April 12, 2000, Cato Institute, http://www
     .cato.org/pub_display.php?pub_id=4736; Seitz, *Global Warming and Ozone Hole Contro-
     versies*; Patrick Michaels, "Apocalypse Machine Blows Up," *Washington Times*, November
     1, 1991, final edition, sec. F; Patrick Michaels, "More Hot Air from the Stratosphere,"
     Washington Times, October 27, 1992, sec. F, 1.

102. S. Fred Singer, "Letters to the Editor: Bad Climate in Ozone Debate," *Wall Street Journal*,
     June 17, 1993, Eastern edition, A1; "Letters to the Editor: Ozone, CFCs and Science Fic-
     tion," *Wall Street Journal*, March 24, 1993, Eastern edition, A15; "Letters to the Editor:
     The Dreaded Ozone Hole," *Wall Street Journal*, March 10, 1992, Eastern edition, A19;
     "Bookshelf: Environmental Fear-mongers Exposed," *Wall Street Journal*, April 28, 1993,
     Eastern edition, A18; "Letters to the Editor: Nobel Politicized Award in Chemistry," *Wall
     Street Journal*, November 3, 1997, Eastern edition, 23.

103. Kent Jeffreys, "Too Many Holes," *Wall Street Journal*, February 11, 1993, Eastern edi-
     tion, 15.

CHAPTER 5

1.   Derek Yach and Aguinaga Bialous, "Junking Science to Promote Tobacco," *American
     Journal of Public Health* 91, no. 11 (November 2001): 1745–48; David A. Kessler, *A Ques-
     tion of Intent: A Great American Battle with a Deadly Industry* (New York: Public Affairs,
     2002); Stanton A. Glantz et al., *The Cigarette Papers* (Berkeley: University of California
     Press, 1996).

2.   U.S. Department of Health and Human Services, Centers for Disease Control and Pre-
     vention, *The Health Consequences of Involuntary Exposure to Tobacco Smoke: A Report of the
     Surgeon General* (Washington, D.C.: U.S. Government Printing Office, 2006), http://
     www.surgeongeneral.gov/library/secondhandsmoke/report/fullreport.pdf; U.S. Depart-
     ment of Health and Human Services, "*The Health Consequences of Involuntary Exposure to
     Tobacco Smoke: A Report of the Surgeon General*," Office of the Surgeon General, http://
     www.surgeongeneral.gov/library/secondhandsmoke/factsheets/factsheet1.html.

3.   Glantz et al., *The Cigarette Papers*, chap. 10.

4.   Ibid., 402–3.

5.   A. N. Koplin, "Anti-Smoking Legislation: The New Jersey Experience," *Journal of Public
     Health Policy* 2, no. 3 (September 1981): 247–55.

6. Ibid.

7. J. R. White and H. F. Froeb, "Small-airways Dysfunction in Nonsmokers Chronically Exposed to Tobacco Smoke," *New England Journal of Medicine* 302, no. 13 (March 27, 1980): 720–23.

8. Glantz et al., *The Cigarette Papers*, 429.

9. This was enormously important, because it meant that if you studied the effect of smoking, you couldn't just compare death rates to nonsmokers in general, you had to compare to nonsmokers who were not exposed to environmental tobacco smoke (ETS). When this was done, it became clear that the effects of smoking were even greater than previously recognized.

10. Mi-Kyung Hong and Lisa Bero, "How the Tobacco Industry Responded to an Influential Study of the Health Effects of Second Hand Smoke," *British Medical Journal* 325 (December 14, 2002): 1413–16; Glantz et al., *The Cigarette Papers*, 413–16.

11. Glantz et al., *The Cigarette Papers*, 414.

12. Quoted in Glantz et al., *The Cigarette Papers*, 415.

13. David Michaels, *Doubt Is Their Product: How Industry's Assault on Science Threatens Your Health* (New York: Oxford University Press, 2008), 80; see also *The Health Consequences of Involuntary Smoking: A Report of the Surgeon General* (Washington, D.C.: U.S. Government Printing Office, 1986), http://profiles.nlm.nih.gov/NN/B/C/P/M/_/ nnbcpm.pdf.

14. Glantz et al., *The Cigarette Papers*, 308. (They did in 1988.)

15. U.S. Department of Health and Human Services, *The Health Consequences of Involuntary Smoking*, 1986.

16. Ibid., viii.

17. National Research Council, Committee on Passive Smoking, Board of Environmental Studies and Toxicology, *Environmental Tobacco Smoke: Measuring Exposures and Assessing Health Effects* (Washington, D.C.: National Academy Press, 1986); U.S. Environmental Protection Agency, Office of Health and Environmental Assessment, *Respiratory Health Effects of Passive Smoking: Lung Cancer and Other Disorders* (Washington, D.C.: 1992).

18. *Ellen Merlo, Vendor Conference Draft*, December 1993, Bates Number (BN): 2040863440, Legacy Tobacco Documents Library.

19. Glantz et al., *The Cigarette Papers*, 366.

20. Ibid., 251, 299–300, 305–13.

21. Glantz et al., *The Cigarette Papers*; Kessler, *A Question of Intent*; Michaels, *Doubt Is Their Product*.

22. Numerous pieces on this in *Bad Science: A Resource Book*, 26 March 1993, BN: 2074143969, Legacy Tobacco Documents Library, for example, Peter Brimelow and Leslie Spencer, "You Can't Get There from Here," *Forbes*, July 6, 1992, 59–64, on pages 120–25 in *Bad Science*. See also articles on pages 181, 217, 225, etc. in *Bad Science*.

23. *Briefing of Ralph Angiuoli Chairman, Executive Committee The Tobacco Institute, Comments: Excise Taxes*, 24 May 1989, BN: TI51541478, Legacy Tobacco Documents Library; *Regional Corporate Affairs*, 1991, 2, BN: 2501146354, Legacy Tobacco Documents Library; Anne Landman, "Beware Secondhand Rhetoric on Cigarette Taxes," PR Watch .org, http://www.prwatch.org/node/8271.

24. Nonsmokers Rights Association, *The Fraser Institute: Economic Think Tank or Front for the*

Tobacco Industry? April 1999, app. A, transcript of original court document, "Proposal for the Organization of the Whitecoat Project," 18, BN: 2065228563, Legacy Tobacco Documents Library; see also *Deposition of Steven Parrish in the* United States of America v. Philip Morris Incorporated et al., CA99-CV-02496, 25 June 2002, BN: PAR-RISHS062502, Legacy Tobacco Documents Library.

25. Merlo, *PM USA Vendor Conference*, Legacy Tobacco Documents Library.

26. *The ETS Program for 1991*, 1990, BN: 2023856052, Legacy Tobacco Documents Library.

27. Ibid.

28. U.S. Environmental Protection Agency, "Fact Sheet: Respiratory Health Effects of Passive Smoking," Smoke-free Homes and Cars Program, January 1993, http://www.epa.gov/smokefree/pubs/etsfs.html.

29. EPA, *Respiratory Health Effects of Passive Smoking*, 1–4.

30. Ibid., 1–6 to 1–7.

31. National Research Council, *Risk Assessment in the Federal Government: Managing the Process* (Washington, D.C.: National Academy Press, 1983).

32. U.S. Environmental Protection Agency, "Risk Assessment Portal," http://www.epa.gov/risk/.

33. EPA, *Respiratory Health Effects of Passive Smoking*, 1–2.

34. Ibid., 2–6.

35. EPA, "Fact Sheet," 1993.

36. *A Review of the Final Version of the Report: "Links between Passive Smoking and Disease: A Best Evidence Synthesis,"* A Report of the Working Group on Passive Smoking, coordinated by Frederick Seitz, 14 April 1989, BN: 512781113, Legacy Tobacco Documents Library.

37. Ibid.

38. Science and Environmental Policy Project, http://www.sepp.org/.

39. Tom Hockaday to Ellen Merlo and others, *Opinion Editorials on Indoor Air Quality and Junk Science*, memorandum, 8 March 1993, BN: 2021178205, Legacy Tobacco Documents Library; S. Fred Singer, *Junk Science at the EPA*, 8 March 1993, BN: 2021178206, Legacy Tobacco Documents Library.

40. Singer, *Junk Science at the EPA*, Legacy Tobacco Documents Library.

41. *Bad Science: A Resource Book*, BN: 2074144197, Legacy Tobacco Documents Library.

42. Ibid.

43. Paul D. Thacker, "Pundit for Hire: Smoked Out," *New Republic*, February 6, 2006, 13–14.

44. *Bad Science: A Resource Book*, BN: 2074144197, Legacy Tobacco Documents Library.

45. Competitive Enterprise Institute, "About CEI," http://cei.org/about.

46. *CEI Science Policy Clips and Highlights, January 1993–April 1994*, BN: 2023585726, Legacy Tobacco Documents Library.

47. Craig L. Fuller to Jim Tozzi, 13 July 1993, BN: 2046597569, Legacy Tobacco Documents Library.

48. Chris Mooney, "Paralysis by Analysis: Jim Tozzi's Regulation to End All Regulation," *Washington Monthly* (May 2004); Chris Mooney, *The Republican War on Science* (New York: Basic Books, 2005); see Mooney's chapter 8 for his discussion of Tozzi's role in the Data Quality Act.

49. Memorandum from James Tozzi to Jim Boland, 29 December 1993, BN: 2024207141, Legacy Tobacco Documents Library.

50. Craig L. Fuller to Tom Borelli, et al., *Subject*: "*Investor's Business Daily*/EPA," 28 January 1993, BN: 2023388137, Legacy Tobacco Documents Library.

51. "Thomas J. Borelli," SourceWatch, http//www.sourcewatch.org/index.php?title=Thomas_J._Borelli.

52. Victor Han to Ellen Merlo, *Subject: Burson/ETS*, memorandum, 22 February 1993, BN: 2023920035, Legacy Tobacco Documents Library.

53. A search on Rush Limbaugh in Legacy Tobacco Documents Library brings up over five hundred documents. These include a document on ETS strategy from Craig Fuller to Jim Boland and others entitled *Getting Rush Limbaugh on the Issue*, 23 January 1993, BN: 2047908408, Legacy Tobacco Documents Library, and a letter from the New York State Association of Wholesale Marketers and Distributors to Limbaugh, dated 13 September 1996, thanking him for his program on the "unreasoned attack by anti-smoking zealots," BN: 621965403, Legacy Tobacco Documents Library.

54. Han to Merlo, *Subject: Burson/ETS*, BN: 2023920035, Legacy Tobacco Documents Library; see also *ETS Media Strategy*, February 1993, BN: 2023920090, Legacy Tobacco Documents Library.

55. Han to Merlo, *Subject: Burson/ETS*, BN: 2023920035, Legacy Tobacco Documents Library.

56. *New Project*, April 1993, BN: 2046662829, Legacy Tobacco Documents Library; *EPA Watch Undertakes "Risk Assessment" on Danger of Showering*, EPA Watch vol.1, no. 2, 16 March 1992, BN: 2021174568, Legacy Tobacco Documents Library.

57. Han to Merlo, *Subject: Burson/ETS*. BN: 2023920035, Legacy Tobacco Documents Library.

58. Committee for a Constructive Tomorrow. "About CFACT," http://www.cfact.org/about/1549/About-CFACT. According to a Greenpeace Web site, Frederick Seitz served on the board of the Committee; see "Factsheet: Frederick Seitz," Exxonsecrets.org, http://www.exxonsecrets.org/html/personfactsheet.php?id=6#srci2; "Biography: Bonner R. Cohen," http://prfamerica.org/biography/Biography-Cohen-Bonner.html.

59. Han to Merlo, *Subject: Burson/ETS*, BN: 2023920035, Legacy Tobacco Documents Library.

60. Ibid.

61. Paul D. Thacker, "The Junkman Climbs to the Top," *Environmental Science and Technology Online News* (May 2005); Thacker, "Smoked Out: Pundit for Hire," 13–14; "Steven J. Milloy: The 'Junkman' Exposed," February 2006, Americans for Nonsmokers' Rights, http://www.no-smoke.org/pdf/stevenmilloy.pdf.

62. APCO vice president Neal Cohen later boasted about this as a general strategy. See Jane Fritsch, "Sometimes, Lobbyists Strive to Keep Public in the Dark," *New York Times*, March 19, 1996, http://query.nytimes.com/gst/fullpage.html?res=9505E1DC1739F93AA25750C0A960958260&sec=&spon=&pagewanted=print.

63. Sheldon Rampton and John Stauber, *Trust Us, We're Experts! How Industry Manipulates Science and Gambles with Your Future* (New York: Tarcher, 2000), 239, 248–49.

64. Sheldon Rampton and John Stauber, "How Big Tobacco Helped Create 'The Junkman,'" *PR Watch* 7, no. 3 (Third Quarter, 2000): 5–15; PR Watch.org, Center for Media and Democracy, http://www.prwatch.org/prissues/2000Q3/junkman.html.

65. Rampton and Stauber, "The Junkman," 5–15.

66. John Lenzi to Ellen Merlo, *Subject: TASSC Update*, memorandum, 13 December 1993, BN: 2046553280, Legacy Tobacco Documents Library.

67. Ibid.; John Lenzi to Vic Han et al., *Subject: TASSC*, 22 February 1994, BN: 2078848225, Legacy Tobacco Documents Library; Tom Hockaday and Neal Cohen to Matt Winokur, *Re: Thoughts on TASSC Europe*, 25 March 1994, BN: 2024233595, Legacy Tobacco Documents Library; see also George Monbiot, "The Denial Industry," *Guardian*, September 19, 2006, http://www.guardian.co.uk/environment/2006/sep/19/ethicalliving.g2.

68. Garrey Carruthers, TASSC chairman and governor of New Mexico, to Dr. Richard Lindzen, *Invitation to Join TASSC*, 12 May 1993, BN: 2046989059, Legacy Tobacco Documents Library.

69. Craig L. Fuller to Michael A. Miles, *January Monthly Report*, 23 February 1994, BN: 2048212857, Legacy Tobacco Documents Library.

70. Steven J. Milloy to Sharon Boyse, *Grant Request from TASSC*, 22 September 1997, BN: 190204008, Legacy Tobacco Documents Library.

71. *Statement of Garrey Carruthers*, 20 December 1995, BN: 2047070949, Legacy Tobacco Documents Library.

72. Mark Dowie, "What's Wrong with the *New York Times* Science Reporting?" *Nation* 267, no. 1 (July 6, 1998): 13–19; "Gina Kolata," SourceWatch, http://www.sourcewatch.org/index.php?title=Gina_Kolata.

73. See discussion in Yach and Bialous, "Junking Science to Promote Tobacco," 1745–48; *Inventory of Comments Received by the Tobacco Institute on the Costs and Benefits of Smoking Restrictions: An Assessment of the Smoke-Free Environment Act of 1993 (H.R. 3434)*, August 1993, BN: 2047232462, Legacy Tobacco Documents Library.

74. *The Tobacco Institute 1995 Proposed Budget*, 11 October 1994, BN: 91082676, Legacy Tobacco Documents Library.

75. S. Fred Singer and Kent Jeffreys, *The EPA and the Science of Environmental Tobacco Smoke*, Alexis de Tocqueville Institution, May 1994, BN: TI31749030, Legacy Tobacco Documents Library; *The Tobacco Institute 1995 Proposed Budget*, Legacy Tobacco Documents Library.

76. Kent Jeffreys, *Who Should Own the Ocean?* (Washington, D.C.: Competitive Enterprise Institute, 1991), 17–18; Kent Jeffreys, "Rescuing the Oceans," in *The True State of the Planet*, ed. Ronald Bailey (New York: Free Press, 1995); "Kent Jeffreys," SourceWatch, http://www.sourcewatch.org/index.php?title=Kent_Jeffreys.

77. *Science, Economics, and Environmental Policy: A Critical Examination*, 11 August 1994, on p. 1, BN: 92756807, Legacy Tobacco Documents Library.

78. Ibid., 7.

79. *Memorandum from Samuel D. Chilcote, Jr. to The Members of the Executive Committee*, 11 August 1994, BN: 980193761, Legacy Tobacco Documents Library.

80. EPA, *Respiratory Health Effects of Passive Smoking*.

81. Letter to William K. Reilly, EPA administrator, from members of the Science Advisory Board, *Subject: Science Advisory Board's Review of the Office of Research and Development Document: Health Effects of Passive Smoking*, 19 April 1991, on p. 2, BN: 2023989358, Legacy Tobacco Documents Library; and on p. 44 of the main report, EPA Science Ad-

visory Board, *Review of Draft Environmental Health Effects Document*, April 1991, EPA-SAB-IAQC-91-007, U.S. Environmental Protection Agency.

82. EPA Science Advisory Board, *Review of Draft*, EPA-SAB-IAQC-91-007, 47–48.

83. Ibid., 45.

84. Ibid., 48.

85. Ibid., 49.

86. EPA Science Advisory Board, *Review of Draft Passive Smoking Health Effects Document. Respiratory Health Effects of Passive Smoking Lung Cancer and Other Disorders*, November 1992, EPA-SAB-IAQC-93-003, on p. 27, BN: 2023989067, Legacy Tobacco Documents Library.

87. Ibid., 1–2.

88. Ibid., 21–22.

89. It is very difficult to find a good, clear explanation of this, because statistical texts are filled with double negatives about rejecting null hypotheses. Type 1 errors, in this lingo, are rejections of the null when the null is actually true. That is to say, there's no effect, but you reject that hypothesis, so you think there is an effect when there isn't one: a false positive. The best recent treatment of this issue, and the related issue of the statistical significance, the confusions surrounding it, and the value judgments embedded in it, is Deirdre McCloskey and Stephen Ziliak, *The Cult of Statistical Significance: How the Standard Error Costs Us Jobs, Justice, and Lives* (Ann Arbor: University of Michigan Press, 2008).

90. It also reflects a long tradition in the history of science that valorizes skepticism as an antidote to religious faith. This is our interpretation of why scientists are so afraid of making type I errors.

91. Valerie J. Easton and John H. McColl, "Type I Error," Statistics Glossary, http://www.stats.gla.ac.uk/steps/glossary/hypothesis_testing.html#1err.

92. "Type I and II Errors," HyperStat Online Contents, http://davidmlane.com/hyperstat/A18652.html.

93. The precautionary principle, widely applied in Europe, is an attempt to remedy this by consciously shifting the burden of proof away from the victim, and asking society to be more cautious when potential harms are recognized. The fact that the United States has not adopted that stance has everything to do with the story told in this book.

94. Ziliak and McCloskey, *The Cult of Statistical Significance*.

95. U.S. Environmental Protection Agency, "Setting the Record Straight: Secondhand Smoke is a Preventable Health Risk," Smoke-Free Homes and Cars Program, June 1994, http://www.epa.gov/smokefree/pubs/strsfs.html.

96. U.S. Environmental Protection Agency, National Center for Environmental Assessment, National Research Council, *Risk Assessment in the Federal Government*, "Guidelines for Carcinogen Risk Assessment (2005)," http://cfpub.epa.gov/ncea/cfm/recordisplay.cfm?deid=116283.

97. Judith Graham, e-mail message to Naomi Oreskes, August 6, 2007. Dr. Graham is referring to EPA, "Guidelines for Carcinogen Risk Assessment (2005)."

98. For a recent review, see D. P. Hayes, "Nutritional Hormesis," *European Journal of Clinical Nutrition* 61 (February 2007): 147–59.

99. This whole section from EPA, "Setting the Record Straight."

100. Ibid.

101. The industry pushed this idea extensively in Europe, through its front organization, the International Center for a Scientific Ecology. A conference in May 1993—organized around the question "Is the concept of linear relationship between dose and effect still a valid model for assessing risk related to low doses of carcinogens?"—included Fred Singer speaking on the Delaney amendment and its consequences on regulation in the United States, and Aaron Wildasky (a political scientist) questioning "Do rodent studies predict human cancers?" See *The International Center for a Scientific Ecology*, BN: 85012622, and also Ron Tully to National Manufacturers Associations, *Subject: International Center for a Scientific Ecology Meeting*, 27 April 1993, BN: 2028385382, Legacy Tobacco Documents Library. In 2008, the Marshall Institute had on its Web site an earlier article, published September 7, 1999, "Paracelsus to Parascience: The Environmental Cancer Distraction," by Bruce N. Ames and Lois Swirsky Gold, arguing the same point for a host of toxins and carcinogens, www.marshall.org/article.php?id=73.

102. Chauncey Starr to George Keyworth, 19 August 1983, and also other letters in William A. Nierenberg (WAN) papers, MC13, 141: 1 Scripps Institution of Oceangraphy, (SIO) Archives; see also Chauncey Starr, "Risk Criteria for Nuclear Power Plants: A Pragmatic Proposal," *Risk Analysis* 1, no. 2 (1981): 113–20; and Chauncey Starr, "Risk Management, Assessment, and Acceptability," *Risk Analysis* 5, no. 2 (1985): 97–102.

103. Emil Mrak speech to Philip Morris Laboratories, "Some Experiences Related to Food Safety," 16 January 1973, Emil M. Mrak Collection D-96, MSS, box 8, Special Collections, Shields Library, University of California, Davis.

104. Ibid.

105. A. R. Feinstein, "Scientific Standards in Epidemiological Studies of the Menace of Daily Life," *Science* 242, no. 4883 (December 2, 1988): 1257–63.

106. "Tobacco Sales Light Up Philip Morris Earnings," *USA Today*, January 26, 1995; see also "Philip Morris's Net More than Tripled in 4th Quarter, Aided by Tobacco Sales," *Wall Street Journal*, January 26, 1995, both clippings in *FYI: Director's Edition*, 1 February 1995, BN: 2041128878, Legacy Tobacco Documents Library.

107. "Five Bargain Blue Chips that Offer Towering Gains," *Money: Wall Street Newsletter*, February 1995, news clipping in *FYI: Director's Edition*, Legacy Tobacco Documents Library.

108. Philip Morris kept track of these positive press clippings; see *FYI: Director's Edition*, Legacy Tobacco Documents Library.

109. "Profiles: FOREST," Tobaccodocuments.org, Tobacco Documents Online, http://tobaccodocuments.org/profiles/forest.html; see also Iain Brotchie, "UK: Scottish Report Exposes Tobacco Tactics," *Tobacco Control* 14, no. 6 (2005): 366.

110. *Air Chief Marshal Sir Christopher Foxley-Norris*, 27 November 1978, BN: 2025024182, Legacy Tobacco Documents Library.

111. *Confidential Conference and Research Proposal*, 1997, BN: 516860591, Legacy Tobacco Documents Library; various other documents, over three thousand in all under search of FOREST, freedom organization in Legacy Tobacco Documents Library.

112. *Confidential Conference and Research Proposal*, 1997.

113. Ibid.

114. "Ralph Harris, Baron Harris of High Cross," Wikipedia, http://en.wikipedia.org/wiki/Ralph_Harris,_Baron_Harris_of_High_Cross. We were unable to find this exact quote

but certainly Lord Harris's writing frequently invoked the metaphor of the "invisible hand"; see J. E. King, "Ralph Harris. Ralph Harris in His Own Words: The Selected Writings of Lord Harris," *History of Economics Review* (Summer 2008), BNET, http://findarticles.com/p/articles/mi_6787/is_48/ai_n31611565/?tag=content;col1; "Lord Harris of High Cross," Obituaries, *Daily Telegraph*, October 2006, telegraph.co.uk, http://www.telegraph.co.uk/news/obituaries/1531862/Lord-Harris-of-High-Cross.html; "Lord Harris of High Cross: Free-Market Thinker Who Served as Director of Institute of Economic Affairs for Three Decades," *Times Online*, October 20, 2006, http://www.timesonline.co.uk/tol/comment/obituaries/article606521.ece. On Harris and the British Institute of Economic Affairs, see Philip Mirowksi and Dieter Plehwe, *The Road from Mont Pèlerin: The Making of the Neoliberal Thought Collective* (Cambridge, Mass.: Harvard University Press, 2009), 45–97.

115. John C. Luik, *Through the Smokescreen of "Science": The Dangers of Politically Corrupted Science for Democratic Public Policy*, on p. 1, BN: 517443033, Legacy Tobacco Documents Library.

116. Ibid., 1.

117. Ibid., 2.

118. Singer and Jeffreys, *The EPA and the Science of Environmental Tobacco Smoke*, 2.

119. William E. Simon, *A Time for Truth* (New York: Reader's Digest Press, 1978), 221.

120. On the Olin Foundation, and its support of right-wing causes, see http://mediatransparency.org/funderprofile.php?funderID=7. This link is now inactive as the Olin Foundation has shut down; see "John M. Olin Foundation," SourceWatch, http://www.sourcewatch.org/index.php?title=John_M._Olin_Foundation; "John M. Olin Foundation," Right Web, http://www.rightweb.irc-online.org/profile/John_M._Olin_Foundation.

121. Russell Seitz, "Making the World Safe for Cigarette Smokers," *Forbes* 160, no. 5 (September 8, 1997): 181, http://www.forbes.com/forbes/1997/0908/6005181a.html.

122. Ibid.

123. Lt. General Daniel Graham to William A. Nierenberg, 27 December 1984, WAN papers, MC13, 43: 17, SIO Archives.

124. Isaiah Berlin, *Liberty: Incorporating Four Essays on Liberty*, ed. Henry Hardy (New York: Oxford University Press, 2002).

125. Luik, *Through the Smokescreen of "Science,"* Legacy Tobacco Documents Library.

126. Ibid., "totalitarian flavour" quote on p. 3, rest of quote on p. 2.

127. Ibid., 3.

128. *Comparative Substance Use III: Pleasure and Quality of Life*, 28 September 1993, BN: 2029104002, Legacy Tobacco Documents Library.

129. *Academic Contact List*, BN: 502563475, Legacy Tobacco Documents Library.

130. Luik, *Through the Smokescreen of "Science,"* Legacy Tobacco Documents Library.

CHAPTER 6

1. John Roach, "2004: The year global warming got respect," *National Geographic News: Reporting Your World Daily*, December 29, 2004, http://news.nationalgeographic.com/news/2004/12/1229_041229_climate_change_consensus.html.

2. Intergovernmental Panel on Climate Change, *Summary for Policy Makers* in *Climate*

Change 2007, the Physical Science Basis, Contribution of Working Group I to the Fourth Assessment Report of the Intergovernmental Panel on Climate Change (Cambridge: Cambridge University Press, 2007), p. 8, http://www.ipcc.ch/pdf/assessment-report/ar4/wg1/ar4-wg1-spm.pdf.

3. Naomi Oreskes, "Behind the Ivory Tower: The Scientific Consensus on Climate Change," *Science* 306, no. 5702 (December 2004): 1686.

4. "Poll: Americans See a Climate Problem," *Time*, March 26, 2006, http://www.time
.com/time/nation/article/0,8599,1176967,00.html. Contrast this with the results of the Intergovernmental Panel on Climate Change Third Assessment Report, which states unequivocally that average global temperatures have risen: *Climate Change 2001, Contribution of Working Groups I, II, and III to the Third Assessment Report of the International Panel on Climate Change* (Cambridge: Cambridge University Press, 2001), http://www.ipcc.ch/ipccreports/tar/vol4/english/index.htm.

5. Gary Langer, "Poll: Public Concern on Warming Gains Intensity: Many See a Change in Weather Patterns," *ABC News*, March 26, 2006, http://abcnews.go.com/Technology/GlobalWarming/story?id=1750492&page=1. For a related poll, see also the Pew Research Center for the People and the Press, "Little Consensus on Global Warming: Partisanship Drives Opinion," July 12, 2006, Survey Reports, http://people-press.org/report/280/little-consensus-on-global-warming. For the Pew results, see "Fewer Americans see Solid Evidence of Global Warming," October 22, 2009, http://pewresearch
.org/pubs/1386/cap-and-trade-global-warming-opinion.

6. James R. Fleming, *The Callendar Effect: The Life and Times of Guy Stewart Callendar (1898–1964), The Scientist Who Established the Carbon Dioxide Theory of Climate Change* (Boston, Mass.: American Meteorological Society, 2007); James R. Fleming, *Historical Perspectives on Climate Change* (New York: Oxford University Press, 1998); Spencer R. Weart, *The Discovery of Global Warming* (Cambridge, Mass.: Harvard University Press, 2008).

7. Roger Revelle et al., "Atmospheric Carbon Dioxide," app. Y.4, in President's Science Advisory Committee, Panel on Environmental Pollution, *Restoring the Quality of Our Environment: Report of the Panel on Environmental Pollution* (Washington, D.C.: The White House, 1965).

8. Ibid., 9.

9. Lyndon B. Johnson, "Special Message to Congress on Conservation and Restoration of Natural Beauty," February 8, 1965, American Presidency Project, http://www.presidency
.ucsb.edu/ws/index.php?pid=27285.

10. Gordon MacDonald et al., *The Long Term Impact of Atmospheric Carbon Dioxide on Climate*, Jason Technical Report JSR-78-07 (Arlington, Va.: SRI International, 1979), 1.

11. Ann K. Finkbeiner, *The Jasons: The Secret History of Science's Postwar Elite* (New York: Viking, 2006). On the personality of the type of physicist who was associated with the Jasons in the early period, see also Myanna Lahsen, "Experiences of Modernity in the Greenhouse: A Cultural Analysis of a Physicist 'Trio' Supporting the Backlash against Global Warming," *Global Environmental Change* 18 (2008): 204–19, http://sciencepolicy.colorado.edu/admin/publication_files/resource-2590-2008
.05.pdf.

12. MacDonald et al., *The Long Term Impact of Atmospheric Carbon Dioxide*, 1.

13. Ibid., iii.

14. Robert M. White, "Oceans and Climate—Introduction," *Oceanus* 21 (1978): 2–3.

15. John S. Perry to Jule Charney, 9 May 1979, MC184, 364: 11, National Academies Archives.

16. Jule Charney et al., *Carbon Dioxide and Climate: A Scientific Assessment, Report of an Ad-Hoc Study Group on Carbon Dioxide and Climate, Woods Hole, Massachusetts, July 23–27, 1979, to the Climate Research Board,* National Research Council (Washington, D.C.: National Academies Press, 1979), 2.

17. Ibid., 2.

18. Verner E. Suomi in Charney et al., *Carbon Dioxide and Climate,* viii.

19. Charney et al., *Carbon Dioxide and Climate,* 10–11.

20. Henry Abarbanel, personal communication with Naomi Oreskes, October 26, 2006.

21. Suomi in Charney et al., *Carbon Dioxide and Climate,* viii.

22. Ibid.

23. Richard Meserve to Verner E. Suomi, 5 October 1979, *Assembly on Mathematical and Physical Sciences, Climate Board: Review Panel on Carbon Dioxide and Climate: General, 1979–1981,* collection, National Academies Archives.

24. Thomas C. Schelling, 18 April 1980, Climate Research Board Collection, National Academy of Sciences, National Academies Archives.

25. Ibid.

26. Ibid., emphasis in original.

27. John S. Perry, "Energy and Climate: Today's Problem, Not Tomorrow's," *Climatic Change* 3, no. 3 (September 1981): 223–25.

28. Ibid., 223–24.

29. Quoted in Perry, "Energy and Climate," 224.

30. Weart, *The Discovery of Global Warming.*

31. Abraham Ribicoff to Philip Handler, 30 October 1979, *Climate Research Board Study Group on Stratospheric Monitoring,* Ad Hoc: Meeting: Agenda, Assembly on Mathematical and Physical Sciences, William A. Nierenberg (WAN) papers, MC13 88: file label "National Academy of Sciences, Energy Security Act Text, August 1979," Scripps Institute of Oceanography (SIO) Archives.

32. John Perry to members of the Climate Research Board, 3 August 1979, National Academy of Sciences, Assembly of Mathematical and Physical Sciences, Climate Board: Review Panel on Carbon Dioxide & Climate, General 1979–1981 collection, National Academies Archives.

33. William Nierenberg to John Perry, 10 August 1979, National Academy of Sciences, Assembly of Mathematical and Physical Sciences, Climate Board: Review Panel on Carbon Dioxide & Climate, General 1979–1981 collection, National Academies Archives.

34. Roger R. Revelle, "Probable Future Changes in Sea Level Resulting from Increased Atmospheric Carbon Dioxide," in William Nierenberg et al., *Changing Climate: Report of the Carbon Dioxide Assessment Committee* (Washington, D.C.: National Academies Press, 1983), 441–442.

35. Ibid.

36. Ibid.

37. Nierenberg et al., *Changing Climate,* 87.

38. Note that the low-end estimate—337 ppm—had already been passed by the year 2000.

39. William D. Nordhaus and Gary W. Yohe, in Nierenberg et al., *Changing Climate*, 151.

40. Thomas C. Schelling, "Climate Change: Implications for Welfare and Policy," in Nierenberg et al., *Changing Climate*, 449.

41. Ibid., 452. In hindsight, Schelling's argument that the problem is not CO_2 but climate change is clearly wrong. Ocean acidification and its effects on marine ecosystems demonstrate that CO_2, per se, is a problem, even if it did not also produce global warming. Changing the chemistry of the atmosphere turns out to change the chemistry of the oceans as well, and this may turn out to be more significant for the biosphere than atmospheric temperature changes.

42. Nierenberg et al., *Changing Climate*, 3.

43. Ibid., 53.

44. Alvin M. Weinberg, "Global Effects of Man's Production of Energy," *Science* 186, no. 4160 (October 18, 1974): 205.

45. Alvin Weinberg, Comments on NRC draft *Report of the Carbon Dioxide Assessment Committee*, July–August 1983, WAN MC13, 86: file label "BASC/CO_2," SIO Archives.

46. Anonymous, *Chapter Reviews*, pp. 5 and 8, WAN papers, MC13, 86: file label "Chapter reviews of Draft CO_2 Assessment Committee Report, July 1983, 1 of 2," SIO Archives.

47. Edward Frieman, personal communication to Naomi Oreskes, March 16, 2007.

48. Philip Shabecoff, "Haste of Global Warming Trend Opposed," *New York Times*, October 21, 1983, Late City edition, sec. A, 1; Stephen Seidel, *Can We Delay a Greenhouse Warming?: The Effectiveness and Feasibility of Options to Slow a Build-Up of Carbon Dioxide in the Atmosphere*, Office of Policy and Resources Management, Office of Policy Analysis, Strategic Studies Staff (Washington, D.C.: U.S. Government Printing Office, 1983); John S. Hoffman et al., *Projecting Future Sea Level Rise: Methodology, Estimates to the Year 2100, and Research Needs*, U.S. Environmental Protection Agency, Office of Policy and Resource Management (Washington, D.C.: U.S. Government Printing Office, 1983).

49. Jay Keyworth to Ed Meese, *OSTP Monthly Report for October 1983*, 28 November 1983, George A. Keyworth Collection, 6: file label "OSTP Monthly Report 1982–84 [1 of 4]," Reagan Presidential Library Archives, Simi Valley, Courtesy of Josh Howe, Stanford University.

50. Shabecoff, "Haste of Global Warming Trend Opposed," 1.

51. *Climate Board, Carbon Dioxide Assessment Committee, Fourth Session, 28–29 September 1981, Washington, D.C.*, WAN papers, MC13, 90: 7, file label "NAS Climate Research Board/CO_2 Committee," SIO Archives.

52. Ibid.

53. John Perry to Carbon Dioxide Assessment Committee, 27 September 1982, 3, WAN papers, MC13, 91: 1, file label "NAS Climate Research Board/CO_2 Committee, Aug–Sep 1982," SIO Archives.

54. Senate Committee on Energy and Natural Resources, *Greenhouse Effect and Global Climate Change: Hearing Before the Committee on Energy and Natural Resources*, 100th Congress, 1st sess., November 9, 1987 (Washington, D.C.: U.S. Government Printing Office), 52.

55. Senate Committee on Energy and Natural Resources, *Greenhouse Effect and Global Climate Change: Hearing Before the Committee on Energy and Natural Resources*, 100th Congress, 1st sess., June 23, 1988, pt. 2 (Washington, D.C.: U.S. Government Printing Office), 1.

56. Ibid., 39.

57. Senate Committee on Energy and Natural Resources, *Greenhouse Effect and Global Climate Change*, 100th Congress, 1st sess., June 23, 1988, pt. 2, 48; and Senate Committee on Energy and Natural Resources, *Greenhouse Effect and Global Climate Change*, 100th Congress, 1st sess., November 9, 1987 (Washington, D.C.: U.S. Government Printing Office), 52.

58. Philip Shabecoff, "Global Warming Has Begun, Expert Tells Senate," *New York Times*, June 24, 1988, sec. A, 1.

59. Richard A. Kerr, "Hansen vs. the World on the Greenhouse Threat," *Science* 244, no. 4908 (June 2, 1989): 1041–43.

60. Bert Bolin, *A History of the Science and Politics of Climate Change: The Role of the Intergovernmental Panel on Climate Change* (Cambridge: Cambridge University Press, 2007), 49.

61. See Bolin, *A History of the Science and Politics of Climate Change*, 50–51; J. T. Houghton, G. J. Jenkins, and J. J. Ephraums, eds., *Climate Change: The IPCC Scientific Assessment* (New York: Cambridge University Press, 1990), iii and v.

62. John Balzar, "Bush Vows 'Zero Tolerance' of Environmental Polluters," *Los Angeles Times*, September 1, 1988, sec. A.

63. Committee on Earth Sciences, *Our Changing Planet: A U.S. Strategy for Global Change Research* (Washington, D.C.: U.S. Government Printing Office, 1989).

64. Senate Committee on Commerce, Science, and Transportation, *National Global Change Research Act of 1989*, 101st Congress, 1st sess., February 22, 1989 (Washington, D.C.: U.S. Government Printing Office, 1989), 1–4.

65. Gus Speth, interview with Naomi Oreskes, August 3, 2007.

66. Robert Jastrow, William Nierenberg, and Frederick Seitz, *Global Warming: What Does the Science Tell Us?* (Washington, D.C.: George C. Marshall Institute, 1989).

67. Leslie Roberts, "Global Warming: Blaming the Sun," *Science* 246, no. 4933 (November 24, 1989): 992–93.

68. Ibid.

69. Ibid.

70. Jastrow et al., *Global Warming: What Does the Science Tell Us?* 30–31, 48–57.

71. Ibid., 56–57; Roberts, "Blaming the Sun," 992–93.

72. James Hansen et al., "Climate Impact of Increasing Atmospheric Carbon Dioxide," *Science* 213, no. 4511 (August 28, 1981): 957–66, diagram on 963.

73. Schneider to Albert Hecht, September 1, 1989, reproduced in Stephen H. Schneider, *Global Warming: Are We Entering the Greenhouse Century?* (New York: Vintage, 1990), 329.

74. Houghton et al., eds., *The IPCC Scientific Assessment*, xi; also see Michael Weisskopf and William Booth, "UN Report Predicts Dire Warming; Break with US Seen in Thatcher Response," *Washington Post*, May 26, 1990, sec. A, 1.

75. Houghton et al., eds., *The IPCC Scientific Assessment*, 63.

76. Bolin, *History of the Science and Politics of Climate Change*, 72; Nierenberg described the Marshall Institute's estimate of climate sensitivity in William Nierenberg, "Global Warming: Look Before We Leap," *New Scientist* (March 9, 1991): 10.

77. Deborah Day, personal communication with Naomi Oreskes, 2008.

78. Bill Kristol to Sam Skinner et al., *Attachment—Chart B*, 23 April 1992, Jeffrey Holmstead, file "Global Warming Implications," OA/ID CF01875, Counsels Office, George H. W. Bush Presidential Library, College Station, Texas.

79. Robert Jastrow to Terry Yosle, 22 February 1991, WAN papers, Accession 2001-01, 60: file label "Marshall Institute Correspondence, 1990–1992," SIO Archives.

80. Roger Revelle, "What Can We Do About Climate Change?" Presented at the AAAS Annual Meeting, New Orleans, 9 February 1990, Revelle Papers, MC6A 165: 9, SIO Archives.

81. Ibid.

82. Ibid.

83. Affidavit of Ms. Christa Beran, *S. Fred Singer v. Justin Lancaster*, Mass., CA93-2219 (August 2, 1993).

84. Walter Munk, personal communication with Naomi Oreskes, January 10, 2005.

85. Affidavit of Ms. Christa Beran, *S. Fred Singer v. Justin Lancaster*, CA93-2219. Most of the relevant materials can be found on Justin Lancaster's Web site, "The Real Truth about the Revelle-Gore Store," The Cosmos Myth, http://home.att.net/~espi/Cosmos_myth.html, including Fred Singer's deposition: http://home.att.net/~S-F-Singer_Deposition.pdf.

86. Houghton et al., eds., *The IPCC Scientific Assessment*, xi; see also Weisskopf and Booth, "UN Report Predicts Dire Warming," 1.

87. S. Fred Singer, "What to Do about Greenhouse Warming," *Environmental Science and Technology* 24, no. 8 (August 1990): 1138–39. Italics in original.

88. Lancaster, "The Real Truth," The Cosmos Myth, and attached galley proofs.

89. Ibid.

90. S. F. Singer, R. Revelle, C. Starr, "What to Do about Greenhouse Warming: Look Before You Leap," *Cosmos* 1, no. 1 (1991): 28–33; republished as S. Fred Singer, Roger Revelle, and Chauncey Starr, "What to Do about Greenhouse Warming: Look Before You Leap," in Richard A. Geyer, ed., *A Global Warming Forum: Scientific, Economic, and Legal Overview* (Boca Raton, Fla.: CRC Press, Inc., 1991), 347–56.

91. Ibid.

92. Singer et al., "What to Do about Greenhouse Warming."

93. Justin Lancaster, interview with Naomi Oreskes, October 20, 2007.

94. Gregg Easterbrook, "Has environmentalism blown it? Green Cassandras," *New Republic* 207, no. 2 (July 6, 1992): 23–25.

95. S. Fred Singer, "Global Warming: Do We Know Enough to Act?" in *Environmental Protection: Regulating for Results*, ed. Kenneth Chilton and Melinda Warren (Boulder, Colo.: Westview Press, 1991), 29–49. The key phrase is repeated on p. 30; the attack on IPCC is on pp. 33–35.

96. George F. Will, "Al Gore's Green Guilt," *Washington Post*, September 3, 1992, final edition, A23.

97. "The 1992 Campaign: In Dispute Quayle and Gore Battle Devolves into a Hand-to-Hand Fight about 4 Issues," *New York Times*, October 14, 1992, sec. A, 19.

98. Carolyn Revelle Hufbauer, "Global Warming: What My Father Really Said," *Washington Post*, September 13, 1992, final edition, sec. C.

99. *Affidavit of Defendant Justin Lancaster, S. Fred Singer v. Justin Lancaster*, Civil Action 93-2219.

100. Roger Revelle, "What Can We Do About Climate Change," *Oceanography* 5, no. 2 (1992): 126–27; Walter H. Munk and Edward Frieman, "Let Roger Revelle Speak for Himself," *Oceanography* 5, no. 2 (1992): 125.

101. Walter Munk, personal communication with Naomi Oreskes, January 10, 2005.

102. S. Fred Singer et al., "Look before You Leap," 347; Roger Revelle, "What Can We Do About Climate Change," 126–27.

103. S. Fred Singer, *Statement Made at Revelle Symposium at Harvard on October 23, 1992,* Biographical Information Files, 22: Roger Revelle, SIO Archives.

104. "Global Warming Lawsuit," February 25, 1994, Living on Earth, http://www.loe.org/shows/shows.htm?programID=94-P13-00008#feature1; Lancaster, "The Real Truth," The Cosmos Myth.

105. Lancaster, "The Real Truth," The Cosmos Myth.

106. Roger Revelle, *The Science of Climate Change and Climate Variability*, 15 November 1990 (revised 20 December 1990), Revelle Papers, MC6A, 165: 11, SIO Archives.

107. United Nations, *United Nations Framework Convention on Climate Change* (1992), http://unfccc.int/resource/docs/convkp/conveng.pdf.

108. George H. W. Bush, "Address to United Nations Conference on Environment and Development in Rio de Janeiro, Brazil," June 12, 1992, http://bulk.resource.org/gpo.gov/papers/1992/1992_vol1_925.pdf; George H. W. Bush, "Address to the United Nations Conference on Environment and Development in Rio de Janeiro, Brazil," June 12, 1992, in: *Public Papers of the Presidents of the United States, George Bush: 1992*, vol. 1 (Washington, D.C.: U.S. Government Printing Office, 1993), 924–25.

109. Benjamin Santer, interview by Erik Conway, February 20, 2009; see also William K. Stevens, *The Change in the Weather: People, Weather, and the Science of Climate* (New York: Delacorte Press, 1999), 218.

110. Santer, interview with Conway, February 20, 2009; Stevens, *The Change in the Weather*, 218–19.

111. K. Hasselmann, "On the Signal-to-Noise Problem in Atmospheric Response Studies," in *Meteorology Over the Tropical Oceans: The Main Papers Presented at a Joint Conference Held 21 to 25 August 1978 in the Rooms of the Royal Society, London*, ed. D. B. Shaw (Bracknell, Berkshire: Royal Society, 1979).

112. Ben Santer, e-mail communication with Naomi Oreskes, October 3, 2009.

113. V. Ramanathan, "The Greenhouse Theory of Climate Change: A Test by an Inadvertent Global Experiment," *Science* 240, no. 4850 (April 15, 1988): 293–99.

114. Ben Santer et al., "Signal-to-Noise Analysis of Time-Dependent Greenhouse Warming Experiments. Part 1: Pattern Analysis," *Climate Dynamics* 9 (1994): 267–85; Ben Santer et al., "Ocean Variability and Its Influence on the Detectability of Greenhouse Warming Signals," *Journal of Geophysical Research* 100, no. C6 (1995): 10693–726;

Ben Santer et al., "Towards the Detection and Attribution of an Anthropogenic Effect on Climate," *Climate Dynamics* 12, no. 2 (December 1995): 77–100; Ben Santer et al., "A Search for Human Influences on the Thermal Structure of the Atmosphere," *Nature* 382, no. 6586 (July 1996): 39–46.

115. Source: IPCC procedures for preparation, review, acceptance, approval, and publication of its reports, Annex 2, document hard copy in possession of Ben Santer. The procedures were subsequently put online, see: International Panel on Climate Change, *Procedures for Preparation, Review, Acceptance, Adoption, Approval and Publication of IPCC Reports, adopted at the 15th session (San Jose, 15–18 April 1999) amended at the 29th session (Paris, 19–21 February 2003) and 21st session (Vienna, 3 and 6–7 November 2003)*, Annex 1, http://www.ipcc.ch/pdf/ipcc-principles/ipcc-principles-appendix-a.pdf.

116. Santer, interview with Conway, February 20, 2009, and Houghton et al., eds., *Climate Change 1995: The Science of Climate Change, A Report of the Intergovernmental Panel on Climate Change* (Cambridge: Cambridge University Press, 1996).

117. Santer et al., "A Search for Human Influences on the Thermal Structure," 39–46. Santer writes, "I checked on this. We submitted our paper to *Nature* in April 1995." Benjamin Santer, e-mail communication with Naomi Oreskes, October 4, 2009.

118. Michael Oppenheimer as quoted in Stevens, *The Change in the Weather*, 226.

119. Stevens, *The Change in the Weather*, 227; and William K. Stevens, "Global Warming Experts Call Human Role Likely," *New York Times*, September 10, 1995.

120. Waldo Jaquith, "Does Virginia Really Have a State Climatologist?" August 10, 2006, cvillenews.com, http://www.cvillenews.com/2006/08/10/state-climatologist/.

121. For example, Patrick J. Michaels, "Climate and the Southern Pine-Beetle in Atlantic Coastal and Piedmont Regions," *Forest Science* 30, no. 1 (March 1, 1984): 143–56; Patrick J. Michaels, "Price, Weather, and 'Acreage Abandonment' in Western Great Plains Wheat Culture, *Journal of Climate and Applied Meteorology* 22, no. 7 (July 1983): 1296–1303.

122. Patrick Michaels, "Apocalypse Machine Blows Up," *Washington Times*, November 1, 1991, final edition, sec. F; Patrick Michaels, "More Hot Air from the Stratosphere," *Washington Times*, October 27, 1992, sec. F.

123. New Hope Environmental Services, http://www.nhes.com/; see discussion in Ross Gelbspan, *The Heat Is On: The High Stakes Battle Over Earth's Threatened Climate* (Reading, Mass.: Addison-Wesley Publishing Company, 1997), 41–43; and Naomi Oreskes, "My Facts Are Better than Your Facts: Spreading Good News about Global Warming," in *How Do Facts Travel?* ed. Mary Morgan and Peter Howlett (Cambridge: Cambridge University Press, in press). According to Gelbspan, Michaels's publication started as *World Climate Review*, then became *World Climate Report*.

124. Oreskes, "My Facts Are Better than Your Facts."

125. House Committee of Science, *Scientific Integrity and Public Trust: The Science Behind Federal Policies and Mandates: Case Study 2—Climate Models and Projections of Potential Impacts of Global Climate Change, Hearing before the Subcommittee on Energy and Environment*, 104th Congress, 1st sess., November 16, 1995 (Washington, D.C.: U.S. Government Printing Office, 1996), 33.

126. Ibid., 1071.

127. Bill Nierenberg to Fred Seitz (handwritten), 27 November 1995, WAN papers, Accession 2001-01, 70: file label "Frederick Seitz, 1994–1995," SIO Archives.

128. Stevens, *The Change in the Weather*, 228.

129. Santer, interview with Conway, February 20, 2009.

130. Ross Gelbspan, *The Heat Is On: The Climate Crisis, the Cover-Up, and the Prescription*, (Reading, Mass.: Basic Books, 1998, updated edition), 38. See also http://www.heatis online.org/contentserver/objecthandlers/index.cfm?id=3872&method=full.

131. Stephen H. Schneider and Paul N. Edwards, "Self-Governance and Peer Review in Science-for-Policy: The Case of the IPCC Second Assessment Report," in *Changing the Atmosphere: Expert Knowledge and Environmental Governance*, ed. Paul N. Edwards and Clark A. Miller (Cambridge, Mass.: MIT Press, 2001), 219–96.

132. Bolin, *History of the Science and Politics of Climate Change*, 113; Stevens, *The Change in the Weather*, 229; Santer, interview with Conway, February 20, 2009.

133. Bolin, 113; Houghton et al., eds., *Climate Change 1995*, 5.

134. Bolin, *History of the Science and Politics of Climate Change*.

135. Santer, interview with Conway, February 20, 2009.

136. S. Fred Singer, "Climate Change and Consensus," *Science* 279, no. 5249 (February 2, 1996): 581–82.

137. T. M. L. Wigley, "Climate Change Report," Letters, *Science* 271, no. 5255 (March 15, 1996): 1481–82.

138. Ibid.

139. S. Fred Singer, "Climate Change Report," Letters, *Science* 271, no. 5255 (March 15, 1996): 1482–83.

140. On the Global Climate Coalition, see Gelbspan, *The Heat Is On*, and Jeremy Leggett, *The Carbon War: Global Warming and the End of the Oil Era* (New York: Routledge, 2001). On Pearlman, see Gelbspan, *The Heat Is On*, 119–20.

141. Stevens, *The Change in the Weather*, 231.

142. Santer, interview with Conway, February 20, 2009.

143. Ibid.

144. Myanna Lahsen, "The Detection and Attribution of Conspiracies: The Controversy over Chapter 8," in *Paranoia within Reason: A Casebook on Conspiracy as Explanation*, ed. G. E. Marcus (Chicago: University of Chicago Press, 1999), 111–36.

145. Frederick Seitz, "A Major Deception on 'Global Warming,'" *Wall Street Journal*, June 12, 1996, A16.

146. Benjamin D. Santer, letter to the editor, *Wall Street Journal*, June 25, 1996; Susan K. Avery et al., "Special Insert: An Open Letter to Ben Santer," UCAR—University Corporation for Atmospheric Research, *Communications Quarterly* (July 25, 1996) and attachment 2, http://www.ucar.edu/communications/quarterly/summer96/insert.html; A15; "Open Letter to Ben Santer" and attachment 2, *Bulletin of the American Meteorological Society* 77 (September 1996): 8, 1961–1962; 1963–1965.

147. Avery et al., "An Open Letter to Ben Santer," and attachment 2; *Bulletin*.

148. Avery et al., "Open Letter," and attachment 3, Bulletin.

149. Avery et al., "Open Letter." 1961–1962; 1963–1965.

150. Ibid., 1961; see Bolin, *History of the Science and Politics of Climate Change*, 129.

151. S. Fred Singer, letter to the editor, *Wall Street Journal*, July 11, 1996, sec. A, 15; see also letters by Frederick Seitz and Hugh Ellsaesser in the same section. On Marshall Scientific Advisory board see Dr. Frederick Seitz, *Global Warming and Ozone Hole Controversies: A*

Challenge to Scientific Judgment (Washington, D.C.: George C. Marshall Institute, 1994), BN: 250135590. On Heartland report see *Heartlander by Mail: Report on December 1995 Activities*, 9 January 1996, BN: 2046851463, Legacy Tobacco Documents Library.

152. Benjamin D. Santer, "Global Warming Critics, Chill Out," *Wall Street Journal*, July 23, 1996, sec. A, 23; see also letter by Bert Bolin and John Houghton in the same section.

153. Gelbspan reprinted this e-mail exchange: Ross Gelbspan, *The Heat Is On*, 230–36.

154. S. Fred Singer, "Disinformation on Global Warming?" *Washington Times*, November 13, 1996, sec. A.

155. S. Fred Singer et al., "Comments on an Open Letter to Ben Santer," *Bulletin of the American Meteorological Society* 78, no. 1 (January 1997): 81–82; S. Fred Singer et al., "Letter to the Bulletin of the AMS," Science and Environmental Policy Project Archives, January 1997, http://www.sepp.org/Archive/controv/IPCCcont/AMSltr.htm.

156. Lahsen, "The Detection and Attribution of Conspiracies," 111–36.

157. Tom M. L. Wigley to William A. Nierenberg, 14 April 1997, WAN papers, Accession 2001-01, 18: file label "EPRI," SIO Archives.

158. Ibid.

159. Wigley to Nierenberg, 24 April 1997, WAN papers, Accession 2001-01, 18: file label "EPRI," SIO Archives. Nierenberg's papers suggest that the paper he was referring to was T. M. L. Wigley, R. Richels, and J. A. Edmonds, "Economic and Environmental Choices in the Stabilization of Atmospheric CO_2 Concentrations," *Nature* 379 (January 18, 1996): 240–43.

160. Klaus Hasselmann to William A. Nierenberg, 18 April 1997, WAN papers, Accession 2001-01, 18: file label "EPRI," SIO Archives.

161. Ibid.

162. Faxed copy of statement in Edward Frieman papers, MC77, 123: 7, SIO Archives; see also Chris Mooney, *The Republican War on Science* (New York: Basic Books, 2005), 62–64.

163. James M. Inhofe, "Climate Change Update: Senate Floor Statement by U.S. Senator James M. Inhofe," January 4, 2005, Floor Speeches, http://inhofe.senate.gov/pressreleases/climateupdate.htm.

164. Interview of Richard Cheney with Jonathan Karl, *ABC News*, broadcast February 23, 2007; transcript at "Exclusive: Cheney on Global Warming," http://abcnews.go.com/Technology/story?id=2898539&page=1.

165. Robert W. Seidel, *Los Alamos and the Making of the Atomic Bomb* (Los Alamos, N.M.: Otowi Press, 1995); Paul Norris Edwards, *The Closed World: Computers and the Politics of Discourse in Cold War America* (Cambridge, Mass.: MIT Press, 1997); Sherry Sontag, Christopher Drew, and Annette Lawrence Drew, *Blind Man's Bluff: The Untold Story of American Submarine Espionage* (New York: Public Affairs, 1999); John P. Craven, *The Silent War: The Cold War Battle Beneath the Sea* (New York: Simon and Schuster, 2001); Peter J. Westwick, *The National Labs: Science in an American System, 1947–1974* (Cambridge, Mass.: Harvard University Press, 2003); Naomi Oreskes, *Science on a Mission: American Oceanography in the Cold War and Beyond* (Chicago: University of Chicago Press, under contract).

166. Eugene Linden, *The Winds of Change: Climate, Weather, and the Destruction of Civilizations* (New York: Simon and Schuster, 2006), 222–23.

167. Maxwell T. Boykoff and Jules M. Boykoff, "Balance as Bias: Global Warming and the US Prestige Press," *Global Environmental Change* 14 (2004): 125–36.

168. Aaron M. McCright and Riley E. Dunlap, "Defeating Kyoto: The Conservative Movement's Impact on U.S. Climate Change Policy," *Social Problems* 50, no. 3 (May 2003): 348–73; *Byrd-Hagel Resolution*, 105th Congress, 1st sess., July 25, 1997, National Center for Public Policy Research, http://www.nationalcenter.org/KyotoSenate.html.

CHAPTER 7

1. Naomi Oreskes, "Science and Public Policy: What's Proof Got to Do with It?" *Environmental Science and Policy* 7, no. 5 (2004): 369–83; Thomas R. Dunlap, *DDT: Scientists, Citizens, and Public Policy* (Princeton, N.J.: Princeton University Press, 1981); Edmund Russell, *War and Nature: Fighting Humans and Insects with Chemicals from World War I to Silent Spring* (Cambridge: Cambridge University Press, 2001); Zuoyue Wang, "Responding to Silent Spring: Scientists, Popular Science Communication, and Environmental Policy in the Kennedy Years," *Science Communication* 19, no. 2 (1997): 141–63; Linda Lear, *Rachel Carson: Witness for Nature* (New York: Henry Holt, 1998).

2. Rachelwaswrong.org, 2009, project of the Competitive Enterprise Institute, http://rachelwaswrong.org/.

3. Roger Bate, "The Rise, Fall, Rise and Imminent Fall of DDT," *Health Policy Outlook* 14 (November 2007): 2–9, AEI Outlook Series, American Enterprise Institute for Public Policy Research, http://www.aei.org/outlook/27063.

4. "DDT Makes a Comeback, featuring Richard Tren," Cato Daily Podcast, September 22, 2006, Cato Institute, http://www.cato.org/dailypodcast/podcast-archive.php?podcast_id=125.

5. Bonner R. Cohen, "Uganda Will Use DDT to Fight Malaria," April 1, 2007, *Environment and Climate News*, Heartland Institute, http://www.heartland.org/policybot/results/20807/Uganda_Will_Use_DDT_to_Fight_Malaria.html.

6. "Global Warming: Was It Ever Really a Crisis?" News, Heartland.org, http://www.heartland.org/events/NewYork09/news.html; Global Warming Facts, Heartland Institute, http://www.globalwarmingheartland.org/; see also Andrew C. Revkin, "Skeptics Dispute Climate Worries and Each Other," *New York Times*, March 8, 2009, http://www.nytimes.com/2009/03/09/science/earth/09climate.html.

7. Edmund Russell, "The Strange Career of DDT: Experts, Federal Capacity, and Environmentalism in WWII," *Technology and Culture* 40, no. 4 (October 1999): 770–96; Russell, *War and Nature*.

8. Russell, "The Strange Career of DDT."

9. "Paul Müller: The Nobel Prize in Physiology or Medicine, 1948," *Nobel Lectures, Physiology or Medicine 1942–1962* (Amsterdam: Elsevier Publishing Company, 1964), Nobelprize.org, http://nobelprize.org/nobel_prizes/medicine/laureates/1948/muller-bio.html.

10. Dunlap, *DDT: Scientists, Citizens, and Public Policy*.

11. Russell, *War and Nature*.

12. *The American Experience: Rachel Carson's Silent Spring*, DVD, produced by Neil Goodwin (WGBH/PBS, 1992); Russell, "The Strange Career of DDT," 770–96; Russell, *War and Nature*.

13. *Silent Spring* was first serialized in the *New Yorker*: "Rachel Carson, A Reporter at Large, 'Silent Spring,'" June 16, 1962, June 23, 1962, June 30, 1962. It was later published as Rachel Carson, *Silent Spring* (Boston, Mass.: Houghton Mifflin, 1962; repr. 1994). Citations are to the 1994 version.

14. Carson, *Silent Spring*, 116.

15. Ibid., 132–33.

16. Wang, "Responding to Silent Spring," 141–163; DDT's development and relationship to chemical warfare is explained in Russell, *War and Nature*.

17. Wang, "Responding to Silent Spring," 141–63; *The American Experience: Rachel Carson's Silent Spring*.

18. Wang, "Responding to Silent Spring," 156.

19. *CBS Reports: The Silent Spring of Rachel Carson*, first broadcast 3 April 1963 by CBS.

20. Russell, "The Strange Career of DDT," 770–96.

21. Russell, *War and Nature*.

22. President's Science Advisory Committee, *Use of Pesticides, A Report of the President's Science Advisory Committee*, May 15, 1963 (Washington, D.C.: U.S. Government Printing Office, 1963), 1–2.

23. PSAC, *Use of Pesticides*, 9; a study in 2007 revealed that women exposed to DDT as young girls in the 1950s and 1960s had a higher risk of breast cancer later in life. Most previous studies had found no cancer risk for DDT because they did not focus on the time of exposure.

24. PSAC, *Use of Pesticides*, 10.

25. Ibid., 4.

26. Dunlap, *DDT: Scientists, Citizens, and Public Policy*; Lear, *Rachel Carson*; Wang, "Responding to Silent Spring"; PSAC, *Use of Pesticides*; President's Science Advisory Committee, *Restoring the Quality of Our Environment, A Report of the Environmental Pollution Panel*, November 1965 (Washington, D.C.: U.S. Government Printing Office, 1965); *Report of Committee on Persistent Pesticides, Division of Biology and Agriculture, National Research Council*, to the U.S. Department of Agriculture, May 1969 (National Academy of Sciences, 1969); E. M. Mrak, *Report of the Secretary's Commission on Pesticides and Their Relationship to Environmental Health*, U.S. Department of Health, Education and Welfare, December 1969 (Washington, D.C.: U.S. Government Printing Office, 1969).

27. This is actually a very complex point. Historian of science Zuoyue Wang notes that at the time the PSAC report was written, it was actually not clear who had the burden of proof. There was, for example, the bizarre legal practice of having a pesticide registered "under protest": a manufacturer could still register and market a pesticide after being refused by the USDA for registration. In its report, PSAC recommended the elimination of "protest" registrations, but it did not shift the burden of proof entirely to the manufacturers (PSAC, *Use of Pesticides*, 15). Rather, it advocated beefed-up resources and research programs on the part of the government to regulate pesticides, primarily to establish a basis for regulation, but implicitly also to meet any challenge in court. Perhaps even more important to PSAC was the *transparency* of the whole regulatory process regardless of the question of burden of proof. Thus it advocated that "all data used as a basis for granting registration and establishing tolerances should be published, thus allowing the hypotheses and the validity and reliability of the data to be

subjected to critical review by the public and the scientific community." Zuoyue Wang, personal communication with Naomi Oreskes, January 19, 2010. Here "data" included both those provided by the manufacturers and the regulatory agencies. So perhaps it is better to suggest not that PSAC shifted the burden of proof from the government to the manufacturers, but rather that it raised the bar of proof, so that reasonable doubt was adequate to deny the registration of a product subject to appeal, and certainly the marketing of any denied product. See also Wang, *In Sputnik's Shadow: The President's Science Advisory Committee and Cold War America* (New Brunswick, N.J.: Rutgers University Press, 2008), 205–7.

28. Besides the legal tradition, PSAC may have been influenced by the medical principle of "first do no harm." Zuoyue Wang notes that the main staff for the PSAC panel were James Hartgering and Peter S. Bing; (both of whom were trained physicians), and the chair of the panel was Colin MacLeod, professor of medicine in NYU's Medical School (see chapter 5). Moreover, most of the PSAC members had a distrust of expedient technological fixes and their advocates, especially from their long struggle over the questions of nuclear arms race. DDT and other pesticides seemed to fit into the same category of such easy fixes and would have, they instinctively knew, unintended consequences. See Wang, *In Sputnik's Shadow.*

29. John C. Whitaker, "Earth Day Recollections: What It Was Like When the Movement Took Off," *EPA Journal* (July–August 1988), U.S. Environmental Protection Agency, http://www.epa.gov/history/topics/earthday/10.htm; NASA Glenn Research Center: Earth Day Committee, http://earthday.grc.nasa.gov/history.html; Gordon MacDonald, "Environment: The Evolution of a Concept," in *Yesterday, Today and Tomorrow: The Harvard Class of 1950 Reflects on the Past and Looks to the Future* (Arlington, Mass.: Travers Press, 2000).

30. MacDonald, "Environment: The Evolution of a Concept."

31. Rachelwaswrong.org.

32. Andrew Kenny by way of Tim Blair, "The Green Terror," A Stitch in Haste Blog, posted June 9, 2005, http://kipesquire.powerblogs.com/posts/1118329320.shtml.

33. Todd Seavey, "The DDT Ban Turns 30—Millions Dead of Malaria Because of Ban, More Deaths Likely," June 1, 2003, American Council on Science and Health, http://www.acsh.org/healthissues/newsID.442/healthissue_detail.asp.

34. Rachelwaswrong.org.

35. Kenny, "The Green Terror."

36. Thomas Sowell, "Intended Consequences," *Jewish World Review*, June 7, 2001, http://www.jewishworldreview.com/cols/sowell060701.asp.

37. "Environmentalists with Blood on Their Hands: Flying Fickle Finger of Fate Award: Rachel Carson," The Maverick Conservative Blog, comment posted February 29, 2008, http://the-maverick-conservative.blogspot.com/2008_02_01_archive.html; Bjørn Lomborg, *The Skeptical Environmentalist: Measuring the Real State of the World* (New York: Cambridge University Press, 2001), 215–16.

38. Angela Logomasini, " 'Silent Spring' was Wrong, Sen. Coburn is Right," Commentary, *Examiner*, May 28, 2007, Examiner.com, http://www.examiner.com/a-751059~Angela_Logomasini___Silent_Spring__was_wrong__Sen__Coburn_is_right.html. Logomasini is director of Risk and Environmental Policy at the Competitive Enterprise Institute and manager of Rachelwaswrong.org.

39. Pete du Pont, "Plus Ça (Climate) Change: The Earth was warming before global warming was cool," February 21, 2007, from the *WSJ* Opinion Archives: Outside the Box, *Wall Street Journal Online*, http://www.opinionjournal.com/columnists/pdupont/?id =110009693.

40. Tina Rosenberg, "What the World Needs Now Is DDT," *New York Times Magazine*, April, 11, 2004, http://www.nytimes.com/2004/04/11/magazine/what-the-world-needs-now-is-ddt.html; John Tierney, "Fateful Voice of a Generation Still Drowns Out Real Science," *New York Times*, June 5, 2007, http://www.nytimes.com/2007/06/05/science/earth/05tier.html?_r=2&8dpc&oref=slogin.

41. Rosenberg, "What the World Needs Now Is DDT."

42. Tierney, "Fateful Voice of a Generation." For the original article, see I. L. Baldwin, "Chemicals and Pests," *Silent Spring* by Rachel Carson (book review), *Science* 137, no. 3535 (September 28, 1962): 1042–43.

43. Centers for Disease Control and Prevention, "Malaria: Vector Control," http://www.cdc .gov/malaria/control_prevention/vector_control.htm.

44. Gordon Patterson, *The Mosquito Crusades: A History of the American Anti-Mosquito Movement from the Reed Commission to the First Earth Day* (New Brunswick, N.J.: Rutgers University Press, 2009), 182.

45. Carson, *Silent Spring*, chap. 16.

46. Conevery Bolton Valenčius, *The Health of the Country: How American Settlers Understood Themselves and Their Land* (New York: Basic Books, 2002).

47. Patterson, *The Mosquito Crusaders*, presents a fascinating view of American efforts to destroy the mosquito pest. Also see Margaret Humphreys, "Kicking a Dying Dog: DDT and the Demise of Malaria in the American South, 1942–1950," *Isis* 87, no. 1 (March 1996), 1–17.

48. Patterson, *The Mosquito Crusaders*, 156.

49. Centers for Disease Control and Prevention, "Malaria: The Panama Canal," http://www.cdc.gov/malaria/history/panama_canal.htm. A general history of the canal effort can be found in David McCullough, *The Path Between the Seas: The Creation of the Panama Canal 1870–1914* (New York: Simon and Schuster, 1977).

50. Centers for Disease Control and Prevention, "Eradication of Malaria in the United States (1947–1951)," http://www.cdc.gov/malaria/history/eradication_us.htm.

51. "Ruckelshaus, Sweeney, and DDT," Jaworowski 2003: A Cornucopia of Misinformation, part 1, Monday Bristlecone Blogging, http://www.someareboojums.org/blog/?p=62. The original source is *In the Matter of Stevens Industries, Inc. et al., I.F&R. Docket Nos. 63 et al. (Consolidated DDT Hearings), Opinion of the Administrator, Decided June 2, 1972*, on p. 26, http://www.someareboojums.org/blog/wp-content/images/ddt/ead.pdf; later published in *Notices, Environmental Protection Agency*, [I.F&R. Docket Nos. 63 et al.]: Consolidated DDT Hearings, *Opinion and Order of the Administrator*, 30 June 1972, *Federal Register* 37, no. 131 (July 7, 1972): 13369–76, on p. 13373, http://www.epa.gov/history/topics/ddt/DDT-Ruckelshaus.pdf.

52. Baldwin, "Chemicals and Pests," 1042.

53. Ibid.

54. Ibid.

55. U.S. Environmental Protection Agency, "DDT," Persistent Bioaccumulative and Toxic

(PBT) Chemical Program, http://www.epa.gov/pbt/pubs/ddt.htm; United States Geological Survey, "DDT," Toxic Substances Hydrology Program, http://toxics.usgs.gov/definitions/ddt.html; Brenda Eskenazi et al., "The Pine River Statement: Human Health Consequences of DDT Use," *Environmental Health Perspectives* 117, no. 9 (September 2009): 1359–67.

56. *California v. Montrose Chemical Corp. of California*, 104 F.3d 1507 (9th Cir. 1997), Lewis and Clark Law School's Environmental Law Online, http://www.elawreview.org/summaries/environmental_quality/hazardous_waste/california_v_montrose_chemical.html.

57. Tina Adler, "Keep the Sprays Away?: Home Pesticides Linked to Childhood Cancers," *Environmental Health Perspectives* 115, no. 12 (December 2007): A594; M. D'Amelio et al., "Paraoxonase Gene Variants Are Associated with Autism in North America, But Not in Italy: Possible Regional Specificity in Gene-Environment Interactions," *Molecular Psychiatry* 10 (November 2005): 1006–16; D. R. Davies et al., "Chronic Organophosphate Induced Neuropsychiatric Disorder (COPIND): Results of Two Postal Questionnaire Surveys," *Journal of Nutritional and Environmental Medicine* 9, no. 2 (1999): 123–34.

58. Walter J. Rogan and Aimin Chen, "Health Risks and Benefits of bis (4-chlorophenyl)-1,1,1-trichloroethane (DDT)," *Lancet* 366, no. 9487 (August 27, 2005): 763–73.

59. Barbara A. Cohn et al., "DDT and Breast Cancer in Young Women: New Data on the Significance of Age at Exposure," *Environmental Health Perspectives* 115, no. 10 (October 2007): 1406–14; Rick Weiss, "Long Hidden Dangers? Early Exposure to DDT May Raise Risk of Breast Cancer," *Washington Post*, October 9, 2007, sec. F, 1, http://www.washingtonpost.com/wp-dyn/content/article/2007/10/05/AR2007100502253.html.

60. Dixy Lee Ray and Lou Guzzo, *Trashing the Planet: How Science Can Help Us Deal with Acid Rain, Depletion of the Ozone, and Nuclear Waste (Among Other Things)* (New York: HarperCollins, 1992), 69.

61. World Health Organization, *Resistance of Vectors and Reservoirs of Disease to Pesticides*, WHO Expert Committee on Insecticides (Geneva: World Health Organization, 1976), 68–69.

62. World Health Organization, "Malaria Situation in SEAR Countries: Sri Lanka," http://www.searo.who.int/EN/Section10/Section21/Section340_4026.htm.

63. World Health Organization, *Resistance of Vectors and Reservoirs of Disease to Pesticides*, 7.

64. Ray and Guzzo, *Trashing the Planet*, 74.

65. Steven J. Milloy to TASSC Members, *Re: Annual Report*, 7 January 1998, Bates Number (BN): 2065254885, Legacy Tobacco Documents Library.

66. J. Gordon Edwards, "DDT: A Case Study in Scientific Fraud," *Journal of American Physicians and Surgeons* 9, no. 3 (Fall 2004): 83–88.

67. Quoted in James Hoare, "Greenpeace, WWF Repudiate Anti-DDT Agenda," April 1, 2005, Environment and Climate News, Heartland Institute, http://www.heartland.org/Article.cfm?artId=16803.

68. On global warming see "Global Warming/Climate," JunkScience.com, http://www.junkscience.com/#GWS; on acid rain see "Cleaner Air Means a Warmer Europe," April 15, 2008, http://junkscience.com/blog_js/2008/04/15/cleaner-air-means-a-warmer

-europe. (He notes that sulphates help cool the air—true, but if he isn't worried about global warming then why does he care about this?) At present this link is inactive but the article is listed in the JunkScience April 2008 archives: http://www.junkscience .com/apr08.html; on ozone see "The 'Ozone Layer'—What's Going On?" JunkScience .com, http://www.junkscience.com/Ozone/ozone_seasonal.html; see also Chris Mooney, "Some Like It Hot," Special Reports: As the World Burns, *Mother Jones* 30, no. 3 (May/June 2005): 36–94, http://www.motherjones.com/environment/2005/05/ some-it-hot.

69. Paul D. Thacker, "Pundit for Hire: Smoked Out," *New Republic*, February 6, 2006, 13–14; Paul D. Thacker, "The Junkman Climbs to the Top," *Environmental Science and Technology Online News* (May 2005); "Steven J. Milloy: The 'Junkman' Exposed," February 2006, Americans for Nonsmokers' Rights, http://www.no-smoke.org/pdf/stevenmilloy .pdf; "Steven J. Milloy," SourceWatch, http://www.sourcewatch.org/index.php?title =Steven_J._Milloy; Steven J. Milloy, *Junk Science Judo: Self-Defense Against Health Scares and Scams* (Washington, D.C.: Cato Institute, 2001); Steven J. Milloy, *Science Without Sense* (Washington, D.C.: 1996).

70. John Berlau, "Rush Limbaugh for the Nobel Peace Prize," May 30, 2007, Competitive Enterprise Institute, http://cei.org/gencon/019,05942.cfm.

71. On the book and AEI inviting Crichton to speak, see Harold Evans, "Crichton's Conspiracy Theory," *BBC News*, http://news.bbc.co.uk/2/hi/uk_news/magazine/4319574 .stm. Folks associated with CEI have also been strong supporters of Crichton; see, for example, the review by Iain Murray, "Science Fiction: Michael Crichton Takes a Novel Approach to Global Warming Alarmism," December 20, 2004, Competitive Enterprise Institute, http://cei.org/gencon/019,04342.cfm.

72. Michael Crichton, *State of Fear* (New York: HarperCollins, 2004), 487.

73. "Welcome to the Heartland Institute," Heartland Institute, http://www.heartland.org/ about/.

74. Deroy Merdock, "DDT Key to Third World's War on Malaria," July 1, 2001, Environment and Climate News, Heartland Institute, http://www.heartland.org/publications/ environment%20climate/article/10415/DDT_Key_to_Third_Worlds_War_on_Malaria .html.

75. On Heartland support for Crichton, see Joseph L. Bast, "Michael Crichton Is Right!" January 1, 2005, News Releases, Heartland Institute, http://www.heartland.org/Article.cfm ?artId=16260. In 2008 they sponsored a conference to continue to insist that climate change is not happening, or if it is, it can be dealt with entirely by letting the free market respond: "The 2008 International Conference on Climate Change, March 2–4, New York, USA," Heartland.org, http://www.heartland.org/NewYork08/newyork08.cfm.

76. "Joseph L. Bast—2008 Resume," January 1, 2008, Heartland Institute, http://www .heartland.org/policybot/results/12825/Joseph_L_Bast_2008_Resum%E9.html.

77. Richard C. Rue, project director of the Heartland Institute, to Roy E. Marder [*sic*], Manager of Industrial Affairs, Philip Morris, 3 August 1993, BN: 2024211094, Legacy Tobacco Documents Library; see also Roy Marden to Thomas Borelli et al., *RE: CA*, 22 April 1997, BN: 2075574228B, Legacy Tobacco Documents Library; for numerous other documents found in search see "search: Heartland Institute," Legacy Tobacco Documents Library, http://legacy.library.ucsf.edu/action/search/advanced?sq[o].f=org

&sq[0].q=heartland+inst&sq[0].op=AND&ps=10&df=er&fd=1&rs=false&ath=true&drf
=ndd&sd=1990&ed=2008&asf=ddu&p=12&ef=true.

78. Roy Marden to Thomas Borelli et al., *RE: CA*, 22 April 1997, BN: 2075574226D, Legacy Tobacco Documents Library.

79. *Fedsuit Actions/Marden*, 26 October 1999, BN: 2077575920A, Legacy Tobacco Documents Library; see also *FET Update*, 28 January 1994, BN: 2046554465, Legacy Tobacco Documents Library.

80. *Policy Payments for Slavit*, 1997, BN: 2078848138, Legacy Tobacco Documents Library.

81. David P. Nicoli to Buffy, 8 March 1994, BN: 2073011685; Merrick Carey, president, Alixis de Tocqueville Institution, to David P. Nicoli, 8 February 1994, BN: 2073011666, Legacy Tobacco Documents Library.

82. In 2008, the Heartland Institute's prospectus listed Roy Marden as a member of its board of directors: "2008 Annual Report," About the Heartland Institute, http://www.heartland.org/about/PDFs/HeartlandProspectus.pdf.

83. Jacques et al., "The Organisation of Denial: Conservative Think Tanks and Environmental Sceptism," *Environmental Politics* 17, no. 3 (June 2008): 349–85.

84. George Orwell, *1984* (New York: Harcourt Brace, 1949).

85. Lt. General Daniel O. Graham to William A. Nierenberg, 27 December 1984, William A. Nierenberg papers, MC13, 43: 17, Scripps Institute of Oceanography Archives.

CONCLUSION

1. Alexis de Tocqueville, *Democracy in America*, ed. J. P. Mayer, trans. George Lawrence (New York: Perennial Classics, 2000), 242.

2. William Ophuls, *Requiem for Modern Politics: The Tragedy of the Enlightenment and the Challenge of the New Millennium* (Boulder, Colo.: Westview Press, 1997). The "all sail" quote derives from the eighteenth-century commentator Thomas Babington Macaulay.

3. Steven Lee Myers and Megan Thee, "Americans Feel Military Is Best at Ending War," *New York Times*, September 10, 2007, http://www.nytimes.com/2007/09/10/washington/10poll.html.

4. Jon A. Krosnick et al., "The Effects of Beliefs About the Health Consequences of Cigarette Smoking on Smoking Onset," *Journal of Communication* 56, no. S1 (August 2006): S18–S37.

5. Anthony Leiserowitz, principal investigator, "American Opinions on Global Warming: A Yale University/Gallup/ClearVision Institute Poll," 2007, http://environment.research.yale.edu/documents/downloads/a-g/AmericansGlobalWarmingReport.pdf.

6. Gail E. Kennedy and Lisa A. Bero, "Print Media Coverage of Research on Passive Smoking, *Tobacco Control* 8 (1999): 254–60.

7. See discussion in chapter 3.

8. "How the Ozone Story Became a Volcano Story," The Donella Meadows Archive: Voice of a Global Citizen, Sustainability Institute, http://www.sustainer.org/dhm_archive/index.php?display_article=vn504ozoneed.

9. Maxwell T. Boykoff and Jules M. Boykoff, "Balance as Bias: Global Warming and the US Prestige Press," *Global Environmental Change* 14 (2004): 125–36. See also Liisa Antilla, "Climate of Scepticism: U.S. Newspaper Coverage of the Science of Climate

Change, *Global Environmental Change* 15 (2005): 338–52, and discussion in Peter J. Jacques et al., "The Organisation of Denial: Conservative Think Tanks and Environmental Scepticism," *Environmental Politics* 17, no. 3 (June 2008): 349–85. On the continuing problem of "false balance," see Chris Mooney, "Sadly, False Balance in the *New York Times*," The Intersection Blogs, Discover Magazine, comment posted February 26, 2009, http://blogs.discovermagazine.com/intersection/2009/02/26/sadly-false-balance-in-the-new-york-times/.

10. We also know that in some cases members of the media were pressured by editors to present "both sides"; see for example, Eugene Linden, *The Winds of Change: Climate, Weather, and the Destruction of Civilizations* (New York: Simon and Schuster, 2006). Of course, there are charlatans on television too, such as the CEO of "WeatherAction"—a Web site that sells weather forecasts up to one year in advance—featured on CNN in the summer of 2007. (Chaos theory explains that it is impossible to predict weather accurately more than a week ahead—hence the ubiquitous five-day forecast. Anyone who claims otherwise is having you on, and if he takes your money, he's a crook.)

11. David Michaels, *Doubt Is Their Product: How Industry's Assault on Science Threatens Your Health* (New York: Oxford University Press, 2008); David Michaels and Celeste Monforton, "Manufacturing Uncertainty: Contested Science and the Protection of the Public's Health and Environment," *American Journal of Public Health* 95, no. S1 (July 2005): S39–S48.

12. *Center for Indoor Air Research, Position Description: Executive Director*, May 1987, Bates Number (BN): 2023555600, Legacy Tobacco Documents Library; *The Tobacco Institute, Inc., Minutes of Meeting of the Executive Committee*, 10 December 1987, BN: TIMN0014390, Legacy Tobacco Documents Library; see also "The Tobacco Institute's Center for Indoor Air Research (CIAR)," TobaccoFreedom.org, http://www.tobaccofreedom.org/issues/documents/ets/cia_center/.

13. Robert Proctor, *The Golden Holocaust* (under review). See also *Center for Indoor Air Research, Position Description*, BN: 2023555600, Legacy Tobacco Documents Library.

14. "Global Warming Petition Project," http://www.petitionproject.org.

15. Arthur B. Robinson and Zachary W. Robinson, "Science Has Spoken: Global Warming Is a Myth," *Wall Street Journal*, December 4, 1997.

16. Arthur Robinson, Sallie Baliunas, Willie Soon, and Zachary Robinson, *Environmental Effects of Increased Atmospheric Carbon Dioxide* (Cave Junction, Ore.: Oregon Institute of Science and Medicine, 1998), http://www.oism.org/pproject/s33p36.htm; later republished as Arthur B. Robinson, Noah E. Robinson, and Willie Soon, "Environmental Effects of Increased Atmospheric Carbon Dioxide," *Journal of American Physicians and Surgeons* 12, no. 3 (Fall 2007): 79–90; see also "Global Warming Petition," Petition Project, http://www.oism.org/pproject/.

17. David Malakoff, "Climate Change: Advocacy Mailing Draws Fire," *Science* 280, no. 5361 (April 10, 1998): 195.

18. Bert Bolin, *A History of the Science and Politics of Climate Change: The Role of the Intergovernmental Panel on Climate Change* (Cambridge: Cambridge University Press, 2007), 155; William K. Stevens, "Science Academy Disputes Attack on Global Warming," *New York Times*, April 22, 1998.

19. S. Fred Singer, "Kyoto Accord Protest Quickening," *Washington Times*, April 22, 1998.

20. "Global Warming Petition," Petition Project; see also Global Warming Petition Project, http://www.petitionproject.org/. We've received many e-mails from people who cite this petition to us as evidence that there is a continuing scientific debate. See also "Letter from Frederick Seitz, Research Review of Global Warming Evidence," Petition Project, http://www.oism.org/pproject/s33p41.htm.

21. *Journal of American Physicians and Surgeons*, Association of American Physicians and Surgeons, http://www.jpands.org/.

22. "Doctors Group: Limbaugh Medical Records Seizure Unlawful," PRNewswire, February 22, 2004, Newsmax.com, http://archive.newsmax.com/archives/articles/2004/2/21/141518.shtml.

23. Michael Fumento, "AIDS—A Heterosexual Epidemic?" *Medical Sentinel* 2, no. 3 (Summer 1997), http://www.haciendapub.com/v2n3.html.

24. Emil Mrak, *Some Experiences Relating to Food Safety*, Philip Morris Laboratories, Richmond, Virginia, 16 January 1973, MSS, 8: Emil M. Mrak Collection D-96, Special Collections, Shields Library, University of California, Davis.

25. Ross Gelbspan, *The Heat Is On: The High Stakes Battle over Earth's Threatened Climate* (Reading, Mass.: Addison-Wesley, 1998); Ross Gelbspan, *Boiling Point: How Politicians, Big Oil and Coal, Journalists, and Activists Are Fueling the Climate Crisis—and What We Can Do to Avert Disaster* (New York: Basic Books, 2004); see also David Adam, "Exxon-Mobil Continuing to Fund Climate Sceptic Groups, Records Show," *Guardian* (UK), July 1, 2009, http://www.guardian.co.uk/environment/2009/jul/01/exxon-mobil-climate-change-sceptics-funding; Chris Mooney, "Some Like It Hot," Special Reports: As the World Burns, *Mother Jones* 30, no. 3 (May/June 2005): 36–94, http://www.motherjones.com/environment/2005/05/some-it-hot; Bill McKibben, "Climate of Denial," Special Reports: As the World Burns, *Mother Jones* 30, no. 3 (May/June 2005): 34–35, http://www.motherjones.com/politics/2005/05/climate-denial; Jeremy Leggett, *The Carbon War: Global Warming and the End of the Oil Era* (London: Penguin, 2000); James Hoggan and Richard Littlemore, *Climate Cover-Up: The Crusade to Deny Global Warming* (Greystone Books, 2009).

26. Chris Mooney, "Some Like It Hot," 36–94.

27. Ibid.

28. William Nierenberg e-mail to Richard Lindzen, 7 September 2000, William A. Nierenberg (WAN) papers, Accession 2001-01, 7: file label "Lindzen, Richard," Scripps Institute of Oceanography (SIO) Archives.

29. "NIPCC Report: Table of Contents," Heartland Institute, http://www.heartland.org/publications/NIPCC%20report. According to Singer's Science and Environmental Policy Project's 2007 IRS form 1099, Heartland paid $143,000 for this work.

30. "Global Warming: Was it Ever Really a Crisis?" News, Heartland.org, http://www.heartland.org/events/NewYork09/news.html; Global Warming Facts, Heartland Institute, http://www.globalwarmingheartland.org/; see also Andrew C. Revkin, "Skeptics Dispute Climate Worries and Each Other," *New York Times*, March 8, 2009, http://www.nytimes.com/2009/03/09/science/earth/09climate.html.

31. *Tobacco Strategy*, March 1994, BN: 2022887066, Tobacco Legacy Documents Library.

32. "Ludwig Von Mises," The Concise Encyclopedia of Economics, Library of Economics and Liberty, http://www.econlib.org/library/Enc/bios/Mises.html.

33. S. Fred Singer, "My Adventures in the Ozone Layer," *National Review* (June 30, 1989): 34–38, quote on 36–37.

34. S. Fred Singer and Kent Jeffreys, *The EPA and the Science of Environmental Tobacco Smoke*, 2, Alexis de Tocqueville Institution, May 1994, BN: TI31749030, Legacy Tobacco Documents Library.

35. Milton Friedman, *Capitalism and Freedom* (Chicago: Chicago University Press, 1962).

36. George Soros, "The Capitalist Threat," *Atlantic Monthly* 279, no. 2 (February 1997): 45–58, http://www.theatlantic.com/issues/97feb/capital/capital.htm.

37. On the influence of market fundamentalism on business schools, see Kelley Holland, "Is It Time to Retrain B-Schools?" *New York Times*, March 15, 2009.

38. Soros, "The Capitalist Threat," 45–58.

39. We are grateful to John Mashey, former chief scientist of Silicon Graphics, for this point.

40. Randall Parker, "An Overview of the Great Depression," *Economic History Encyclopedia*, EH.Net, http://eh.net/encyclopedia/article/parker.depression; Wendy Wall, *Inventing the "American Way": The Politics of Consensus from the New Deal to the Civil Rights Movement* (New York: Oxford University Press, 2008), 20.

41. Wall, *Inventing the "American Way."*

42. Friedman, *Capitalism and Freedom*, xiii, "preface 1982" in 2002 edition.

43. UCSD 25th Anniversary Oral Histories, RSS 52, 1: 8, file label "Revelle, Roger," 147, University of California, San Diego, Mandeville Special Collections Library.

44. Quoted in Andrew Revkin, "Let's Be Sensible on Global Warming," *Christian Science Monitor*, June 30, 1992; George Will, "Chicken Little: The Persistence of Eco-Pessimism," *Washington Post*, May 31, 1992, C7.

45. Dixy Lee Ray, "Global Warming and Other Environmental Myths: The Economic Consequences of Fact vs. Media Perception," An Address to the Progress Foundation International Economic Conference, September 21, 1992, Hotel Savory Baur en Ville, Zurich, Switzerland, quote on p. 1. Of course some environmentalists did reject the concept of progress, but that in no way proved that global warming was a myth.

46. Ibid., 4.

47. Dixy Lee Ray, "Science and the Environment," Acton Institute, http://www.acton.org/publications/randl/rl_interview_52.php; also discussed in Jacques et al., "The Organisation of Denial," 349–85.

48. On Smith, see Fred L. Smith, Jr., President and Founder, Competitive Enterprise Institute, http://cei.org/people/fred-l-smith-jr; on Ray's affiliation with Smith, see Jacques et al., "The Organisation of Denial," 349–85.

49. S. Fred Singer, "Earth Summit Will Shackle Planet, Not Save It," *Wall Street Journal* February 19, 1992, A14, found in *Bad Science: A Resource Book*, 26 March 1993, BN: 2074143969, 49, Legacy Tobacco Documents Library.

50. Patrick J. Michaels, "Give Industry a Bigger Science Role," *Roanoke Times and World News*, December 29, 1992, A7, quote found in *Bad Science: A Resource Book*, 152, Legacy Tobacco Documents Library.

51. "About Consumer Distorts," Consumer Distorts: The Consumer Reports Watchdog, http://www.junkscience.com/consumer/consumer_about.html; see discussion in Sheldon Rampton and John Stauber, "How Big Tobacco Helped Create 'The Junkman,'" PR

Watch 7, no. 3 (3rd Quarter, 2000), PR Watch.org, http://www.prwatch.org/prwissues/2000Q3/junkman.html.

52. Patrick J. Michaels, "Cap and-Trade Is Dead. Long Live Cap-and-Trade," September 18, 2009, Cato Institute, http://www.cato.org/pub_display.php?pub_id=10558.

53. Nicholas Wade, "The Editorial Notebook; Mr. Darman and Green Vegetables," *New York Times*, May 14, 1990, A16.

54. Jacques et al., "The Organisation of Denial," 349–85.

55. Nicholas Stern, *Stern Review: The Economics of Climate Change, Executive Summary*, i, http://www.hm-treasury.gov.uk/d/Executive_Summary.pdf.

56. Charles Krauthammer, "The New Socialism," *Washington Post*, December 11, 2009, http://www.washingtonpost.com/wp-dyn/content/article/2009/12/10/AR2009121003163.html.

57. James Gustave Speth, *The Bridge at the Edge of the World: Capitalism, the Environment, and Crossing from Crisis to Sustainability* (New Haven, Conn.: Yale University Press, 2008), xi.

58. Ibid., 9.

59. John Perry to Carbon Dioxide Assessment Committee, 27 September 1982, 3, WAN papers, MC13, 91: 1, file label "NAS Climate Research Board/CO_2 Committee, Aug–Sep 1982, SIO Archives.

60. On the father of the green revolution, who died in 2009, see Justin Gillis, "Norman Borlaug, Plant Scientist Who Fought Famine, Dies at 95," *New York Times*, September 13, 2009, http://www.nytimes.com/2009/09/14/business/energy-environment/14borlaug.html. Some obituaries noted that after the initial major successes, further advances became harder to achieve, in part because of the social challenges in regions with poor infrastructure, unstable labor markets, an uneducated work force, etc. The spectacular gains that were realized in Asia were not reproduced in Africa.

61. Most historians would hold that there have been major ups and downs, and that while it is true that many people have far more material comforts than they once did, it is still unclear whether the human condition, overall, has improved. One of us (N.O.) puts it this way: We live longer and have more things, to be sure, but are we happier? No one knows.

62. Ed Regis, "The Doomslayer," Wired.com, http://www.wired.com/wired/archive/5.02/ffsimon_pr.html.

63. Julian L. Simon and Herman Kahn, eds., *The Resourceful Earth: A Response to Global 2000* (New York: Blackwell, 1984), 1.

64. Julian L. Simon, ed., *The State of Humanity* (Cambridge, Mass.: Wiley-Blackwell, 1995).

65. Dixy Lee Ray and Lou Guzzo, *Trashing the Planet: How Science Can Help Us Deal with Acid Rain, Depletion of the Ozone, and Nuclear Waste (Among Other Things)* (New York: HarperPerennial, 1992); originally published by Regenry Gateway, 1990.

66. S. Fred Singer, "Will the World Come to a Horrible End?" *Science* 170, no. 3954 (October 9, 1970): 125.

67. Ibid.

68. Ibid.

69. "Board of Advisors," About the Institute, Independent Institute, http://www.independent.org/aboutus/advisors.asp.

70. "Biography," Bjørn Lomborg, http://www.lomborg.com/about/biography/.

71. John Mashey, "Lomborg and Playing the Long Game," The Way Things Break, comment

posted January 8, 2009, http://thingsbreak.wordpress.com/2009/01/08/lomborg-long -game/. The quote is from Julian Simon's book *The Ultimate Resource* (Princeton, N.J.: Princeton University Press, 1996). The ultimate resource is human ingenuity.

72. Stuart Pimm and Jeff Harvey, "No Need to Worry About the Future," Book Review: The Skeptical Environmentalist, *Nature* 414 (November 8, 2001): 149–50, http://www .nature.com/nature/journal/v414/n6860/full/414149a0.html. On the resonances between Lomborg's thought, Julian Simon, and various free market–oriented think tanks, see Mashey, "Lomborg and Playing the Long Game," and also Danish biologist Kåre Fog, "When Lomborg Became a So-Called 'Skeptical Environmentalist,'" Lomborg-Errors, The Lomborg Story, http://www.lomborg-errors.dk/lomborgstory2.htm.

73. Pimm and Harvey, "No Need to Worry About the Future," 149–50.

74. John Rennie, editor in chief, et al., "Misleading Math About the Earth: Science Defends Itself Against *The Skeptical Environmentalist,*" *Scientific American* 286, no. 1 (January 2002): 61–71, http://www.scientificamerican.com/article.cfm?id=misleading-math-about-the.

75. Lone Frank, "Scholarly Conduct: *Skeptical Environmentalist* labeled 'Dishonest,'" *Science* 299 (January 17, 2003): 326, http://www.sciencemag.org/cgi/reprint/299/5605/ 326b.pdf; Andrew C. Revkin, "Environment and Science: Danes Rebuke a 'Skeptic,'" *New York Times,* January 8, 2003, http://www.nytimes.com/2003/01/08/world/ environment-and-science-danes-rebuke-a-skeptic.html. One Danish biologist has dedicated an entire Web site to "Lomborg's Errors": Kåre Fog, This is the Lomborg-Errors Web Site, http://www.lomborg-errors.dk/.

76. See "Bjørn Lomborg: Danish Writer Cleared of 'Scientific Dishonesty,'" December 18, 2003, Center for the Defense of Free Enterprise, http://www.eskimo.com/~rarnold/ lomborg_cleared.htm; *Bjørn Lomborgs Klage Over Udvalgene Vedrørende Videnskabelig Uredeligheds (UVVU) Afgørelse af 6. Januar 2003,* December 17, 2003, http://www.dr .dk/nyheder/htm/baggrund/tema2003/striden%20om%20lomborg/images/87.pdf; see also *Årsberetning 2005: Udvalgene Vedrørende Videnskabelig Uredelighed,* November 2006, esp. p. 27, http://www.fi.dk/publikationer/2006/aarsberetning-2005-udvalgene videnskabelig-uredelighed/aarsberetning-2005-udvalgene-vedrorende-videnskabeligu. pdf. We thank Professor Kasper Eskildsen of Roskilde University for translating the relevant portions for us. Danish scientists also broadly complained about the overturning of the findings; see Alison Abbott, "Social Scientists Call for Overturning of Dishonesty Committee," *Nature* 421, no. 6924 (Febuary 13, 2003): 681.

77. See for example Nicholas Stern, *Stern Review, Executive Summary,* as well as M. L. Perry et al., eds., *Contribution of Working Group II to the Fourth Assessment Report of the Intergovernmental Panel on Climate Change* (Cambridge: Cambridge University Press, 2007), http://www.ipcc.ch/publications_and_data/publications_ipcc_fourth_assessment_ report_wg2_report_impacts_adaptation_and_vulnerability.htm.

78. Naomi Oreskes and Bjørn Lomborg, "Er klimaskeptikere nu domt ude?" [Have climate skeptics been ruled out of bounds?] A debate with Bjørn Lomborg, moderated by Morten Jastrup, *Politiken* (Copenhagen), December 5, 2004, p. 2, sec. 4.

79. Bjørn Lomborg, *The Skeptical Environmentalist: Measuring the Real State of the World* (New York: Cambridge University Press, 2001), 11.

80. Naomi Oreskes, "Science and Public Policy: What's Proof Got to Do with It?" *Environmental Science and Policy* 7, no. 5 (October 2004): 369–383.

81. "Bjørn Lomborg: Danish Writer Cleared of 'Scientific Dishonesty,'" Center for the Defense of Free Enterprise.

82. Mashey, "Lomborg and Playing the Long Game."

83. "The 2010 Sir John M. Templeton Fellowships Essay Contest," Academic Programs, Independent Institute, http://www.independent.org/students/essay/.

84. Friedman, *Capitalism and Freedom*, 4.

85. David A. Hounshell, *From the American System to Mass Production, 1800–1932: The Development of Manufacturing Technology in the United States* (Baltimore, Md.: Johns Hopkins University Press, 1984), 4–5; and especially Merritt Roe Smith, *Harpers Ferry Armory and the New Technology: The Challenge of Change* (Ithaca, N.Y.: Cornell University Press, 1977).

86. On the history of Silicon Valley, see Christophe Lécuyer, *Making Silicon Valley: Innovation and the Growth of High Tech, 1930–1970* (Cambridge, Mass.: MIT Press, 2007); Steve Blank, "The Secret History of Silicon Valley," Lecture, Google TechTalks, December 18, 2007, http://www.youtube.com/watch?v=hFSPHfZQpIQ.

87. Janet Abbate, *Inventing the Internet* (Cambridge, Mass.: MIT Press, 1999), 43–82. See also Thomas C. Greene, "Net Builders Kahn, Cerf Recognise Al Gore," October 2, 2000, *Register*, http://www.theregister.co.uk/2000/10/02/net_builders_kahn_cerf_recognise/. On this post, two scientists who played leading roles in the development of the Internet, Vint Cerf and Bob Kahn, acknowledge Al Gore's role in providing intellectual leadership, and helping to secure the passage of the High Performance Computing and Communications Act in 1991. In their words, "The fact of the matter is that Gore was talking about and promoting the Internet long before most people were listening."

88. Roger E. Bilstein, *Flight in America: From the Wrights to the Astronauts* (Baltimore, Md.: Johns Hopkins University Press, revised ed., 1994), illuminates the government's complex role in aviation development.

89. John Mashey, e-mail communication with authors, October 1, 2009. Mashey, an early contributor to the UNIX operating system, serves on the board of the Computer History Museum in Silicon Valley.

90. David E. Nye, *Electrifying America: Social Meanings of a New Technology, 1880–1940* (Cambridge, Mass.: MIT Press, 1990), 287–335.

91. "Stephen H. Schneider," http://stephenschneider.stanford.edu/; Stephen H. Schneider, "Mediarology," February 2007, http://stephenschneider.stanford.edu/Mediarology/MediarologyFrameset.html; Stephen H. Schneider, *Science as a Contact Sport* (Washington, D.C.: National Geographic, 2009), 203–232.

92. Excerpted from Naomi Oreskes, "The Scientific Consensus on Climate Change: How Do We Know We're Not Wrong?" in *Climate Change: What It Means for Us, Our Children, and Our Grandchildren*, ed. Joseph F. DiMento and Pamela Doughman (Cambridge, Mass.: MIT Press, 2007), 65–99.

93. Stephen Schneider makes a similar point; see Stephen H. Schneider, "Mediarology." Schneider expressed his concern that scientists may at times need to focus on the scare stories, or no one will pay attention. This has been frequently taken out of context to suggest that he thinks it is acceptable for scientists to be dishonest if the exigency of the problem warrants it. The full quote is:

On the one hand, as scientists we are ethically bound to the scientific method, in effect promising to tell the truth, the whole truth, and nothing but—which means that we must include all the doubts, the caveats, the ifs, ands, and buts. On the other hand, we are not just scientists but human beings as well. And like most people we'd like to see the world a better place, which in this context translates into our working to reduce the risk of potentially disastrous climatic change. To do that we need to get some broadbased support, to capture the public's imagination. That, of course, entails getting loads of media coverage. So we have to offer up scary scenarios, make simplified, dramatic statements, and make little mention of any doubts we might have. This "double ethical bind" we frequently find ourselves in cannot be solved by any formula. Each of us has to decide what the right balance is between being effective and being honest. I hope that means being both.

Quoted in interview for *Discover Magazine* (October 1989): 45–48; for the original, together with Schneider's commentary on its misrepresentation, see also Stephen H. Schneider, "Don't Bet All Environmental Changes Will Be Beneficial," American Physical Society, *APS News* 5, no. 8 (August/September 1996).

94. David Ignatius, "A Bid to Chill Thinking: Behind Joe Barton's Assault on Climate Scientists," *Washington Post*, July 22, 2005, http://www.washingtonpost.com/wp-dyn/content/article/2005/07/21/AR2005072102186.html.

95. "First 'Four Pillars,'" Senate Floor Statement by U.S. Senator James M. Inhofe (R-Okla.), April 8, 2005, Floor Speeches, http://inhofe.senate.gov/pressreleases/pillar.htm.

96. Antonio Regaldo, "In Climate Debate, the 'Hockey Stick' Leads to a Face-Off," *Wall Street Journal*, February 14, 2005, A1; David Appell, "Behind the Hockey Stick," *Scientific American* 992, no. 3 (March 2005): 34–35, http://www.scientificamerican.com/article.cfm?id=behind-the-hockey-stick; Michael Le Page, "Climate Myths: The 'Hockey Stick' Graph has Been Proven Wrong," *New Scientist*, September 2009, http://www.newscientist.com/article/dn11646; Gerald North, Chair, et al., *Surface Temperature Reconstructions for the Last 2000 Years*, Board on Atmospheric Sciences and Climate (Washington, D.C.: National Academies Press, 2006); Ben Santer, "The MSU Debate, Climate Auditing, and Freedom of Information," presented at AEED Seminar, Lawrence Livermore National Lab, December 10, 2008, Doc. No. LLNL-PRES-409614.

97. Stefan Ramstorf, personal communication with Naomi Oreskes, March 2009, Copenhagen, Denmark, at conference on "Climate Change: Global Risks, Challenges and Decisions," held in Copenhagen, March 10–12, 2009.

98. Confidential communication with a leading oceanographer by Naomi Oreskes, March 2009, Copenhagen, Denmark, "Climate Change: Global Risks, Challenges and Decisions."

99. Kåre Fog, "The First Debate in Denmark," Lomborg-Errors, The Lomborg Story, http://www.lomborg-errors.dk/lomborgstory3.htm.

100. Edward Frieman, personal communication with Naomi Oreskes, March 16, 2007.

101. Robert J. Samuelson, "Global Warming's Real Inconvenient Truth," *Washington Post*, July 5, 2006, http://www.washingtonpost.com/wp-dyn/content/article/2006/07/04/AR2006070400789.html; idem., "A Different View of Global Warming," http://www.msnbc.com/id/20226462/site/newsweek/print/1/displaymode/1098.

EPILOGUE

1.   Ronald N. Giere et al., *Understanding Scientific Reasoning*, 5th ed. (Belmont, Calif.: Thomson Wadsworth, 2006), 299.

2.   Giere et al., *Understanding Scientific Reasoning*, 299.

3.   There are a few articles in peer-reviewed journals, but Singer is not the first author. See for example David H. Douglass, Benjamin D. Pearson, and S. Fred Singer, "Altitude Dependence of Atmospheric Temperature Trends: Climate Models Versus Observation," *Geophysical Research Letters* 31 (July 9, 2004): L13208.

4.   Fred Singer, "Warming Theories Need Warning Labels," *Bulletin of Atomic Scientists* 48, no. 5 (June 1992): 34–39. Available online at http://books.google.com/books?id =kQsAAAAAMBAJ&pg=PA34&lpg=PA34&dq=Singer+Warming+Theorieseed+Warning +Labels&source=bl&ots=qoFV8Bn732&sig=gPI4434dsCtZKiId8262nqaFZXk&hl=en& ei=CubMSrvqJoH8tQPo7eiVAQ&sa=X&oi=book_result&ct=result&resnum=1#v =onepage&q=&f=false. He also reprises his old acid rain complaint: "Does it make sense to waste \$100 billion a year on what is still a phantom threat?" Singer, "Warning Labels," 39

5.   See S. Fred Singer, "World Demand for Oil," in *The Resourceful Earth: A Response to Global 2000*, ed. Julian L. Simon and Herman Kahn (New York: Blackwell, 1984), 339–60, and citations therein.

6.   See, for example, John J. Reilly, "Richard Feynman and Isaac Asimov on Spelling Reform," *Journal of the Simplified Spelling Society* 25 (1999/1): 31–32, http://www .spellingsociety.org/journals/j25/feynman.php#feynman.

7.   Michael Smithson, "Toward a Social Theory of Ignorance," *Journal for the Theory of Social Behaviour* 15, no. 2 (1985): 151–72; see also Michael Smithson, "Social Theories of Ignorance," in *Agnotology: The Making and Unmaking of Ignorance*, ed. Robert Proctor and Londa Schiebinger (Stanford, Calif.: Stanford University Press, 2008), 209–29.

8.   For a full discussion of this issue, see Robert Evans "Demarcation Socialized: Constructing Boundaries and Recognizing Difference," *Science, Technology, & Human Values* 30, No. 1: 3–16.

9.   Austin Bradford Hill, "The Environment and Disease: Association or Causation?" *Proceedings of the Royal Society of Medicine* 58, no. 5 (May 1965): 295–300.

10. Carolyn Symon, Lelani Arris, and Bill Heel, *Arctic Climate Impact Assessment* (Cambridge: Cambridge University Press, 2005).

11. William Shakespeare, *Macbeth*, Act V, Scene V.

12. S. Green, *Smoking Associated Disease and Causality*, n.d., BN: 1192.02, Legacy Tobacco Documents Library. An earlier, slightly different version was written in 1976; see S. Green, *Cigarette Smoking and Causal Relationships*, 27 October 1976, BN: 2231.08, Legacy Tobacco Documents Library. S. Green is discussed in David A. Kessler's *A Question of Intent: A Great American Battle with a Deadly Industry* (New York: Public Affairs, 2002), 228.

13. Steve LaRue, "Early Action Urged in Fight on Acid Rain," *San Diego Union*, local sec., August 8, 1984, B2.

Index

A Note on the Authors

Naomi Oreskes is a professor of history and science studies at the University of California, San Diego. Her study "Beyond the Ivory Tower," published in *Science*, was a milestone in the fight against global warming denial and was cited by Al Gore in *An Inconvenient Truth*. **Erik M. Conway** has published four previous books, including *Atmospheric Science at NASA: A History*. *Merchants of Doubt* is their first book together.